Python Deep Learning

Understand how deep neural networks work and apply them to real-world tasks

Ivan Vasilev

BIRMINGHAM—MUMBAI

Python Deep Learning

Associate Group Product Manager: Niranjan Naikwadi

Associate Publishing Product Manager: Nitin Nainani

Book Project Manager: Hemangi Lotlikar

Senior Editor: Tiksha Lad

Technical Editor: Rahul Limbachiya

Copy Editor: Safis Editing

Proofreader: Safis Editing

Indexer: Subalakshmi Govindhan

Production Designer: Shankar Kalbhor

DevRel Marketing Coordinator: Vinishka Kalra

First published: April 2017

Second edition: January 2019

Third edition: November 2023

Production reference: 1271023

Published by

Packt Publishing Ltd.

Grosvenor House

11 St Paul's Square

Birmingham

B3 1RB, UK.

ISBN 978-1-83763-850-5

www.packtpub.com

Contributors

About the author

Ivan Vasilev started working on the first open source Java deep learning library with GPU support in 2013. The library was acquired by a German company, with whom he continued its development. He has also worked as a machine learning engineer and researcher in medical image classification and segmentation with deep neural networks. Since 2017, he has focused on financial machine learning. He co-founded an algorithmic trading company, where he's the lead engineer.

He holds an MSc in artificial intelligence from Sofia University St. Kliment Ohridski and has written two previous books on the same topic.

About the reviewer

Krishnan Raghavan is an IT professional with over 20+ years of experience in the areas of software development and delivery excellence, across multiple domains and technologies, ranging from C++ to Java, Python, Angular, Golang, and data warehousing.

When not working, Krishnan likes to spend time with his wife and daughter, besides reading fiction, nonfiction, and technical books and participating in hackathons. Krishnan tries to give back to the community by being part of a GDG, the Pune volunteer group, helping the team to organize events.

You can connect with Krishnan at `mailtokrishnan@gmail.com` or via LinkedIn at `www.linkedin.com/in/krishnan-Raghavan`.

I would like to thank my wife, Anita, and daughter, Ananya, for giving me the time and space to review this book.

Table of Contents

3

Deep Learning Fundamentals 65

Part 2: Deep Neural Networks for Computer Vision

4

Computer Vision with Convolutional Networks 93

5

Advanced Computer Vision Applications 135

Part 3: Natural Language Processing and Transformers

6

Natural Language Processing and Recurrent Neural Networks 173

7

The Attention Mechanism and Transformers 211

8

Exploring Large Language Models in Depth   245

9

Advanced Applications of Large Language Models   277

Part 4: Developing and Deploying Deep Neural Networks

10

Preface

The book will start from the theoretical foundations of deep **neural networks** (**NN**), and it will delve into the most popular network architectures – transformers, transformer-based **large language models** (**LLMs**), and convolutional networks. It will introduce these models in the context of various computer vision and **natural language processing** (**NLP**) examples, including state-of-the-art applications such as text-to-image generation and chatbots.

Each chapter is organized with a comprehensive theoretical introduction to the topic as its main body. This is followed by coding examples that serve to validate the presented theory, providing readers with practical hands-on experience. The examples are executed using PyTorch, Keras, or Hugging Face Transformers.

Who this book is for

This book is for individuals already familiar with programming – software developers/engineers, students, data scientists, data analysts, machine learning engineers, statisticians, and anyone interested in deep learning who has Python programming experience. It is designed for people with minimal prior deep learning knowledge, employing clear and straightforward language throughout.

What this book covers

Chapter 1, Machine Learning – an Introduction, discusses the basic machine learning paradigms. It will explore various machine learning algorithms and introduce the first NN, implemented with PyTorch.

Chapter 2, Neural Networks, starts by introducing the mathematical branches related to NNs – linear algebra, probability, and differential calculus. It will focus on the building blocks and structure of NNs. It will also discuss how to train NNs with gradient descent and backpropagation.

Chapter 3, Deep Learning Fundamentals, introduces the basic paradigms of deep learning. It will make the transition from classic networks to deep NNs. It will outline the challenges of developing and using deep networks, and it will discuss how to solve them.

Chapter 4, Computer Vision with Convolutional Networks, introduces convolutional networks – the main network architecture for computer vision applications. It will discuss in detail their properties and building blocks. It will also introduce the most popular convolutional network models in use today.

Chapter 5, Advanced Computer Vision Applications, discusses applying convolutional networks for advanced computer vision tasks – object detection and image segmentation. It will also explore using NNs to generate new images.

Chapter 6, Natural Language Processing and Recurrent Neural Networks, introduces the main paradigms and data processing pipeline of NLP. It will also explore recurrent NNs and their two most popular variants – long short-term memory and gated recurrent units.

Chapter 7, The Attention Mechanism and Transformers, introduces one of the most significant recent deep learning advances – the attention mechanism and the transformer model based around it.

Chapter 8, Exploring Large Language Models in Depth, introduces transformer-based LLMs. It will discuss their properties and what makes them different than other NN models. It will also introduce the Hugging Face Transformers library.

Chapter 9, Advanced Applications of Large Language Models, discusses using LLMs for computer vision tasks. It will focus on classic tasks such as image classification and object detection, but it will also explore state-of-the-art applications such as text-to-image generation. It will introduce the LangChain framework for LLM-driven application development.

Chapter 10, Machine Learning Operations (MLOps), will introduce various libraries and techniques for easier development and production deployment of NN models.

To get the most out of this book

Many code examples in the book require the presence of a GPU. Don't worry if you don't have one. To avoid any hardware limitations, all code examples are available as Jupyter notebooks, executed on Google Colab. So, even if your hardware is not sufficient to run the examples, you can still run them under Colab.

Software/hardware covered in the book	Operating system requirements
PyTorch 2.0.1	Windows, macOS, or Linux
TensorFlow 2.13	Windows (legacy support), macOS, or Linux
Hugging Face Transformers 4.33	Windows, macOS, or Linux

Some code examples in the book may use additional packages not listed in the table. You can see the full list (with versions) in the requirements.txt file in the book's GitHub repo.

If you are using the digital version of this book, we advise you to type the code yourself or access the code from the book's GitHub repository (a link is available in the next section). Doing so will help you avoid any potential errors related to the copying and pasting of code.

Download the example code files

You can download the example code files for this book from GitHub at `https://github.com/PacktPublishing/Python-Deep-Learning-Third-Edition/`. If there's an update to the code, it will be updated in the GitHub repository.

We also have other code bundles from our rich catalog of books and videos available at `https://github.com/PacktPublishing/`. Check them out!

Conventions used

There are a number of text conventions used throughout this book.

`Code in text`: Indicates code words in text, database table names, folder names, filenames, file extensions, pathnames, dummy URLs, user input, and Twitter handles. Here is an example: "Use `opencv-python` to read the RGB image located at `image_file_path`."

A block of code is set as follows:

```
def build_fe_model():
    """"Create feature extraction model from the pre-trained model
ResNet50V2"""

    # create the pre-trained part of the network, excluding FC layers
    base_model = tf.keras.applications.MobileNetV3Small(
```

Any command-line input or output is written as follows:

```
import tensorflow.keras
```

When we wish to draw your attention to a particular part of a code block, the relevant lines or items are set in bold:

```
import io
image = Image.open(io.BytesIO(response.content))
image.show()
```

> **Tips or important notes**
> Appear like this.

Get in touch

Feedback from our readers is always welcome.

General feedback: If you have questions about any aspect of this book, email us at `customercare@packtpub.com` and mention the book title in the subject of your message.

Errata: Although we have taken every care to ensure the accuracy of our content, mistakes do happen. If you have found a mistake in this book, we would be grateful if you would report this to us. Please visit www.`packtpub.com/support/errata` and fill in the form.

Piracy: If you come across any illegal copies of our works in any form on the internet, we would be grateful if you would provide us with the location address or website name. Please contact us at `copyright@packt.com` with a link to the material.

If you are interested in becoming an author: If there is a topic that you have expertise in and you are interested in either writing or contributing to a book, please visit `authors.packtpub.com`.

Share Your Thoughts

Once you've read *Python Deep Learning, Third Edition*, we'd love to hear your thoughts! Scan the QR code below to go straight to the Amazon review page for this book and share your feedback.

`https://packt.link/r/1837638500`

Your review is important to us and the tech community and will help us make sure we're delivering excellent quality content.

Download a free PDF copy of this book

Thanks for purchasing this book!

Do you like to read on the go but are unable to carry your print books everywhere?

Is your eBook purchase not compatible with the device of your choice?

Don't worry, now with every Packt book you get a DRM-free PDF version of that book at no cost.

Read anywhere, any place, on any device. Search, copy, and paste code from your favorite technical books directly into your application.

The perks don't stop there, you can get exclusive access to discounts, newsletters, and great free content in your inbox daily

Follow these simple steps to get the benefits:

1. Scan the QR code or visit the link below

https://packt.link/free-ebook/9781837638505

2. Submit your proof of purchase
3. That's it! We'll send your free PDF and other benefits to your email directly

Part 1:
Introduction
to Neural Networks

We'll start this part by introducing you to basic machine learning theory and concepts. Then, we'll follow with a thorough introduction to neural networks – a special type of machine learning algorithm.

We'll discuss the mathematical principles behind them and learn how to train them. Finally, we'll make the transition from shallow to deep networks.

This part has the following chapters:

- *Chapter 1, Machine Learning – an Introduction*
- *Chapter 2, Neural Networks*
- *Chapter 3, Deep Learning Fundamentals*

1

Machine Learning – an Introduction

Machine learning (**ML**) techniques are being applied in a variety of fields, and data scientists are being sought after in many different industries. With ML, we identify the processes through which we gain knowledge that is not readily apparent from data to make decisions. Applications of ML techniques may vary greatly and are found in disciplines as diverse as medicine, finance, and advertising.

In this chapter, we'll present different ML approaches, techniques, and some of their applications to real-world problems, and we'll also introduce one of the major open source packages available in Python for ML, PyTorch. This will lay the foundation for later chapters in which we'll focus on a particular type of ML approach using **neural networks** (**NNs**). In particular, we will focus on **deep learning** (**DL**). DL makes use of more advanced NNs than those used previously. This is not only a result of recent developments in the theory but also advancements in computer hardware. This chapter will summarize what ML is and what it can do, preparing you to better understand how DL differentiates itself from popular traditional ML techniques.

In this chapter, we're going to cover the following main topics:

- Introduction to ML
- Different ML approaches
- Neural networks
- Introduction to PyTorch

Technical requirements

We'll implement the example in this chapter using Python and PyTorch. If you don't have an environment set up with these tools, fret not – the example is available as a Jupyter notebook on Google Colab. You can find the code examples in the book's GitHub repository: `https://github.com/PacktPublishing/Python-Deep-Learning-Third-Edition/tree/main/Chapter01`.

Introduction to ML

ML is often associated with terms such as **big data** and **artificial intelligence** (**AI**). However, both are quite different from ML. To understand what ML is and why it's useful, it's important to understand what big data is and how ML applies to it.

Big data is a term used to describe huge datasets that are created as the result of large increases in data that is gathered and stored. For example, this may be through cameras, sensors, or internet social sites.

How much data do we create daily?

It's estimated that Google alone processes over 20 petabytes of information per day, and this number is only going to increase. A few years ago, Forbes estimated that every day, 2.5 quintillion bytes of data are created and that 90% of all the data in the world has been created in the last two years.

(https://www.forbes.com/sites/bernardmarr/2018/05/21/how-much-data-do-we-create-every-day-the-mind-blowing-stats-everyone-should-read/)

Humans alone are unable to grasp, let alone analyze, such huge amounts of data, and ML techniques are used to make sense of these very large datasets. ML is the tool that's used for large-scale data processing. It is well suited to complex datasets that have huge numbers of variables and features. One of the strengths of many ML techniques, and DL in particular, is that they perform best when used on large datasets, thus improving their analytic and predictive power. In other words, ML techniques, and DL NNs in particular, learn best when they can access large datasets where they can discover patterns and regularities hidden in the data.

On the other hand, ML's predictive ability can be successfully adapted to AI systems. ML can be thought of as the brain of an AI system. AI can be defined (though this definition may not be unique) as a system that can interact with its environment. Also, AI machines are endowed with sensors that enable them to know the environment they are in and tools with which they can relate to the environment. Therefore, ML is the brain that allows the machine to analyze the data ingested through its sensors to formulate an appropriate answer. A simple example is Siri on an iPhone. Siri hears the command through its microphone and outputs an answer through its speakers or its display, but to do so, it needs to understand what it's being told. Similarly, driverless cars will be equipped with cameras, GPS systems, sonars, and LiDAR, but all this information needs to be processed to provide a correct answer. This may include whether to accelerate, brake, or turn. ML is the information-processing method that leads to the answer.

We've explained what ML is, but what about DL? For now, let's just say that DL is a subfield of ML. DL methods share some special common features. The most popular representatives of such methods are deep NNs.

Different ML approaches

As we have seen, the term ML is used in a very general way and refers to the general techniques that are used to extrapolate patterns from large sets, or it is the ability to make predictions on new data based on what is learned by analyzing available known data. ML techniques can roughly be divided into two core classes, while one more class is often added. Here are the classes:

- Supervised learning
- Unsupervised learning
- Reinforcement learning

Let's take a closer look.

Supervised learning

Supervised learning algorithms are a class of ML algorithms that use previously labeled data to learn its features, so they can classify similar but unlabeled data. Let's use an example to understand this concept better.

Let's assume that a user receives many emails every day, some of which are important business emails and some of which are unsolicited junk emails, also known as **spam**. A supervised machine algorithm will be presented with a large body of emails that have already been labeled by a teacher as spam or not spam (this is called **training data**). For each sample, the machine will try to predict whether the email is spam or not, and it will compare the prediction with the original target label. If the prediction differs from the target, the machine will adjust its internal parameters in such a way that the next time it encounters this sample, it will classify it correctly. Conversely, if the prediction is correct, the parameters will stay the same. The more training data we feed to the algorithm, the better it becomes (this rule has caveats, as we'll see next).

In the example we used, the emails had only two classes (spam or not spam), but the same principles apply to tasks with arbitrary numbers of classes (or categories). For example, Gmail, the free email service by Google, allows the user to select up to five categories, which are labeled as follows:

- **Primary**: Includes person-to-person conversations
- **Promotions**: Includes marketing emails, offers, and discounts
- **Social**: Includes messages from social networks and media-sharing sites
- **Updates**: Includes bills, bank statements, and receipts
- **Forums**: Includes messages from online groups and mailing lists

To summarize, the ML task, which maps a set of input values to a finite number of classes, is called **classification**.

In some cases, the outcome may not necessarily be discrete, and we may not have a finite number of classes to classify our data into. For example, we may try to predict the life expectancy of a group of people based on their predetermined health parameters. In this case, the outcome is a numerical value, and we don't talk about classification but rather **regression**.

One way to think of supervised learning is to imagine we are building a function, f, defined over a dataset, which comprises information organized by **features**. In the case of email classification, the features can be specific words that may appear more frequently than others in spam emails. The use of explicit sex-related words will most likely identify a spam email rather than a business/work email. On the contrary, words such as *meeting*, *business*, or *presentation* are more likely to describe a work email. If we have access to metadata, we may also use the sender's information as a feature. Each email will then have an associated set of features, and each feature will have a value (in this case, how many times the specific word is present in the email's body). The ML algorithm will then seek to map those values to a discrete range that represents the set of classes, or a real value in the case of regression. The definition of the f function is as follows:

$$f:\text{space of features} \rightarrow \text{classes} = (\text{discrete values or real values})$$

In later chapters, we'll see several examples of either classification or regression problems. One such problem we'll discuss is classifying handwritten digits of the **Modified National Institute of Standards and Technology** (**MNIST**) database (http://yann.lecun.com/exdb/mnist/). When given a set of images representing 0 to 9, the ML algorithm will try to classify each image in one of the 10 classes, wherein each class corresponds to one of the 10 digits. Each image is 28×28 (= 784) pixels in size. If we think of each pixel as one feature, then the algorithm will use a 784-dimensional feature space to classify the digits.

The following figure depicts the handwritten digits from the MNIST dataset:

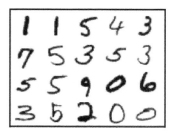

Figure 1.1 – An example of handwritten digits from the MNIST dataset

In the next sections, we'll talk about some of the most popular classical supervised algorithms. The following is by no means an exhaustive list or a thorough description of each ML method. We recommend referring to the book *Python Machine Learning*, by Sebastian Raschka (https://www.packtpub.

`com/product/python-machine-learning-third-edition/9781789955750`). It's
a simple review meant to provide you with a flavor of the different ML techniques in Python.

Linear and logistic regression

A **regression** algorithm is a type of supervised algorithm that uses features of the input data to predict
a numeric value, such as the cost of a house, given certain features, such as size, age, number of
bathrooms, number of floors, and location. Regression analysis tries to find the value of the parameters
for the function that best fits an input dataset.

In a **linear regression** algorithm, the goal is to minimize a **cost function** by finding appropriate
parameters for the function over the input data that best approximates the target values. A cost
function is a function of the error – that is, how far we are from getting a correct result. A popular
cost function is the **mean squared error** (**MSE**), where we take the square of the difference between
the expected value and the predicted result. The sum of all the input examples gives us the error of
the algorithm and represents the cost function.

Say we have a 100-square-meter house that was built 25 years ago with three bathrooms and two
floors. Let's also assume that the city is divided into 10 different neighborhoods, which we'll denote
with integers from 1 to 10, and say this house is located in the area denoted by 7. We can parameterize
this house with a five-dimensional vector, $\mathbf{x} = (x_1, x_2, x_3, x_4, x_5) = (100, 25, 3, 2, 7)$. Say that we also
know that this house has an estimated value of \$100,000 (in today's world, this would be enough for
just a tiny shack near the North Pole, but let's pretend). What we want is to create a function, f, such
that $f(\mathbf{x}) = 100000$.

A note of encouragement

Don't worry If you don't fully understand some of the terms in this section. We'll discuss vectors,
cost functions, linear regression, and gradient descent in more detail in *Chapter 2*. We will also
see that training NNs and linear/logistic regressions have a lot in common. For now, you can
think of a vector as an array. We'll denote vectors with boldface font – for example, $\mathbf{x}$. We'll
denote the vector elements with italic font and subscript – for example, x_i.

In linear regression, this means finding a vector of weights, $\mathbf{w} = (w_1, w_2, w_3, w_4, w_5)$, such that the dot
product of the vectors, $\mathbf{x} \cdot \mathbf{w} = 100000$, would be $100 w_1 + 25 w_2 + 3 w_3 + 2 w_4 + 7 w_5 = 100000$ or
$\sum x_i w_i = 100000$. If we had 1,000 houses, we could repeat the same process for every house, and
ideally, we would like to find a single vector, $\mathbf{w}$, that can predict the correct value that is close enough
for every house. The most common way to train a linear regression model can be seen in the following
pseudocode block:

```
Initialize the vector w with some random values
repeat:
    E = 0 # initialize the cost function E with 0
    for every pair (x⁽ⁱ⁾, t⁽ⁱ⁾) of the training set:
```

```
    E += (x⁽ⁱ⁾ · w - t⁽ⁱ⁾)² # here t⁽ⁱ⁾ is the real house price
MSE = ──────E──────── # Mean Square Error
      total number of samples
    use gradient descent to update the weights w based on MSE until MSE
falls below threshold
```

First, we iterate over the training data to compute the cost function, MSE. Once we know the value of MSE, we'll use the **gradient descent** algorithm to update the weights of the vector, **w**. To do this, we'll calculate the derivatives of the cost function concerning each weight, w_i. In this way, we'll know how the cost function changes (increase or decrease) concerning w_i. Then we'll update that weight's value accordingly.

Previously, we demonstrated how to solve a regression problem with linear regression. Now, let's take a classification task: trying to determine whether a house is overvalued or undervalued. In this case, the target data would be categorical [1, 0] – 1 for overvalued and 0 for undervalued. The price of the house will be an input parameter instead of the target value as before. To solve the task, we'll use logistic regression. This is similar to linear regression but with one difference: in linear regression, the output is $x \cdot w$. However, here, the output will be a special logistic function (https://en.wikipedia. org/wiki/Logistic_function), $\sigma(x \cdot w)$. This will squash the value of $x \cdot w$ in the (0:1) interval. You can think of the logistic function as a probability, and the closer the result is to 1, the more chance there is that the house is overvalued, and vice versa. Training is the same as with linear regression, but the output of the function is in the (0:1) interval, and the labels are either 0 or 1.

Logistic regression is not a classification algorithm, but we can turn it into one. We just have to introduce a rule that determines the class based on the logistic function's output. For example, we can say that a house is overvalued if the value of $\sigma(x \cdot w) < 0.5$ and undervalued otherwise.

> **Multivariate regression**
>
> The regression examples in this section have a single numerical output. A regression analysis can have more than one output. We'll refer to such analysis as **multivariate regression**.

Support vector machines

A **support vector machine** (**SVM**) is a supervised ML algorithm that's mainly used for classification. It is the most popular member of the kernel method class of algorithms. An SVM tries to find a hyperplane, which separates the samples in the dataset.

> **Hyperplanes**
>
> A hyperplane is a plane in a high-dimensional space. For example, a hyperplane in a one-dimensional space is a point, and in a two-dimensional space, it would just be a line. In three-dimensional space, the hyperplane would be a plane, and we can't visualize the hyperplane in four-dimensional space, but we know that it exists.

We can think of classification as the process of trying to find a hyperplane that will separate different groups of data points. Once we have defined our features, every sample (in our case, an email) in the dataset can be thought of as a point in the multidimensional space of features. One dimension of that space represents all the possible values of one feature. The coordinates of a point (a sample) are the specific values of each feature for that sample. The ML algorithm task will be to draw a hyperplane to separate points with different classes. In our case, the hyperplane would separate spam from non-spam emails.

In the following diagram, at the top and bottom, we can see two classes of points (red and blue) that are in a two-dimensional feature space (the x and y axes). If both the x and y values of a point are below 5, then the point is blue. In all other cases, the point is red. In this case, the classes are **linearly separable**, meaning we can separate them with a hyperplane. Conversely, the classes in the image on the right are linearly inseparable:

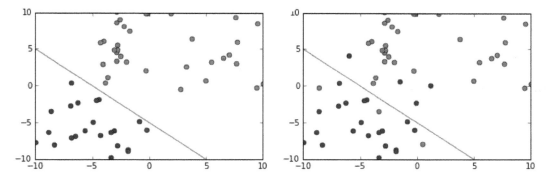

Figure 1.2 – A linearly separable set of points (left) and a linearly inseparable set (right)

The SVM tries to find a hyperplane that maximizes the distance between itself and the points. In other words, from all possible hyperplanes that can separate the samples, the SVM finds the one that has the maximum distance from all points. In addition, SVMs can deal with data that is not linearly separable. There are two methods for this: introducing soft margins or using the **kernel trick**.

Soft margins work by allowing a few misclassified elements while retaining the most predictive ability of the algorithm. In practice, it's better not to overfit the ML model, and we could do so by relaxing some of the SVM hypotheses.

The kernel trick solves the same problem differently. Imagine that we have a two-dimensional feature space, but the classes are linearly inseparable. The kernel trick uses a kernel function that transforms the data by adding more dimensions to it. In our case, after the transformation, the data will be three-dimensional. The linearly inseparable classes in the two-dimensional space will become linearly separable in the three dimensions and our problem is solved:

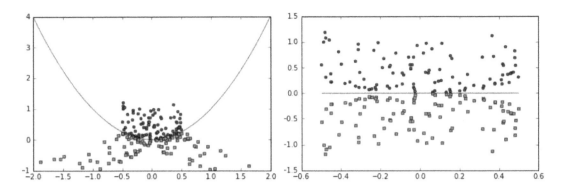

Figure 1.3 – A non-linearly separable set before the kernel was applied (left) and the same dataset after the kernel has been applied, and the data can be linearly separated (right)

Lets move to the last one in our list, decision trees.

Decision trees

Another popular supervised algorithm is the decision tree, which creates a classifier in the form of a tree. It is composed of decision nodes, where tests on specific attributes are performed, and leaf nodes, which indicate the value of the target attribute. To classify a new sample, we start at the root of the tree and navigate down the nodes until we reach a leaf.

A classic application of this algorithm is the Iris flower dataset (`http://archive.ics. uci.edu/ml/datasets/Iris`), which contains data from 50 samples of three types of irises (**Iris Setosa**, **Iris Virginica**, and **Iris Versicolor**). Ronald Fisher, who created the dataset, measured four different features of these flowers:

- The length of their sepals
- The width of their sepals
- The length of their petals
- The width of their petals

Based on the different combinations of these features, it's possible to create a decision tree to decide which species each flower belongs to. In the following diagram, we have defined a decision tree that will correctly classify almost all the flowers using only two of these features, the petal length and width:

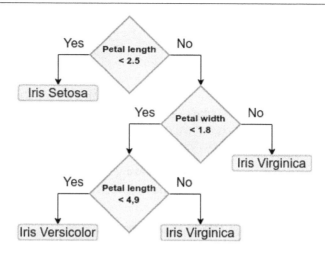

Figure 1.4 – A decision tree for classifying the Iris dataset

To classify a new sample, we start at the root note of the tree (petal length). If the sample satisfies the condition, we go left to the leaf, representing the Iris Setosa class. If not, we go right to a new node (petal width). This process continues until we reach a leaf.

In recent years, decision trees have seen two major improvements. The first is **random forests**, which is an ensemble method that combines the predictions of multiple trees. The second is a class of algorithms called **gradient boosting**, which creates multiple sequential decision trees, where each tree tries to improve the errors made by the previous tree. Thanks to these improvements, decision trees have become very popular when working with certain types of data. For example, they are one of the most popular algorithms used in Kaggle competitions.

Unsupervised learning

The second class of ML algorithms is unsupervised learning. Here, we don't label the data beforehand; instead, we let the algorithm come to its conclusion. One of the advantages of unsupervised learning algorithms over supervised ones is that we don't need labeled data. Producing labels for supervised algorithms can be costly and slow. One way to solve this issue is to modify the supervised algorithm so that it uses less labeled data; there are different techniques for this. But another approach is to use an algorithm, which doesn't need labels in the first place. In this section, we'll discuss some of these unsupervised algorithms.

Clustering

One of the most common, and perhaps simplest, examples of unsupervised learning is clustering. This is a technique that attempts to separate the data into subsets.

To illustrate this, let's view the spam-or-not-spam email classification as an unsupervised learning problem. In the supervised case, for each email, we had a set of features and a label (spam or not spam). Here, we'll use the same set of features, but the emails will not be labeled. Instead, we'll ask the algorithm, when given the set of features, to put each sample in one of two separate groups (or clusters). Then, the algorithm will try to combine the samples in such a way that the intraclass similarity (which is the similarity between samples in the same cluster) is high and the similarity between different clusters is low. Different clustering algorithms use different metrics to measure similarity. For some more advanced algorithms, you don't have to specify the number of clusters.

The following graph shows how a set of points can be classified to form three subsets:

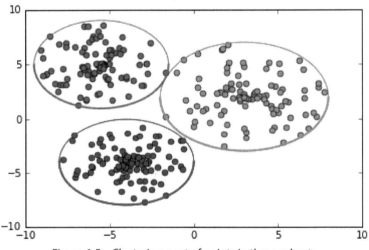

Figure 1.5 – Clustering a set of points in three subsets

K-means

K-means is a clustering algorithm that groups the elements of a dataset into k distinct clusters (hence the k in the name). Here's how it works:

1. Choose k random points, called **centroids**, from the feature space, which will represent the center of each of the k clusters.

2. Assign each sample of the dataset (that is, each point in the feature space) to the cluster with the closest centroid.

3. For each cluster, we recomputed new centroids by taking the mean values of all the points in the cluster.

4. With the new centroids, we repeat *Steps 2* and *3* until the stopping criteria are met.

The preceding method is sensitive to the initial choice of random centroids, and it may be a good idea to repeat it with different initial choices. It's also possible for some centroids to not be close to any of the points in the dataset, reducing the number of clusters down from k. Finally, it's worth mentioning that if we used k-means with $k=3$ on the Iris dataset, we may get different distributions of the samples compared to the distribution of the decision tree that we'd introduced. Once more, this highlights how important it is to carefully choose and use the correct ML method for each problem.

Now, let's discuss a practical example that uses k-means clustering. Let's say a pizza delivery place wants to open four new franchises in a city, and they need to choose the locations for the sites. We can solve this problem with k-means:

1. Find the locations where pizza is ordered most often; these will be our data points.

2. Choose four random points where the sites will be located.

3. By using k-means clustering, we can identify the four best locations that minimize the distance to each delivery place:

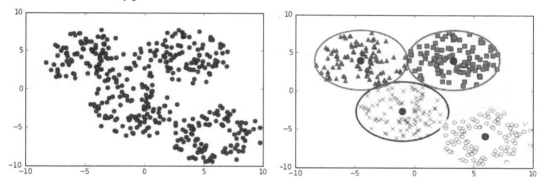

Figure 1.6 – The distribution of points where pizza is delivered most often (left); the round points indicate where the new franchises should be located and their corresponding delivery areas (right)

Self-supervised learning

Self-supervised learning refers to a combination of problems and datasets, which allow us to *automatically generate* (that is, without human intervention) labeled data from the dataset. Once we have these labels, we can train a supervised algorithm to solve our task. To understand this concept better, let's discuss some use cases:

*   **Time series forecasting**: Imagine that we have to predict the future value of a time series based on its most recent historical values. Examples of this include stock (and nowadays crypto) price prediction and weather forecasting. To generate a labeled data sample, let's take a window with length k of the historical data that ends at past moment t. We'll take the historical values in the range $[t - k; t]$ and we'll use them as input for the supervised algorithm. We'll also take the historical value at moment $t + 1$ and we'll use it as the label for the given input sample. We can apply this division to the rest of the historical values and generate a labeled training dataset automatically.

- **Natural language processing** (**NLP**): Similar to time series, the natural text represents a sequence of words (or tokens). We can train an NLP algorithm to predict the next word based on the preceding k words in a similar manner to a time series. However, the natural text does not carry the same strict past/future division as time series do. Because of this, we can use the whole context around the target word as input – that is, words that come both before and after the target word in the consequence, instead of the preceding words only. As we'll see in *Chapter 6*, this technique is foundational to contemporary NLP algorithms.

- **Autoencoders**: This is a special type of NN that tries to reproduce its input. In other words, the target value (label) of an autoencoder is equal to the input data, $y_i = x_i$, where i is the sample index. We can formally say that it tries to learn an identity function (a function that repeats its input). Since our labels are just input data, the autoencoder is an unsupervised algorithm. You might be wondering what the point of an algorithm that tries to predict its input is. The autoencoder is split into two parts – an encoder and a decoder. First, the encoder tries to compress the input data into a vector with a smaller size than the input itself. Next, the decoder tries to reproduce the original input based on this smaller internal state vector. By setting this limitation, the autoencoder is forced to extract only the most significant features of the input data. The goal of the autoencoder is to learn a representation of the data that is more efficient or compact than the original representation, while still retaining as much of the original information as possible.

Another interesting application of self-supervised learning is in generative models, as opposed to discriminative models. Let's discuss the difference between the two. Given input data, a discriminative model will map it to a certain label (in other words, classification or regression). A typical example is the classification of MNIST images in 1 of 10-digit classes, where the NN maps input data features (pixel intensities) to the digit label. We can also say this in another way: a discriminative model gives us the probability of y (class), given x (input) – $P(Y|X = x)$. In the case of MNIST, this is the probability of the digit when given the pixel intensities of the image.

On the other hand, a generative model learns how classes are distributed. You can think of it as the opposite of what the discriminative model does. Instead of predicting the class probability, y, given certain input features, it tries to predict the probability of the input features when given a class, $y - P(X|Y = y)$. For example, a generative model will be able to create an image of a handwritten digit when given the digit class. Since we only have 10 classes, it will be able to generate just 10 images. However, we've only used this example to illustrate this concept. In reality, the class could be an arbitrary tensor of values, and the model would be able to generate an unlimited number of images with different features. If you don't understand this now, don't worry; we'll discuss this topic again in *Chapter 5*. In *Chapter 8* and *Chapter 9*, we'll discuss how transformers (a new type of NN) have been used to produce some impressive generative models. They have gained popularity both in the research community and the mainstream public because of the attractive results they produce. Two of the most popular visual models are **Stable Diffusion** (https://github.com/CompVis/stable-diffusion), by Stability AI (https://stability.ai/), and **DALL-E** (https://openai.com/dall-e-2/), by OpenAI, which can create photorealistic or artistic images from a natural language description. When prompted with the text *Musician frog playing on a guitar*, Stable Diffusion produces the following figure:

Figure 1.7 – Stable Diffusion output for the prompt Musician frog playing on a guitar

Another interesting generative model is OpenAI's **ChatGPT** (GPT stands for **Generative Pre-trained Transformer**), which (as its name suggests) acts as a smart chatbot. ChatGPT can answer follow-up questions, admit mistakes, challenge incorrect premises, and reject inappropriate requests.

Reinforcement learning

The third class of ML techniques is called **reinforcement learning** (**RL**). We will illustrate this with one of the most popular applications of RL: teaching machines how to play games. The machine (or agent) interacts with the game (or environment). The goal of the agent is to win the game. To do this, the agent takes actions that can change the environment's state. The environment reacts to the agent's actions and provides it with reward (or penalty) signals that help the agent to decide its next action. Winning the game would provide the biggest reward. In formal terms, the goal of the agent is to maximize the total rewards it receives throughout the game:

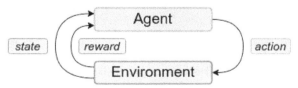

Figure 1.8 – The interaction between different elements of an RL system

Let's imagine a game of chess as an RL problem. Here, the environment would include the chessboard, along with the locations of the pieces. The goal of our agent is to beat the opponent. The agent will then receive a reward when they capture the opponent's piece, and they will win the biggest reward if they checkmate the opponent. Conversely, if the opponent captures a piece or checkmates the agent, the reward will be negative. However, as part of their larger strategies, the players will have to make moves that neither capture a piece nor checkmate the other's king. The agent won't receive any reward then. If this was a supervised learning problem, we would have to provide a label or a reward for each move. This is not the case with RL. In the RL framework, the agent will improvise with a trial-and-error approach to decide its next actions.

Let's take another example, in which sometimes we have to sacrifice a pawn to achieve a more important goal (such as a better position on the chessboard). In such situations, our humble agent has to be smart enough to take a short-term loss as a long-term gain. In an even more extreme case, imagine we had the bad luck of playing against Ding Liren, the current world chess champion. Surely, the agent will lose in this case. However, how would we know which moves were wrong and led to the agent's loss? Chess belongs to a class of problems where the game should be considered in its entirety to reach a successful solution, rather than just looking at the immediate consequences of each action. RL will give us the framework that will help the agent to navigate and learn in this complex environment.

An interesting problem arises from this newfound freedom to take action. Imagine that the agent has learned one successful chess-playing strategy (or **policy**, in RL terms). After some games, the opponent might guess what that policy is and manage to beat us. The agent will now face a dilemma with the following decisions: either to follow the current policy and risk becoming predictable, or to experiment with new moves that will surprise the opponent, but also carry the risk of turning out even worse. In general terms, the agent uses a policy that gives them a certain reward, but their ultimate goal is to maximize the total reward. A modified policy might be more rewarding, and the agent will be ineffective if they don't try to find such a policy. One of the challenges of RL is the trade-off between exploitation (following the current policy) and exploration (trying new moves).

So far, we've used only games as examples; however, many problems can fall into the RL domain. For example, you can think of autonomous vehicle driving as an RL problem. The vehicle can get positive rewards if it stays within its lane and observes the traffic rules. It will gain negative rewards if it crashes. Another interesting recent application of RL is in managing stock portfolios. The goal of the agent would be to maximize the portfolio value. The reward is directly derived from the value of the stocks in the portfolio.

> **On the absence of RL in this edition**
>
> The second edition of this book had two chapters on RL. In this edition, we'll omit those chapters, and we'll discuss transformers and their applications instead. On one hand, RL is a promising field, but at present, training RL models is slow, and their practical applications are limited. Because of this, RL research is mostly concentrated in well-funded commercial companies and academic institutions. On the other hand, transformers have represented the next big step in the field of ML, in the same way as GPU-trained deep networks sparked interest in the field in the 2009-2012 period.

Q-learning

Q-learning is an off-policy temporal-difference RL algorithm. What a mouthful! But fear not; let's not worry about what all this means, and instead just see how the algorithm works. To do this, we'll use the game of chess we introduced in the previous section. As a reminder, the board's configuration (the locations of the pieces) is the current state of the environment. Here, the agents can take actions, a, by moving pieces, thus changing the state into a new one. We'll represent a game of chess as a graph where the different board configurations are the graph's vertices, and the possible moves from each configuration are the edges. To make a move, the agent follows the edge from the current state, s, to a new state, s'. The basic Q-learning algorithm uses a **Q-table** to help the agent decide which moves to make. The Q-table contains one row for each board configuration, while the columns of the table are all possible actions that the agent can take (the moves). A table cell, $q(s, a)$, contains the cumulative expected reward, called a **Q-value**. This is the potential total reward that the agent will receive for the remainder of the game if they take an action, a, from their current state, s. At the beginning, the Q-table is initialized with an arbitrary value. With that knowledge, let's see how Q-learning works:

```
Initialize the Q table with some arbitrary value
for each episode:
    Observe the initial state s
    for each step of the episode:
        Select new action a using Q-table based policy
        Observe reward r and go to the new state s'
        Use Bellman eq to update q(s, a) in the Q-table
    until we reach a terminal state for the episode
```

An episode starts with a random initial state and finishes when we reach the terminal state. In our case, one episode would be one full game of chess.

The question that arises is, how does the agent's policy determine what will be the next action? To do so, the policy has to consider the Q-values of all the possible actions from the current state. The higher the Q-value, the more attractive the action is. However, the policy will sometimes ignore the Q-table (exploitation of the existing knowledge) and choose another random action to find higher potential rewards (exploration). In the beginning, the agent will take random actions because the Q-table doesn't contain much information (a trial-and-error approach). As time progresses and the Q-table is gradually filled, the agent will become more informed in interacting with the environment.

We update $q(s, a)$ after each new action by using the **Bellman equation**. The Bellman equation is beyond the scope of this introduction, but it's enough to know that the updated value, $q(s, a)$, is based on the newly received reward, r, as well as the maximum possible Q-value, $q^*(s', a')$, of the new state, s'.

This example was intended to help you understand the basic workings of Q-learning, but you might have noticed an issue with this. We store the combination of all possible board configurations and moves in the Q-table. This would make the table huge and impossible to fit in today's computer memory. Fortunately, there is a solution for this: we can replace the Q-table with an NN, which will tell the agent what the optimal action is in each state. In recent years, this development has allowed RL algorithms to achieve superhuman performance on tasks such as the games of Go, Dota 2, Doom, and StarCraft.

Components of an ML solution

So far, we've discussed three major classes of ML algorithms. However, to solve an ML problem, we'll need a system in which the ML algorithm is only part of it. The most important aspects of such a system are as follows:

- **Learner**: This algorithm is used with its learning philosophy. The choice of this algorithm is determined by the problem we're trying to solve since different problems can be better suited for certain ML algorithms.

- **Training data**: This is the raw dataset that we are interested in. This can be labeled or unlabeled. It's important to have enough sample data for the learner to understand the structure of the problem.

- **Representation**: This is how we express the data in terms of the chosen features so that we can feed it to the learner. For example, to classify handwritten images of digits, we'll represent the image as a two-dimensional array of values, where each cell will contain the color value of one pixel. A good choice of representation of the data is important for achieving better results.

- **Goal**: This represents the reason to learn from the data for the problem at hand. This is strictly related to the target and helps us define how and what the learner should use and what representation to use. For example, the goal may be to clean our mailboxes of unwanted emails, and this goal defines what the target of our learner is. In this case, it is the detection of spam emails.

- **Target**: This represents what is being learned as well as the final output. The target can be a classification of unlabeled data, a representation of input data according to hidden patterns or characteristics, a simulator for future predictions, or a response to an outside stimulus or strategy (in the case of RL).

It can never be emphasized enough: any ML algorithm can only achieve an approximation of the target and not a perfect numerical description. ML algorithms are not exact mathematical solutions to problems – they are just approximations. In the *Supervised learning section*, we defined learning as a function from the space of features (the input) into a range of classes. Later, we'll see how certain ML algorithms, such as NNs, can approximate any function to any degree, in theory. This is called the universal approximation theorem (`https://en.wikipedia.org/wiki/Universal_approximation_theorem`), but it does not imply that we can get a precise solution to our problem. In addition, solutions to the problem can be better achieved by better understanding the training data.

Typically, a problem that can be solved with classic ML techniques may require a thorough understanding and processing of the training data before deployment. The steps to solve an ML problem are as follows:

1. **Data collection**: This involves gathering as much data as possible. In the case of supervised learning, this also includes correct labeling.

2. **Data processing**: This involves cleaning the data, such as removing redundant or highly correlated features, or even filling in missing data, and understanding the features that define the training data.

3. **Creation of the test case**: Usually, the data can be divided into three sets:

- **Training set**: We use this set to train the ML algorithm. In most cases, we'll train the algorithm by iterating the whole training set more than once. We'll refer to the number of full training set iterations as **epochs**.

- **Validation set**: We use this set to evaluate the accuracy of the algorithm with unknown data during training. We'll train the algorithm for some time on the training set and then we'll use the validation set to check its performance. If we are not satisfied with the result, we can tune the hyperparameters of the algorithm and repeat the process. The validation set can also help us determine when to stop the training. We'll learn more about this later in this section.

- **Test set**: When we finish tuning the algorithm with the training or validation cycle, we'll use the test set only once for a final evaluation. The test set is similar to the validation set in the sense that the algorithm hasn't used it during training. However, when we strive to improve the algorithm on the validation data, we may inadvertently introduce bias, which can skew the results in favor of the validation set and not reflect the actual performance. Because we use the test only once, this will provide a more objective measurement of the algorithm.

> **Note**
>
> One of the reasons for the success of DL algorithms is that they usually require less data processing than classic methods. For a classic algorithm, you would have to apply different data processing and extract different features for each problem. With DL, you can apply the same data processing pipeline for most tasks. With DL, you can be more productive, and you don't need as much domain knowledge for the task at hand compared to the classic ML algorithms.

There are many valid reasons to create testing and validation datasets. As mentioned previously, ML techniques can only produce an approximation of the desired result. Often, we can only include a finite and limited number of variables, and there may be many variables that are outside of our control. If we only used a single dataset, our model may end up memorizing the data and producing an extremely high accuracy value on the data it has memorized. However, this result may not be reproducible on other similar but unknown datasets. One of the key goals of ML algorithms is their ability to generalize. This is why we create both a validation set used for tuning our model selection during training and a final test set only used at the end of the process to confirm the validity of the selected algorithm.

To understand the importance of selecting valid features and to avoid memorizing the data, which is also referred to as **overfitting** in the literature (we'll use this term from now on), let's use a joke taken from an XKND comic as an example (http://xkcd.com/1122):

> *"Up until 1996, no democratic US presidential candidate who was an incumbent and with no combat experience had ever beaten anyone whose first name was worth more in Scrabble."*

It's apparent that such a rule is meaningless, but it underscores the importance of selecting valid features, and how the question, "How much is a name worth in Scrabble?" can bear any relevance while selecting a US president. Also, this example doesn't have any predictive power over unknown data. We'll call this overfitting, which refers to making predictions that fit the data at hand perfectly but don't generalize to larger datasets. Overfitting is the process of trying to make sense of what we'll call noise (information that does not have any real meaning) and trying to fit the model to small perturbations.

To explain this further, let's try to use ML to predict the trajectory of a ball thrown from the ground up into the air (not perpendicularly) until it reaches the ground again. Physics teaches us that the trajectory is shaped like a parabola. We also expect that a good ML algorithm observing thousands of such throws would come up with a parabola as a solution. However, if we were to zoom into the ball and observe the smallest fluctuations in the air due to turbulence, we might notice that the ball does not hold a steady trajectory but may be subject to small perturbations, which in this case is the noise. An ML algorithm that tries to model these small perturbations would fail to see the big picture and produce a result that is not satisfactory. In other words, overfitting is the process that makes the ML algorithm see the trees but forget about the forest:

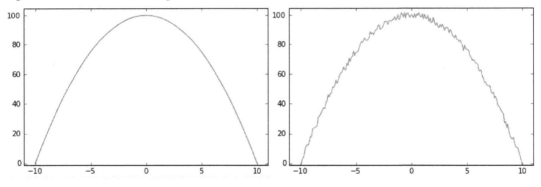

Figure 1.9 – A good prediction model (left) and a bad (overfitted) prediction
model, with the trajectory of a ball thrown from the ground (right)

This is why we separate the training data from the validation and test data; if the algorithm's accuracy over the test data was not similar to the training data accuracy, that would be a good indication that the model overfits. We need to make sure that we don't make the opposite error either – that is, underfitting the model. In practice, though, if we aim to make our prediction model as accurate as possible on our training data, underfitting is much less of a risk, and care is taken to avoid overfitting.

The following figure depicts underfitting:

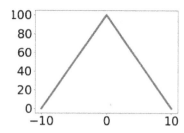

Figure 1.10 – Underfitting can be a problem as well

Neural networks

We introduced some of the popular classical ML algorithms in the previous sections. In this section, we'll talk about NNs, which are the focus of this book.

The first example of an NN is called the perceptron, and this was invented by Frank Rosenblatt in 1957. The perceptron is a classification algorithm that is very similar to logistic regression. Similar to logistic regression, it has a vector of weights, $\mathbf{w}$, and its output is a function, $f(\mathbf{x} \cdot \mathbf{w})$, of the dot product, $\mathbf{x} \cdot \mathbf{w}$ (or $f(\sum_i w_i x_i)$), of the weights and input.

The only difference is that f is a simple step function – that is, if $\mathbf{x} \cdot \mathbf{w} > 0$, then $f(\mathbf{x} \cdot \mathbf{w}) = 1$, or else $f(\mathbf{x} \cdot \mathbf{w}) = 0$, wherein we apply a similar logistic regression rule over the output of the logistic function. The perceptron is an example of a simple one-layer neural feedforward network:

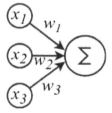

Figure 1.11 – A simple perceptron with three inputs and one output

The perceptron was very promising, but it was soon discovered that it has serious limitations as it only works for linearly separable classes. In 1969, Marvin Minsky and Seymour Paper demonstrated that it could not learn even a simple logical function such as XOR. This led to a significant decline in the interest in perceptrons.

However, other NNs can solve this problem. A classic multilayer perceptron has multiple interconnected perceptrons, such as units that are organized in different sequential layers (input layer, one or more hidden layers, and an output layer). Each unit of a layer is connected to all units of the next layer. First, the information is presented to the input layer, then we use it to compute the output (or activation), y_i, for each unit of the first hidden layer. We propagate forward, with the output as input for the next layers in the network (hence feedforward), and so on until we reach the output. The most common way to train NNs is by using gradient descent in combination with backpropagation. We'll discuss this in detail in *Chapter 2*.

The following diagram depicts an NN with one hidden layer:

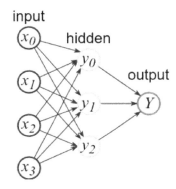

Figure 1.12 – An NN with one hidden layer

Think of the hidden layers as an abstract representation of the input data. This is the way the NN understands the features of the data with its internal logic. However, NNs are non-interpretable models. This means that if we observe the y_i activations of the hidden layer, we wouldn't be able to understand them. For us, they are just a vector of numerical values. We need the output layer to bridge the gap between the network's representation and the actual data we're interested in. You can think of this as a translator; we use it to understand the network's logic, and at the same time, we can convert it into the actual target values that we are interested in.

The universal approximation theorem tells us that a feedforward network with one hidden layer can represent any function. It's good to know that there are no theoretical limits on networks with one hidden layer, but in practice, we can achieve limited success with such architectures. In *Chapter 3*, we'll discuss how to achieve better performance with deep NNs, and their advantages over the shallow ones. For now, let's apply our knowledge by solving a simple classification task with an NN.

Introducing PyTorch

In this section, we'll introduce PyTorch – an open source Python DL framework that was developed primarily by Facebook that has been gaining momentum recently. It provides **graphics processing unit** (**GPU**) accelerated multidimensional array (or tensor) operations, and computational graphs, which we can use to build NNs. Throughout this book, we'll use PyTorch and Keras, and we'll talk about these libraries and compare them in detail in *Chapter 3*.

Let's create a simple NN that will classify the Iris flower dataset. The steps are as follows:

1. Start by loading the dataset:

```
import pandas as pd
dataset = pd.read_csv('https://archive.ics.uci.edu/ml/machine-
learning-databases/iris/iris.data', names=['sepal_length',
'sepal_width', 'petal_length', 'petal_width', 'species'])

dataset['species'] = pd.Categorical(dataset['species']).codes
dataset = dataset.sample(frac=1, random_state=1234)

train_input = dataset.values[:120, :4]
train_target = dataset.values[:120, 4]
test_input = dataset.values[120:, :4]
test_target = dataset.values[120:, 4]
```

The preceding code is boilerplate code that downloads the Iris dataset's CSV file and then loads it into a **pandas DataFrame** called `dataset`. Then, we shuffle the DataFrame's rows and split the code into NumPy arrays, `train_input`/`train_target` (flower properties/flower class), for the training data and `test_input`/`test_target` for the test data. We'll use 120 samples for training and 30 for testing. If you are not familiar with pandas, think of this as an advanced version of NumPy.

2. Next, define our first NN. We'll use a feedforward network with one hidden layer with five units, a **ReLU** activation function (this is just another type of activation, defined simply as $f(x) = max(0, x)$), and an output layer with three units. The output layer has three units, and each unit corresponds to one of the three classes of Iris flowers. The following is the PyTorch definition of the network:

```
import torch
torch.manual_seed(1234)

hidden_units = 5
net = torch.nn.Sequential(
    torch.nn.Linear(4, hidden_units),
    torch.nn.ReLU(),
    torch.nn.Linear(hidden_units, 3)
)
```

We'll use **one-hot encoding** for the target data. This means that each class of the flower will be represented as an array (`Iris Setosa = [1, 0, 0]`, `Iris Versicolor = [0, 1, 0]`, and `Iris Virginica = [0, 0, 1]`), and one element of the array will be the target for one unit of the output layer. When the network classifies a new sample, we'll determine the class by taking the unit with the highest activation value. `torch.manual_seed(1234)` enables us to use the same random data seed every time for the reproducibility of results.

3. Choose the `loss` function:

    ```
    criterion = torch.nn.CrossEntropyLoss()
    ```

    With the `criterion` variable, we define the loss function as **cross-entropy loss**. The loss function will measure how different the output of the network is compared to the target data.

4. Define the **stochastic gradient descent** (**SGD**) optimizer (a variation of the gradient descent algorithm) with a **learning rate** of 0.1 and a **momentum** of 0.9 (we'll discuss SGD and its parameters in *Chapter 2*):

    ```
    optimizer = torch.optim.SGD(net.parameters(), lr=0.1,
            momentum=0.9)
    ```

5. Train the network:

    ```
    epochs = 50

    for epoch in range(epochs):
        inputs = torch.autograd.Variable(
            torch.Tensor(train_input).float())
        targets = torch.autograd.Variable(
            torch.Tensor(train_target).long())

        optimizer.zero_grad()
        out = net(inputs)
        loss = criterion(out, targets)
        loss.backward()
        optimizer.step()

        if epoch == 0 or (epoch + 1) % 10 == 0:
            print('Epoch %d Loss: %.4f' % (epoch + 1,
            loss.item()))
    ```

    We'll run the training for 50 epochs, which means that we'll iterate 50 times over the training dataset:

    I.   Create the `torch` variables from the NumPy array – that is, `train_input` and `train_target`.

II.   Zero the gradients of the optimizer to prevent accumulation from the previous iterations. We feed the training data to the NN, net(inputs), and we compute the loss function's criterion(out, targets) between the network output and the target data.

III.   Propagate the loss value back through the network. We're doing this so that we can calculate how each network weight affects the loss function.

IV.   The optimizer updates the weights of the network in a way that will reduce the future loss function's values.

When we run the training, the output will be as follows:

```
Epoch 1 Loss: 1.2181
Epoch 10 Loss: 0.6745
Epoch 20 Loss: 0.2447
Epoch 30 Loss: 0.1397
Epoch 40 Loss: 0.1001
Epoch 50 Loss: 0.0855
```

The following graph shows how the loss function decreases with each epoch. This shows how the network gradually learns the training data:

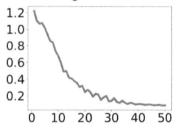

Figure 1.13 – The loss function decreases with each epoch

6.   Let's see what the final accuracy of our model is:

```
import numpy as np
inputs = torch.autograd.Variable(torch.Tensor(test_input).
float())
targets = torch.autograd.Variable(torch.Tensor(test_target).
long())

optimizer.zero_grad()
out = net(inputs)
_, predicted = torch.max(out.data, 1)

error_count = test_target.size - np.count_nonzero((targets ==
predicted).numpy())
print('Errors: %d; Accuracy: %d%%' % (error_count, 100 * torch.
sum(targets == predicted) / test_target.size))
```

We do this by feeding the test set to the network and computing the error manually. The output is as follows:

```
Errors: 0; Accuracy: 100%
```

We were able to classify all 30 test samples correctly.

We must also try different hyperparameters of the network and see how the accuracy and loss functions work. You could try changing the number of units in the hidden layer, the number of epochs we train in the network, as well as the learning rate.

Summary

This chapter covered what ML is and why it's so important. We talked about the main classes of ML techniques and some of the most popular classic ML algorithms. We also introduced a particular type of ML algorithm, called NNs, which is the basis for DL. Then, we looked at a coding example where we used a popular ML library to solve a particular classification problem.

In the next chapter, we'll cover NNs in more detail and explore their theoretical justifications.

2
Neural Networks

In *Chapter 1*, we introduced a number of basic **machine learning** (**ML**) concepts and techniques. We went through the main ML paradigms, as well as some popular classic ML algorithms, and we finished on **neural networks** (**NN**). In this chapter, we will formally introduce what NNs are, discuss their mathematical foundations, describe in detail how their building blocks work, see how we can stack many layers to create a deep feedforward NN, and then learn how to train them.

In this chapter, we will cover the following main topics:

- The need for NNs
- The math of NNs
- An introduction to NNs
- Training NNs

> **The link between NNs and the human brain**
>
> Initially, NNs were inspired by the biological brain (hence the name). Over time, however, we've stopped trying to emulate how the brain works, and instead, we're focused on finding the correct configurations for specific tasks, including computer vision, natural language processing, and speech recognition. You can think of it this way – for a long time, we were inspired by the flight of birds, but, in the end, we created airplanes, which are quite different. We are still far from matching the potential of the brain. Perhaps the machine learning algorithms of the future will resemble the brain more, but that's not the case now. Hence, for the rest of this book, we won't try to create analogies between the brain and NNs. To follow this train of thought, we'll call the smallest building NN building blocks **units**, instead of neurons, as they were originally known.

Technical requirements

We'll implement the example in this chapter using Python. If you don't have an environment set up with these tools, fret not – the example is available as a Jupyter notebook on Google Colab. You can find the code examples in the book's GitHub repository: `https://github.com/PacktPublishing/Python-Deep-Learning-Third-Edition/tree/main/Chapter02`.

The need for NNs

NNs have been around for many years, and they've gone through several periods, during which they've fallen in and out of favor. However, recently, they have steadily gained ground over many other competing machine learning algorithms. This resurgence is due to computers getting faster, the use of **graphical processing units** (**GPUs**) versus the most traditional use of **central processing units** (**CPUs**), better algorithms and NN design, and increasingly larger datasets, which we'll look at in this book. To get an idea of their success, let's look at the ImageNet Large Scale Visual Recognition Challenge (`http://image-net.org/challenges/LSVRC/`, or just **ImageNet**). The participants train their algorithms using the ImageNet database. It contains more than 1 million high-resolution color images in over 1,000 categories (one category may be images of cars, another of people, trees, and so on). One of the tasks in the challenge is to classify unknown images into these categories. In 2011, the winner achieved a top-five accuracy of 74.2%. In 2012, Alex Krizhevsky and his team entered the competition with a convolutional network (a special type of deep network). That year, they won with a top-five accuracy of 84.7%. Since then, the winning algorithms have always been NNs, and the current top-five accuracy is around 99%. However, deep learning algorithms have excelled in other areas – for example, both Google Now and Apple's Siri assistants rely on deep networks for speech recognition and Google's use of deep learning for their translation engines. Recent image and text generation systems such as Stability AI's Stable Diffusion and OpenAI's DALL-E and ChatGPT are implemented with NNs.

We'll talk about these exciting advances in the following chapters, but for now, we'll focus on the mathematical foundations of NNs. To help us with this task, we'll use simple networks with one or two layers. You can think of these as toy examples that are not deep networks, but understanding how they work is important. Here's why:

- Knowing the theory of NNs will help you understand the rest of the book because a large majority of NNs in use today share common principles. Understanding simple networks means that you'll understand deep networks too.

- Having some fundamental knowledge is always good. It will help you a lot when you face some new material (even material not included in this book).

I hope these arguments will convince you of the importance of this chapter. As a small consolation, we'll talk about deep learning in depth (pun intended) in *Chapter 3*.

The math of NNs

In the following few sections, we'll discuss the mathematical principles of NNs. This way, we'll be able to explain NNs through these very principles in a fundamental and structured way.

Linear algebra

Linear algebra deals with objects such as vectors and matrices, linear transformations, and linear equations such as $a_1 x_1 + a_2 x_2 + \ldots + a_n x_n + b = 0$.

Linear algebra identifies the following mathematical objects:

- **Scalar**: A single number.

- **Vector**: A one-dimensional array of numbers (also known as components or **scalars**), where each element has an index. We can denote vectors either with a superscript arrow ($\vec{x}$) or in bold (**x**), but we'll mostly use the bold notation throughout the book. The following is an example of a vector:

$$\mathbf{x} = \vec{x} = [x_1, x_2 \ldots x_n]$$

We can represent a n-dimensional vector as the coordinates of a point in an n-dimensional Euclidean space, $\mathbb{R}^n$. Think of Euclidean space as a coordinate system – the vector starts at the center of that system, and each of the vector's elements represents the coordinate of the point along its corresponding coordinate axis. The following figure shows a vector in a three-dimensional coordinate system, $\mathbb{R}^3$:

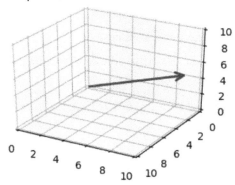

Figure 2.1 – Vector representation in a three-dimensional space

The figure can also help us define two additional properties of the vector:

- **Magnitude** (or **length**):

$$|\mathbf{x}| = \sqrt{x_1^2 + x_2^2 + \ldots + x_n^2}$$

Think of the magnitude as a generalization of the Pythagorean theorem for an n-dimensional space.

- **Direction**: The angle between the vector and each axis of the vector space.

- **Matrix**: A two-dimensional array of scalars, where each element is identified by a row and a column. We'll denote a matrix with a bold capital letter – for example, **A**. Conversely, we'll denote the matrix elements with the small matrix letter, with the row and column as subscript indices – for example, a_{ij}. We can see an example of the matrix notation in the following formula:

$$\mathbf{A} = \begin{bmatrix} a_{11} & a_{12} & \cdots & a_{1n} \\ a_{21} & a_{22} & \cdots & a_{2n} \\ \vdots & & \ddots & \vdots \\ a_{m1} & a_{m2} & \cdots & a_{mn} \end{bmatrix}$$

  We can represent a vector as either a single-row $1 \times n$ matrix (**row matrix**) or a single-column $n \times 1$ matrix (**column matrix**). Transformed like this, the vector can participate in different matrix operations.

- **Tensor**: The term *tensor* originates from mathematics and physics, where it existed long before we started using it in ML. Its definition in these fields differs from the one in ML. Fortunately, the tensor in the context of ML is just a multi-dimensional array with the following properties:

  - **Rank**: The number of array dimensions. Vectors and matrices are special cases of tensors. A tensor of rank 0 is a scalar, a tensor of rank 1 is a vector, and a tensor of rank 2 is a matrix. There is no limit on the number of dimensions, and some types of NNs can use tensors of rank 4 or more.

  - **Shape**: The size of each of the tensor's dimensions.

  - **Data type** of the tensor values. In practice, the data types include 16-, 32-, and 64-bit floats and 8-, 16-, 32-, and 64-bit integers.

The tensor is the main data structure of libraries such as PyTorch, Keras, and TensorFlow.

The nature of tensors

You can find a detailed discussion on the nature of tensors here: `https://stats.stackexchange.com/questions/198061/why-the-sudden-fascination-with-tensors`. You can also compare this with the PyTorch (`https://pytorch.org/docs/stable/tensors.html`) and TensorFlow (`https://www.tensorflow.org/guide/tensor`) and tensor definitions.

Now that we've introduced vectors, matrices, and tensors, let's continue with some of the linear operations they can participate in.

Vector and matrix operations

We'll focus on the operations that relate to NNs, starting with vectors:

- **Vector addition**: Adds two or more n-dimensional vectors, **a** and **b** (and so on) to a new vector:

$$\mathbf{a} + \mathbf{b} = \left[a_1 + b_1, a_2 + b_2 ... a_n + b_n\right]$$

- **Dot (or scalar) product**: Combines two n-dimensional vectors, **a** and **b**, into a **scalar value**:

$$\mathbf{a} \cdot \mathbf{b} = |\mathbf{a}||\mathbf{b}|cos\theta$$

Here, the angle between the two vectors is θ, and $|\mathbf{a}|$ and $|\mathbf{b}|$ are their magnitudes. For example, if the vectors are *two*-dimensional and their components are $a_1, b_1, a_2, b_2 ... a_n, b_n$, the preceding formula becomes the following:

$$\mathbf{a} \cdot \mathbf{b} = a_1 b_1 + a_2 b_2 + ... + a_n b_n$$

The following diagram illustrates the dot product of **a** and **b**:

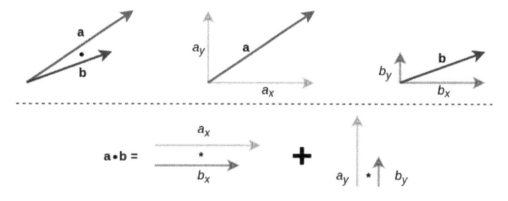

Figure 2.2 – The dot product of vectors – top: vector components,
and bottom: the dot product of the two vectors

We can think of the dot product as a similarity measure between the two vectors, where the angle θ indicates how similar they are. If θ is small (the vectors have similar directions), then their dot product will be higher because $cos\theta$ will converge toward 1. In this context, we can define a **cosine similarity** between two vectors as follows:

$$cos\theta = \frac{\mathbf{a} \cdot \mathbf{b}}{|\mathbf{a}||\mathbf{b}|}$$

- **Cross (or vector) product**: A combination of two vectors, **a** and **b**, in a new vector, which is perpendicular to both initial vectors. The magnitude of the output vector is equal to the following:

$$\mathbf{a} \times \mathbf{b} = |\mathbf{a}||\mathbf{b}|sin\theta$$

We can see an example of a cross product of two-dimensional vectors in the following diagram:

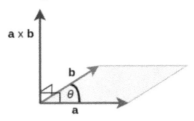

Figure 2.3 – A cross product of two two-dimensional vectors

The output vector is perpendicular (or **normal**) to the plane, which contains the input vectors. The output vector magnitude is equal to the area of a parallelogram with sides, the **a** and **b** vectors (denoted with light blue in the preceding diagram).

Now, let's focus on the matrix operations:

- **Matrix transpose**: Flip the matrix along its main diagonal, represented by the set of all matrix elements, a_{ij}, where $i=j$. We'll denote the transpose operation with T in superscript form. The cell a_{ji} of **A** is equal to the cell a_{ij} of $\mathbf{A}^\mathsf{T}$:

$$[\mathbf{A}^\mathsf{T}]_{ij} = \mathbf{A}_{ji}$$

The transpose of an $m{\times}n$ matrix is an $n{\times}m$ matrix, as we can see with the following examples:

$$\mathbf{A} = \begin{bmatrix} a_{11} & a_{12} & a_{13} \\ a_{21} & a_{22} & a_{23} \\ a_{31} & a_{32} & a_{33} \end{bmatrix} \Rightarrow \mathbf{A}^\mathsf{T} = \begin{bmatrix} a_{11} & a_{21} & a_{31} \\ a_{12} & a_{22} & a_{32} \\ a_{13} & a_{23} & a_{33} \end{bmatrix}$$

$$\mathbf{A} = \begin{bmatrix} a_{11} & a_{12} & a_{13} \\ a_{21} & a_{22} & a_{23} \end{bmatrix} \Rightarrow \mathbf{A}^\mathsf{T} = \begin{bmatrix} a_{11} & a_{21} \\ a_{12} & a_{22} \\ a_{13} & a_{23} \end{bmatrix}$$

$$\mathbf{A} = \begin{bmatrix} a_{11} & a_{12} & a_{13} \end{bmatrix} \Rightarrow \mathbf{A}^\mathsf{T} = \begin{bmatrix} a_{11} \\ a_{12} \\ a_{13} \end{bmatrix}$$

- **Matrix-scalar multiplication**: Multiplication of a matrix, **A**, by a scalar, y, into a new matrix with the same size as the original:

$$\mathbf{A}y = \begin{bmatrix} a_{11} & a_{12} \\ a_{21} & a_{22} \end{bmatrix} y = \begin{bmatrix} a_{11} \times y & a_{12} \times y \\ a_{21} \times y & a_{22} \times y \end{bmatrix}$$

- **Matrix-matrix addition**: Element-wise addition of two or more matrices, **A** and **B** (and so on), into a new matrix. All input matrices must be the same size:

$$\mathbf{A} + \mathbf{B} = \begin{bmatrix} a_{11} & a_{12} \\ a_{21} & a_{22} \end{bmatrix} + \begin{bmatrix} b_{11} & b_{12} \\ b_{21} & b_{22} \end{bmatrix} = \begin{bmatrix} a_{11} + b_{11} & a_{12} + b_{12} \\ a_{21} + b_{21} & a_{22} + b_{22} \end{bmatrix}$$

- **Matrix-vector multiplication**: Multiplication of a matrix, **A**, by a vector, **x**, into a new vector:

$$\mathbf{Ax} = \begin{bmatrix} a_{11} & a_{12} \\ a_{21} & a_{22} \\ a_{31} & a_{32} \end{bmatrix} \begin{bmatrix} x_1 \\ x_2 \end{bmatrix} = \begin{bmatrix} a_{11}x_1 + a_{12}x_2 \\ a_{21}x_1 + a_{22}x_2 \\ a_{31}x_1 + a_{32}x_2 \end{bmatrix}$$

$$\begin{bmatrix} a_{11} & a_{12} \end{bmatrix} \begin{bmatrix} x_1 \\ x_2 \end{bmatrix} = \begin{bmatrix} a_{11}x_1 + a_{12}x_2 \end{bmatrix}$$

$$\mathbf{Ax} = \begin{bmatrix} 1 & 2 \\ 3 & 4 \end{bmatrix} \begin{bmatrix} 5 \\ 6 \end{bmatrix} = \begin{bmatrix} 1 \times 5 + 2 \times 6 \\ 3 \times 5 + 4 \times 6 \end{bmatrix} = \begin{bmatrix} 17 \\ 39 \end{bmatrix}$$

The number of matrix columns must be equal to the vector size. The result of an $m \times n$ matrix, multiplied by an n-dimensional vector, is an m-dimensional vector. We can assume that the rows of the matrix are n-dimensional vectors. Then, each value of the output vector is the dot product between the corresponding matrix row vector and **x**.

- **Matrix multiplication**: A binary operation, which represents the multiplication of two matrices, **A** and **B**, into a single output matrix. We can think of it as multiple matrix-vector multiplications, where each column of the second matrix is a vector:

$$\mathbf{AB} = \begin{bmatrix} a_{11} & a_{12} & a_{13} \\ a_{21} & a_{22} & a_{23} \end{bmatrix} \begin{bmatrix} b_{11} & b_{12} \\ b_{21} & b_{22} \\ b_{31} & b_{32} \end{bmatrix} = \begin{bmatrix} a_{11}b_{11} + a_{12}b_{21} + a_{13}b_{31} & a_{11}b_{12} + a_{12}b_{22} + a_{13}b_{32} \\ a_{21}b_{11} + a_{22}b_{21} + a_{23}b_{31} & a_{21}b_{12} + a_{22}b_{22} + a_{23}b_{32} \end{bmatrix}$$

$$\mathbf{AB} = \begin{bmatrix} 1 & 2 & 3 \\ 4 & 5 & 6 \end{bmatrix} \begin{bmatrix} 1 & 2 \\ 3 & 4 \\ 5 & 6 \end{bmatrix} = \begin{bmatrix} 1 + 6 + 15 & 2 + 8 + 18 \\ 4 + 15 + 30 & 8 + 20 + 36 \end{bmatrix} = \begin{bmatrix} 22 & 28 \\ 49 & 64 \end{bmatrix}$$

If we represent the two vectors as matrices, their dot product, $\mathbf{a} \cdot \mathbf{b} = \mathbf{a}\mathbf{b}^\top$, is equivalent to matrix-matrix multiplication.

You can now breathe a sigh of relief because we've concluded our introduction to linear algebra. Not all is well though, as we'll focus on probability theory next.

An introduction to probability

In this section, we'll introduce some basic concepts of probability theory. They will help us later in the book when we discuss NN training algorithms and natural language processing.

We'll start with the concept of a **statistical experiment**, which has the following properties:

- It is composed of multiple independent trials
- The outcome of each trial is determined by chance (it is **non-deterministic**)
- It has multiple possible outcomes, known as **events**
- We know in advance all possible experiment outcomes

Examples of statistical experiments include a coin toss with two possible outcomes (heads or tails) and a dice roll with six possible outcomes (1 to 6).

The likelihood that some event, e, would occur is known as **probability** $P(e)$. It is a value in the range of $[0, 1]$. $P(e) = 0.5$ indicates a 50–50 chance that the event will occur, $P(e) = 0$ indicates that the event cannot occur, and $P(e) = 1$ indicates that it will always occur.

We can approach probability in two ways:

- **Theoretical**: All events are equally likely to occur and the probability of the event (outcome) we're interested in is as follows:

$$P(e) = \frac{\text{Number of successful outcomes}}{\text{Total number of outcomes}}$$

  The theoretical probability of the two possible outcomes in the coin toss example is $P(\text{heads}) = P(\text{tails}) = 1/2$. In the dice roll example, we have $P(\text{each side of the dice}) = 1/6$.

- **Empirical**: This is the number of times an event, e, occurs in relation to all trials:

$$P(e) = \frac{\text{Number of times } e \text{ occurs}}{\text{Total number of trials}}$$

  The empirical result of the experiment may show that the events aren't equally likely. For example, if we toss a coin 100 times and observe heads 47 times, the empirical probability for heads is $P(\text{heads}) = 47 / 100 = 0.47$. The law of large numbers tells us that we'll calculate the probability more accurately with a higher number of trials.

Now, we'll discuss probability in the context of sets.

Probability and sets

In this section, we'll introduce sets and their properties. We'll also see how to apply these properties in probability theory. Let's start with some definitions:

- **Sample space**: The **set** (get it?) of all possible events (outcomes) of an experiment. We'll denote it with a capital letter. Like Python, we'll list all events in the sample space with {}. For example, the sample spaces of coin toss and dice roll events are $S_c = \{\text{heads, tails}\}$ and $S_d = \{1, 2, 3, 4, 5, 6\}$ respectively.

- **Sample point**: A single event (for example, tails) of the sample space.

- **Event**: A single sample point or a combination (**subset**) of sample points of the sample space. For example, a combined event is for the dice to land on an odd number, $S_o = \{1,3,5\}$.

Let's assume that we have a set (sample space), S = {1, 2, 3, 4, 5}, and two subsets (combined events), A = {1, 2, 3} and B = {3, 4, 5}. We'll use them to define the following set operations:

- **Intersection**: A set of elements that exist in both A and B:

$$A \cap B = \{3\}$$

If the intersection of A and B is an empty set {}, they are **disjoint**.

- **Complement**: A set of all elements that aren't included in A or B:

$$A' = \{4,5\} \qquad B' = \{1,2\}$$

- **Union**: A set of all elements that exist in either A or B:

$$A \cup B = \{1,2,3,4,5\}$$

The following Venn diagrams illustrate these operations:

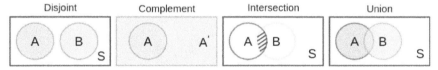

Figure 2.4 – Venn diagrams of the possible set relationships

Now, let's see how to transfer the set operations in the probability domain. We'll work with independent events – that is, the occurrence of one event doesn't affect the probability of the occurrence of another. For example, the outcomes of the different coin tosses are independent of one another. With that, let's define the set operations in terms of probability and events:

- **The intersection of two events**: A set of sample points that exist in both events. The probability of the intersection is called **joint probability**:

$$P(A \cap B)=P(A)\times P(B)$$

Let's say that we want to compute the probability of a card being simultaneously spades and an ace (more poetically, the ace of spades). The probability for spades is P(spades) = 13/52 = ¼, and the probability of an ace is P(ace) = 4/52 = 1/13. The joint probability of the two is P(ace, spades) = (1/13) × (1/4) = 1/52. We can intuitively validate this result because the ace of spades is a unique card, and its probability would be 1/52. Since we draw a single card, the two events occur at the same time and are independent. Had they occurred successively – for example, two card draws, where one is a black and the other is an ace – we would fall in the domain of conditional probability.

The probability of the occurrence of a single event, P(A), is also known as **marginal probability** (as opposed to joint probability).

- **Disjoint (or mutually exclusive) events**: Two or more events that don't share any outcomes. In other words, their respective sample space subsets are disjoint. For example, the events of odd or even dice rolls are disjoint. The following is true for disjoint events:

  - The joint probability for these events to occur together is $P(A \cap B) = 0$

  - The sum of the probabilities of disjoint events is $\sum P(\text{disjoint events}) \leq 1$

- **Jointly exhaustive events**: The subsets of such events contain the whole sample space between themselves. For example, events A = {1, 2, 3} and B = {4, 5, 6} are jointly exhaustive because, together, they cover the whole sample space S = {1, 2, 3, 4, 5, 6}. The following is true for the probability of jointly exhaustive events:

$$\sum P(\text{jointly exhaustive events}) = 1$$

- **Complement events**: Two or more events that are disjoint and jointly exhaustive at the same time. For example, odd and even dice roll events are complementary.

- **Union of events**: A set of events coming from either A or B (not necessarily in both). The probability of this union is:

$$P(A \cup B) = P(A) + P(B) - P(A \cap B)$$

So far, we have discussed independent events. Now, let's see what happens if the events are not independent.

Conditional probability and the Bayes rule

If event A occurs before B and the occurrence of A changes the probability of the occurrence of B, then the two events are dependent. To understand this, let's imagine that we draw consecutive cards from the deck. When the deck is full, the probability of drawing a spade is P(spade) = 13/52 = 0.25. However, once we've drawn the first spade, the probability of picking a spade on the second turn changes. Now, we only have 51 cards and 1 less spade. The probability of the second draw is called **conditional probability**, P(B|A). This is the probability of event B (the second draw of a spade) to occur, given that event A (the first draw of a spade) has occurred. The probability of picking a spade on the second draw is P(spade2|spade1) = 12/51 = 0.235.

We can extend the joint probability formula (introduced in the preceding section) for dependent events:

$$P(A \cap B) = P(A) \times P(B|A)$$

However, this formula is just a special case for two events. We can extend it even further for multiple events, $A_1, A_2 \ldots A_n$. This new generic formula is known as the **chain rule of probability**:

$$P(A_n \cap \ldots \cap A_1) = P(A_n | A_{n-1} \cap \ldots \cap A_1) \times P(A_{n-1} \cap \ldots \cap A_1)$$

For example, the chain rule for three events is as follows:

$$P\left(A_3 \cap A_2 \cap A_1\right) = P\left(A_3 | A_2 \cap A_1\right) \times P\left(A_2 \cap A_1\right) = P\left(A_3 | A_2 \cap A_1\right) \times P\left(A_2 | A_1\right) \times P\left(A_1\right)$$

We can use this property to derive the formula for the conditional probability itself:

$$P(B|A) = \frac{P(A \cap B)}{P(A)}$$

Let's discuss the intuition behind this:

- **P(A ∩ B)** indicates that we're only interested in the occurrences of B, if A has already occurred – that is, we're interested in the joint occurrence of the events, hence the joint probability.

- **P(A)** indicates that we're interested only in the subset of outcomes when event A has occurred. We already know that A has occurred, and therefore, we restrict our observations to these outcomes.

The following is true for dependent events:

$$P(A \cap B) = P(A) \times P(B|A)$$

$$P(A \cap B) = P(B) \times P(A|B)$$

We can use this rule to replace the value of P(A∩B) in the conditional probability formula to derive what is known as the **Bayes rule**:

$$P(A \cap B) = P(A) \times P(B|A) = P(B) \times P(A|B) \quad \Leftrightarrow \quad P(B|A) = \frac{P(A \cap B)}{P(A)} = \frac{P(B) \times P(A|B)}{P(A)}$$

The Bayes rule makes it possible to compute the conditional probability, P(B|A), if we know the opposite conditional probability, P(A|B). P(A) and P(B|A) are known as **prior probability and posterior probability**, respectively.

We can illustrate the Bayes rule with a classic example from the realm of medical testing. A patient is administered a medical test for a disease, which comes out positive. Most tests have a sensitivity value, which is the percentage chance of the test being positive when administered to people with a particular disease. Using this information, we'll apply the Bayes rule to compute the actual probability of the patient having the disease, given that the test is positive. We get the following:

$$P(\text{has disease} | \text{test} = positive) = \frac{P(\text{has disease}) \times P(\text{test} = positive | \text{has disease})}{P(\text{test} = positive)}$$

We can think of P(has disease) as the probability of the disease in the general population.

Now, we'll make some assumptions about the disease and the sensitivity of the test:

- The test is 98% sensitive – that is, it will detect only 98% of all positive cases: P(test=positive|has disease) = 0.98

- Two per cent of the people under 50 have this kind of disease: P(has disease) = 0.02

- The test is positive for 3.9% of the population when administered to people under 50: P(test=positive) = 0.039

We can ask the following question: if a test is 98% sensitive and a 45-year-old person is administered the test, which turns out to be positive, what is the probability that they have the disease? We can calculate it with the Bayes rule:

$$P\big(\text{has disease}\,|\,\text{test} = positive\big) = \frac{P(\text{has disease}) \times P(\text{test} = positive\,|\,\text{has disease})}{P(\text{test} = positive)} = \frac{0.02 \times 0.98}{0.039} = 0.5$$

This example can serve as an introduction to the next section, where we'll introduce the confusion matrix.

Confusion matrix

The **confusion matrix** is used to evaluate the performance of a binary classification algorithm, similar to the medical test we introduced in the *Conditional probability and the Bayes rule* section:

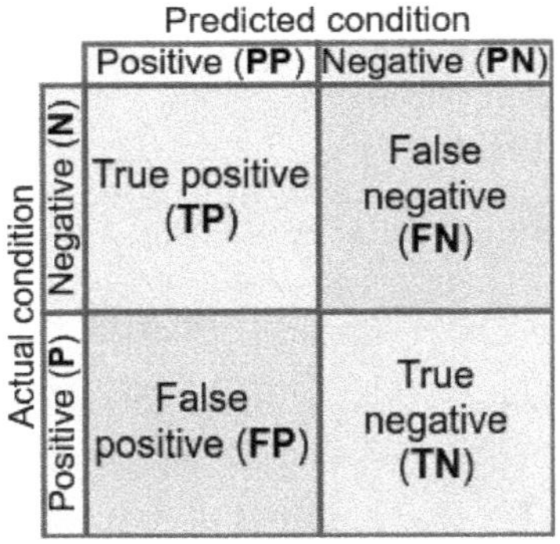

	Predicted condition	
	Positive (PP)	**Negative (PN)**
Actual condition — Negative (**N**)	True positive (**TP**)	False negative (**FN**)
Actual condition — Positive (**P**)	False positive (**FP**)	True negative (**TN**)

Figure 2.5 – Confusion matrix

The relationship between the actual condition (**P** and **N**) and the predicted outcome (**PP** and **PN**) allows us to place the prediction in one of four categories:

- **True positive (TP)**: The actual and predicted values are both true
- **True negative (TN)**: The actual and predicted values are both false
- **False positive (FP)**: The actual value is negative, but the classification algorithm has predicted a positive value
- **False negative (FN)**: The actual value is positive, but the algorithm has predicted a negative value

Based on these categories, we'll introduce some measures that evaluate different aspects of the performance of the algorithm:

- **Accuracy** = $\frac{TP+TN}{TP+FP+FN+TN}$: The fraction of correct predictions among all cases.
- **Precision** = $\frac{TP}{TP+FP}$: The fraction of positive predictions that were actually correct.
- **Recall** (or **sensitivity**) = $\frac{TP}{TP+FN}$: The fraction of actual positive cases that were predicted correctly.

- **Specificity** = $\frac{TN}{TN+FP}$: The fraction of actual negative cases that were predicted correctly.

- **F1 score** = $2 \cdot \frac{precision \times recall}{precision+recall}$: Represents the balance between precision and recall. Because of the multiplication of the two measures, the F1 score will have a high value when both measures have high values.

In the following section, we'll discuss the field of differential calculus, which will help us to train NNs.

Differential calculus

We can think of an ML algorithm as a mathematical function with inputs and parameters (which is the case for NNs). Our goal is to adjust these parameters in a way that will allow the ML function to closely approximate some other target function. To do this, we need to know how the output of the ML function changes when we change some of its parameters (called weights). Fortunately, differential calculus can help us here – it deals with the rate of change of a function with respect to a variable (parameter) that the function depends on. To understand how this works, we'll start with a function, $f(x)$, with a single parameter, x, which has the following graph:

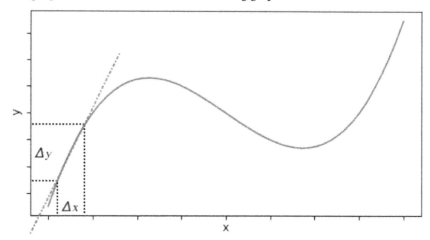

Figure 2.6 – A sample graph of a function, f(x), with a single parameter, x. The function graph is denoted with an uninterrupted blue line; the slope is denoted with an interrupted red line

We can approximate how $f(x)$ changes with respect to x for any value of x by calculating the slope of the function for that value. If the slope is positive, the function increases, and it decreases if the slope is negative. The steepness of the slope indicates the rate of change of the function for that value. We can calculate the slope with the following formula:

$$\text{slope} = \frac{\Delta y}{\Delta x} = \frac{f(x + \Delta x) - f(x)}{\Delta x}$$

The idea here is simple – we calculate the difference between the two values of f at x and $x+\Delta x$ (Δx is a very small value) – $\Delta y = f(x + \Delta x) - f(x)$. The ratio between Δy and Δx gives us the slope. But why is Δx required to be small? If Δx is too big, the part of the function graph between x and $x+\Delta x$ may change significantly, and the slope measurement would be inaccurate. When Δx converges to 0, we'll assume that our slope approximates the actual slope at a single point of the graph. In this case, we call the slope the **first derivative** of $f(x)$. We can express this in mathematical terms via the following equation:

$$f'(x) \; = \; \frac{dy}{dx} \; = \; \lim_{\Delta x \to 0} \frac{f(x + \Delta x) - f(x)}{\Delta x}$$

Here, $\lim_{\Delta x \to 0}$ is the mathematical concept of the limit (Δx approaches 0), and $f'(x)$ and dy/dx are Lagrange's and Leibniz's notations for derivatives, respectively. The process of finding the derivative of f is called **differentiation**. The following diagram shows slopes at different values of x:

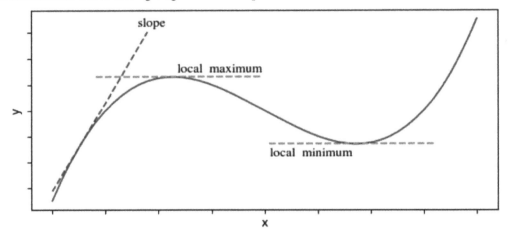

Figure 2.7 – Slopes at different values of x

The points, where f neither increases nor decreases as we change x, are called **saddle points**. The values of f at the saddle points are called the **local minimum** and the **local maximum**. Conversely, the slopes of f at the saddle points are 0.

So far, we have discussed a function with a single parameter, x. Now, let's focus on a function with multiple parameters, $f(x_1, x_2 \ldots x_n)$. The derivative of f with respect to any of the parameters, x_i, is called a **partial derivative** and is denoted by $\partial f/\partial x$. To compute the partial derivative, we will assume that all the other parameters, $x_j \neq x_i$, are constants. We'll denote the partial derivatives of the components of a vector with $\nabla = \left(\frac{\partial}{\partial x_1}, \ldots, \frac{\partial}{\partial x_n} \right)$.

Finally, let's discuss some useful differentiation rules:

- **Chain rule**: f and g are functions and $h(x) = f(g(x))$. The derivative of f with respect to x for any x is as follows:

$$h'(x) = f'(g(x))g'(x) \text{ or } \frac{dh}{dx} = \frac{d}{dx}[f(g(x))] = \frac{d}{dg(x)}[f(g(x))]\frac{d}{dx}[g(x)]$$

- **Sum rule**: f and g are some functions and $h(x) = f(x) + g(x)$. The sum rule states the following:

$$h(x) = f(x) + g(x) \implies h'(x) = (f(x) + g(x))' = f'(x) + g'(x)$$

- **Common functions**:

  - $x' = 1$

  - $(ax)' = a$, where a is a scalar

  - $a' = 0$, where a is a scalar

  - $x^2 = 2x$

  - $(e^x)' = e^x$

The mathematical foundations of NNs and NNs themselves form a kind of knowledge hierarchy. Think of the topics we discussed in *The math of NNs* section as the building blocks of NNs. They represent an important step toward a comprehensive understanding of NNs, which will help us throughout this book and beyond. Now, we have the necessary preparation to learn full-fledged NNs.

An introduction to NNs

We can describe NNs as a mathematical model for information processing. As discussed in *Chapter 1*, this is a good way to describe any ML algorithm, but in this chapter, it has a specific meaning in the context of NNs. An NN is not a fixed program but rather a model, a system that processes information, or inputs. The characteristics of an NN are as follows:

- Information processing occurs in its simplest form, over simple elements called **units**

- Units are connected, and they exchange signals between them through connection links

- Connection links between units can be stronger or weaker, and this determines how information is processed

- Each unit has an internal state that is determined by all the incoming connections from other units

- Each unit has a different **activation function** that is calculated on its state and determines its output signal

A more general description of an NN would be as a computational graph of mathematical operations, but we will learn more about that later.

We can identify two main characteristics of an NN:

- **Neural net architecture**: This describes the set of connections – namely, feedforward, recurrent, multi- or single-layered, and so on – between the units, the number of layers, and the number of units in each layer.

- **Learning**: This describes what is commonly defined as training. The most common but not exclusive way to train an NN is with **gradient descent** (**GD**) and **backpropagation** (**BP**).

We'll start our discussion from the smallest building block of the NN – the unit.

Units – the smallest NN building block

Units are mathematical functions that can be defined as follows:

$$y = f\left(\sum_{i=1}^{n} x_i w_i + b\right)$$

Here, we do the following:

1. We compute the weighted sum $\sum x_i w_i + b$ (also known as an activation value). Let's focus on the components of this sum:

   - The inputs x_i are numerical values that represent either the outputs of other units of the network, or the values of the input data itself

   - The weights w_i are numerical values that represent either the strength of the inputs or the strength of the connections between the units

   - The weight b is a special weight called **bias**, which represents an always-on input unit with a value of 1

   Alternatively, we can substitute x_i and w_i with their vector representations, where $\mathbf{x} = \vec{x} = [x_1, x_2, …, x_n]$ and $\mathbf{w} = \vec{w} = [w_1, w_2, …, w_n]$. Here, the formula will use the dot product of the two vectors:

$$y = f(\mathbf{x} \cdot \mathbf{w} + b) = f(\vec{x} \cdot \vec{w} + b)$$

2. The sum $\sum x_i w_i + b$ serves as input to the **activation function** f (also known as **transfer function**). The output of f is a single **numerical value**, which represents the output of the unit itself. The activation function has the following properties:

   - **Non-linear**: f is the source of non-linearity in an NN – if the NN was entirely linear, it would only be able to approximate other linear functions

   - **Differentiable**: This makes it possible to train the network with GD and BP

Don't worry if you don't understand everything – we'll discuss activation functions in detail later in the chapter.

The following diagram (left) shows a unit:

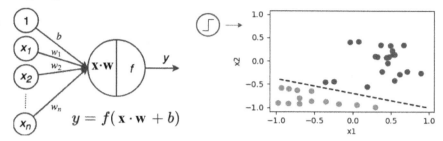

Figure 2.8 – Left: a unit and its equivalent formula, and right: a geometric representation of a perceptron

The input vector $\mathbf{x}$ will be perpendicular to the weight vector $\mathbf{w}$ if $\mathbf{x} \cdot \mathbf{w} = 0$. Therefore, all $\mathbf{x}$ vectors, where $\mathbf{x} \cdot \mathbf{w} = 0$, define a hyperplane in vector space $\mathbb{R}^n$, where n is the dimension of $\mathbf{x}$. In the case of two-dimensional input (x_1, x_2), we can represent the hyperplane as a line. This could be illustrated with the perceptron (or **binary classifier**) – a unit with a **threshold activation function**, $f(a) = 1$ if $a \geq 0$ else 0, that classifies its input in one of the two classes. The geometric representation of the perceptron with two inputs (x_1, x_2) is a line (or **decision boundary**) separating the two classes (to the right in the preceding diagram).

In the preceding diagram, we can also see that the role of the bias, b, is to allow the hyperplane to shift away from the center of the coordinate system. If we don't use bias, the unit will have limited representation power.

The unit is a generalization of several algorithms we introduced in *Chapter 1*:

- A unit with an identity activation function $f(x) = x$ is equivalent to **multiple linear regression**
- A unit with a **sigmoid activation** function is equivalent to logistic regression
- A unit with a threshold activation function is equivalent to a perceptron

We already know from *Chapter 1* that the perceptron (hence the unit) only works with linearly separable classes, and now we know that is because it defines a hyperplane. This imposes a serious limitation on the unit because it cannot classify linearly inseparable problems – even simple ones such as **exclusive or (XOR)**. To overcome this limitation, we'll need to organize the units in an NN. However, before we discuss full-fledged NNs, we'll focus on the next NN building block – the layers.

Layers as operations

An NN can have an indefinite number of units, which are organized in interconnected layers. A layer has the following properties:

- It combines the scalar outputs of multiple units in a single output vector. A unit can convey limited information because its output is a scalar. By combining the unit outputs, instead of a single activation, we can now consider the vector in its entirety. This way, we can convey a lot more information, not only because the vector has multiple values but also because the relative ratios between them carry additional meaning.

- The units of one layer can be connected to the units of other layers, but not to other units of the same layer. Because of this, we can parallelize the computation of the outputs of all units in a single layer (thereby increasing the computational speed). This ability is one of the major reasons for the success of DL in recent years.

- We can generalize multivariate regression to a layer, as opposed to only linear or logistic regression to a single unit. In other words, we can approximate multiple values with a layer as opposed to a single value with a unit. This happens in the case of classification output, where each output unit represents the probability the input belongs to a certain class.

In classical NNs (that is, NNs before DL, when they were just one of many ML algorithms), the primary type of layer is the **fully connected** (**FC**) layer. In this layer, every unit receives weighted input from all the components of the input vector, **x**. This can represent either the output of another layer in the network or a sample of the input dataset. Let's assume that the size of the input vector is m, and that the FC layer has n units and an activation function, f, which is the same for all the units. Each of the n units will have m weights – one for each of the m inputs. The following is a formula we can use for the output of a single unit, j, of an FC layer. It's the same as the formula we defined in the *Units – the smallest NN building block* section, but we'll include the unit index here:

$$y_j = f\left(\sum_{i=1}^{m} x_i w_{ij} + b_j\right)$$

Here, w_{ij} is the weight between the j-th layer unit and the i-th value of the input vector, **x**. We can represent the weights connecting the elements of **x** to the units as an $m \times n$ matrix, **W**. Each column of **W** represents the weight vector of all the inputs to a single unit of the layer. In this case, the output vector of the layer, **y**, is the result of matrix-vector multiplication.

We can also combine multiple input sample vectors, $\mathbf{x}^{(i)}$, in an input matrix, **X**, where each input data vector, $\mathbf{x}^{(i)}$, is represented by a row in **X**. The matrix itself is referred to as a **batch**. Then, we'll simultaneously compute all output vectors, $\mathbf{y}^{(i)}$, corresponding to the input samples, $\mathbf{x}^{(i)}$. In this case, we will have matrix-matrix multiplication, **XW**, and the layer output is also a matrix, **Y**.

The following diagram shows an example of an FC layer, as well as its equivalent formulas in the batch and single sample scenarios:

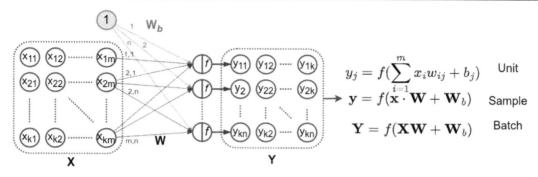

Figure 2.9 – An FC layer with vector/matrix inputs and outputs and its equivalent formulas

We have explicitly separated the bias and input weight matrices, but in practice, the underlying implementation may use a shared weight matrix and append an additional row of 1s to the input data.

So far, we represented the input data samples as vectors, which we can combine in a matrix. However, the input data can have more dimensions. For example, we can represent an RGB image with three dimensions – three two-dimensional channels (one channel for each color). To combine multiple images in a batch, we'll need a fourth dimension. In such cases, we can use input/output **tensors** instead of matrices.

We'll also use different types of layers to process multidimensional data. One such type is the convolutional layer, which we'll discuss in *Chapter 4*. We have many other layer types, such as attention, pooling, and so on. Some of the layers have trainable weights (FC, attention, convolutional), while others don't (pooling). We can also use the terms functions or operations interchangeably with the layer. For example, in TensorFlow and PyTorch, the FC layer we just described is a combination of two sequential operations. First, we perform the weighted sum of the weights and inputs, and then we feed the result as an input to the activation function operation. In practice (that is, when working with DL libraries), the basic building block of an NN is not the unit but an operation that takes one or more tensors as input and outputs one or more tensors:

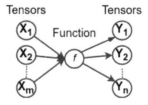

Figure 2.10 – A function (or operation) with input and output tensors

Finally, we have all the necessary information to discuss NNs in their full glory.

Multi-layer NNs

As we have mentioned several times, single-layer neural nets can only classify linearly separable classes. However, there is nothing that prevents us from introducing more layers between the input and the output. These extra layers are called hidden layers. The following diagram demonstrates a three-layer fully connected NN with two hidden layers:

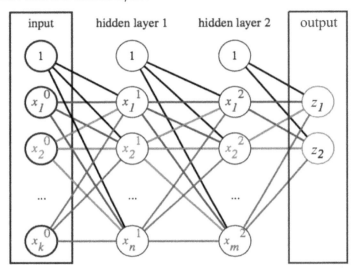

Figure 2.11 – Multi-layer feed-forward network

The input layer has k input units, the first hidden layer has n hidden units, and the second hidden layer has m hidden units. On top is the always-on bias unit. The output, in this example, is the two units, z_1 and z_2, where each unit represents one of two possible classes. The output unit with the highest activation value will determine the NN's class prediction for the given input sample. Each of the hidden units has a non-linear activation function, and the outputs have a special activation function called **softmax**, which we'll discuss in the *Activation functions* section. A unit from one layer is connected to all units from the previous and following layers (hence fully connected). Each connection has its own weight, w, which is not depicted for reasons of simplicity.

As we mentioned in *Chapter 1*, we can think of the hidden layers as the NN's internal representation of the input data. This is the way the NN understands the input sample with its own internal logic. However, this internal representation is non-interpretable by humans. To bridge the gap between the network's representation and the actual data we're interested in, we need the output layer. You can think of this as a translator; we use it to understand the network's logic, and at the same time, we can convert it to the actual target values that we are interested in.

However, we are not limited to single-path networks with sequential layers, as shown in the preceding diagram. The layers (or operations in general) form **directed acyclic graphs**. In such a graph, the information cannot pass twice through the same layer (no loops) and it flows in only one direction, from the input to the output. The network in the preceding diagram is just a special case of a graph whose layers are connected sequentially. The following diagram also depicts a valid NN with two input layers, a single output layer, and randomly interconnected hidden layers. The layers are represented as operations $f^{(i)}(\mathbf{x})$ (i is an index that helps us differentiate between multiple operations):

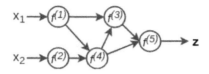

Figure 2.12 – NN as a graph of operations

> **Recurrent networks**
>
> There is a special class of NNs called **recurrent networks**, which represent a **directed cyclic graph** (they can have loops). We'll discuss them in detail in *Chapter 6*.

In this section, we introduced the most basic type of NN – that is, the unit – gradually expanded it to layers, and then generalized it as a graph of operations. We can also think of it in another way. The operations have precise mathematical definitions. Therefore, the NN, as a composition of functions is also a mathematical function, where the input data represents the function arguments, and the set of network weights, θ (a set of all weight matrices, $\mathbf{W}$), are its parameters. We'll denote it with either $f_\theta(\mathbf{x})$ or $f(\theta, \mathbf{x})$. Let's assume that, when an operation receives input from more than one source (input data or other operations), we use the elementwise sum to combine the multiple input tensors. Then, we can represent the NN as a series of nested functions/operations. The equivalent formula for the feed-forward network on the left is as follows:

$$f_\theta^{(f)}(\mathbf{x}) \;=\; f^{(5)}\Big(f^{(3)}\big(f^{(1)}(x_1) + f^{(4)}\big(f^{(1)}(x_1) + f^{(2)}(x_2)\big)\big) + f^{(4)}\big(f^{(1)}(x_1) + f^{(2)}(x_2)\big)\Big)$$

Now that we're familiar with the full NN architecture, let's discuss the different types of activation functions.

Activation functions

We now know that multi-layer networks can classify linearly inseparable classes, but to do this, they need to satisfy one more condition. If the units don't have activation functions, their output would be the weighted sum of the inputs, $\sum x_i w_i + b$, which is a linear function. Then, the entire NN – that is, a composition of units – becomes a composition of linear functions, which is also a linear function. This means that even if we add hidden layers, the network will still be equivalent to a simple linear regression model, with all its limitations. To turn the network into a non-linear function, we'll use non-linear activation functions for the units. Usually, all units in the same layer have the same activation function, but different activation functions. We'll start with three popular activation functions. The first two are from the *classic* period of NNs, while the third is contemporary:

- **Sigmoid**: Its output is bounded between 0 and 1 and can be interpreted stochastically as the probability of the unit being active. Because of these properties, the sigmoid was the most popular activation function for a long time. However, it also has some less desirable properties (more on that later), which led to its decline in use. The following diagram shows the sigmoid function, its derivative, and their graphs (the derivative will be useful when we discuss BP):

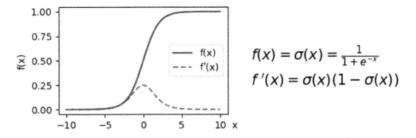

$$f(x) = \sigma(x) = \frac{1}{1+e^{-x}}$$
$$f'(x) = \sigma(x)(1 - \sigma(x))$$

Figure 2.13 – The sigmoid activation function

- **Hyperbolic tangent (tanh)**: The name speaks for itself. The principal difference with the sigmoid is that the tanh is in the (-1, 1) range. The following diagram shows the tanh function, its derivative, and their graphs:

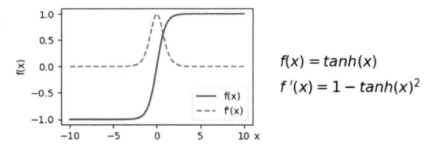

$$f(x) = tanh(x)$$
$$f'(x) = 1 - tanh(x)^2$$

Figure 2.14 – The hyperbolic tangent activation function

- **Rectified Linear Unit (ReLU)**: This is the new kid on the block (that is, compared to the *veterans*). ReLU was first successfully used in 2011 (see *Deep Sparse Rectifier Neural Networks* at http://proceedings.mlr.press/v15/glorot11a/glorot11a.pdf). The following diagram shows the ReLU function, its derivative, and their graphs:

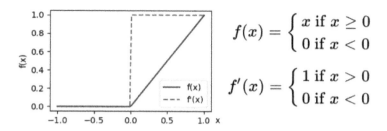

$$f(x) = \begin{cases} x \text{ if } x \geq 0 \\ 0 \text{ if } x < 0 \end{cases}$$

$$f'(x) = \begin{cases} 1 \text{ if } x > 0 \\ 0 \text{ if } x < 0 \end{cases}$$

Figure 2.15 – The ReLU activation function

As we can see, the ReLU repeats its input when x > 0 and stays at 0 otherwise. This activation has several important advantages over sigmoid and tanh, which make it possible to train NNs with more hidden layers (that is, deeper networks). We'll discuss these advantages, as well as other types of activation functions, in *Chapter 3*.

In the next section, we'll demonstrate how NNs can approximate any function.

The universal approximation theorem

In the *Multi-layer NNs* section, we defined the NN as a function, $f_\theta(\mathbf{x})$, where $\mathbf{x}$ is the input data (most often a vector or a tensor) and θ is the NN weights. Conversely, the training dataset is a collection of input samples and labels, which represents another, real-world, function $g(\mathbf{x})$. The NN function $f_\theta(\mathbf{x})$ **approximates** the function $g(\mathbf{x})$:

$$f_\theta(\mathbf{x}) \approx g(\mathbf{x})$$

The universal approximation theorem states that any continuous function on compact subsets of $\mathbb{R}^n$ can be approximated to an arbitrary degree of accuracy by a feedforward NN, with at least one hidden layer with a finite number of units and a non-linear activation. This is significant because it tells us that there are no theoretical insurmountable limitations in terms of NNs. In practice, an NN with a single hidden layer will perform poorly in many tasks, but at least we can aspire to a bright future with all-powerful NNs. We can understand the universal approximation theorem with an intuitive example.

> **Note**
>
> The idea for the following example was inspired by Michael A. Nielsen's book *Neural Networks and Deep Learning* (http://neuralnetworksanddeeplearning.com/).

We'll design an NN with one hidden layer that approximates the **boxcar function** (shown on the right in the following diagram). This is a type of step function, which is 0 across all input values, except in a narrow range, where it is equal to a constant value, *A*. A series of **translated** step functions can approximate any continuous function on a compact subset of $\mathbb{R}$, as shown in the left figure of the following diagram:

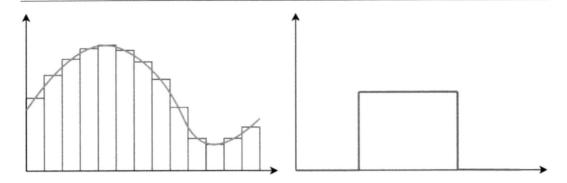

Figure 2.16 – Left: a continuous function approximation with a series
of step functions, and right: a single boxcar step function

We'll start building our boxcar NN with a single unit, with a single scalar input, *x*, and sigmoid activation. The following figure shows the unit, as well as its equivalent formula:

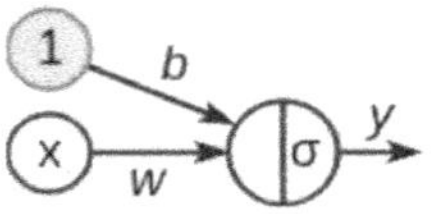

$$f(x) = \frac{1}{1 + exp(-(wx + b))}$$

Figure 2.17 – A unit with single input and sigmoid activation

In the following diagrams, we can see the unit output for different values of *b* and *w* with inputs, *x*, in the range [-10: 10]:

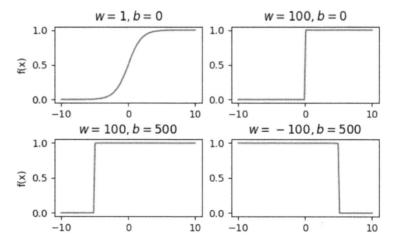

Figure 2.18 – The unit output based on different values of w and b.
The network input, x, is represented on the x axis

We can see that the weight, w, determines the steepness of the sigmoid function. We can also see that the formula, $t = -b/w$, determines the translation of the function along the x axis. Let's cover the different graphs in the preceding diagram:

- **Top-left**: Regular sigmoid

- **Top-right**: A large weight, w, amplifies the input, x, to a point where the sigmoid output resembles threshold activation

- **Bottom-left**: The bias, b, translates the unit activation along the x axis

- **Bottom-right**: We can simultaneously translate the activation along the x axis with the bias, b, and reverse the activation with a negative value of the weight, w

We can intuitively see that the unit can implement all the pieces of the box function. However, to create a full box function, we'll have to combine two such units in an NN with one hidden layer. The following diagram shows the NN architecture, along with the weights and biases of the units, as well as the box function that's produced by the network:

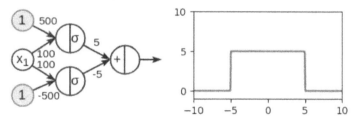

Figure 2.19 – A box function approximation NN

Here's how it works:

- $x < -5$: the NN output is 0.

- $x \geq 5$: The top unit activates for the upper step of the function and stays active for all values of $x \geq 5$.

- $x \geq 10$: The bottom unit activates for the bottom step of the function and stays active for all $x \geq 10$. The outputs of the two hidden units cancel each other out because of the weights in the output layer, which are the same but with opposite signs.

- The weights of the output layer determine the constant value, $A = 5$, of the boxcar function.

The output of this network is 5 in the [-5:5] interval and 0 for all other inputs. We can approximate additional boxes by adding more units to the hidden layer in a similar manner.

Now that we're familiar with the structure of an NN, let's focus on their training process.

Training NNs

The NN function $f_\theta(\mathbf{x})$ **approximates** the function $g(\mathbf{x})$: $f_\theta(\mathbf{x}) \approx g(\mathbf{x})$. The goal of the training is to find parameters, θ, such that $f_\theta(\mathbf{x})$ will best approximate $g(\mathbf{x})$. First, we'll see how to do that for a single-layer network, using an optimization algorithm called GD. Then, we'll extend it to a deep feedforward network with the help of BP.

> **Note**
>
> We should note that an NN and its training algorithm are two separate things. This means we can adjust the weights of a network in some way other than GD and BP, but this is the most popular and efficient way to do so and is, ostensibly, the only way that is currently used in practice.

GD

For the purposes of this section, we'll train a simple NN using the **mean square error** (**MSE**) cost function. It measures the difference (known as **error**) between the network output and the training data labels of all training samples:

$$J(\theta) = \frac{1}{2n} \sum_{i=1}^{n} \left(f_\theta(\mathbf{x}^{(i)}) - t^{(i)} \right)^2$$

At first, this might look scary, but fear not! Behind the scenes, it's very simple and straightforward mathematics (I know that sounds even scarier!). Let's discuss its components:

- $f_\theta(\mathbf{x}^{(i)})$: The output of the NN, where θ is the set of all network weights. For the remainder of this section, we'll denote the individual weights with θ_j (unlike the w notation in the rest of the sections).

- n: The total number of samples in the training set.

- $\mathbf{x}^{(i)}$: The vector representation of a training sample, where the superscript i indicates the i-th sample of the dataset. We use superscript because $\mathbf{x}^{(i)}$ is a vector, and the subscript is reserved for each of the vector components. For example, $x_j^{(i)}$ is the j-th component of the i-th training sample.

- $t^{(i)}$: The label associated with the training sample, $\mathbf{x}^{(i)}$.

> **Note**
>
> In this example, we will use MSE, but in practice, there are different types of cost functions. We'll discuss them in *Chapter 3*.

First, GD computes the derivative (gradient) of $J(\theta)$ with respect to all the network weights. The gradient gives us an indication of how $J(\theta)$ changes with respect to each weight. Then, the algorithm uses this information to update the weights in a way that will minimize $J(\theta)$ in future occurrences of the same input/target pairs. The goal is to gradually reach the **global minimum** of the cost function, where the gradient is 0. The following is a visualization of GD for MSE with respect to a single NN weight:

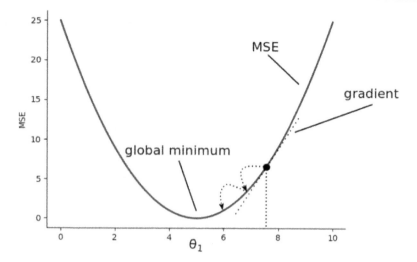

Figure 2.20 – An MSE diagram

Let's go over the execution of GD step by step:

1. Initialize the network weights, θ, with random values.

2. Repeat until the cost function, $J(\theta)$, falls below a certain threshold:

   - **Forward pass**: Compute the MSE $J(\theta)$ cost function for all the samples of the training set using the preceding formula

   - **Backward pass**: Compute the partial derivatives (gradients) of $J(\theta)$ with respect to all the network weights θ_j, using the chain rule:

$$\frac{\partial J(\theta)}{\partial \theta_j} = \frac{\partial \frac{1}{2n}\sum_i^n \left(f_\theta(\mathbf{x}^{(i)}) - t^{(i)}\right)^2}{\partial \theta_j} = \frac{1}{2n}\sum_i^n \frac{\partial \left(f_\theta(\mathbf{x}^{(i)}) - t^{(i)}\right)^2}{\partial \theta_j} = \frac{1}{n}\sum_i^n \frac{\partial f_\theta(\mathbf{x}^{(i)})}{\partial \theta_j}\left[f_\theta(\mathbf{x}^{(i)}) - t^{(i)}\right]$$

Let's analyze the partial derivative $\partial J(\theta)/\partial \theta_j$. J is a function of θ_j by being a function of the network output. Therefore, it is also a function of the NN function itself – that is, $J(f(\theta))$. Then, by following the chain rule, we get the following: $\frac{\partial J(\theta)}{\partial \theta_j} = \frac{\partial J(f(\theta))}{\partial \theta_j} = \frac{\partial J(f(\theta))}{\partial f(\theta)}\frac{\partial f(\theta)}{\partial \theta_j}$

   - Use these gradient values to update each of the network weights:

$$\theta_j \leftarrow \theta_j - \eta\frac{\partial J(\theta)}{\partial \theta_j} = \theta_j - \eta\frac{1}{n}\sum_i^n \frac{\partial f_\theta(\mathbf{x}^{(i)})}{\partial \theta_j}\left[f_\theta(\mathbf{x}^{(i)}) - t^{(i)}\right]$$

Here, η is the **learning rate**, which determines the step size at which the optimizer makes updates to the weights during training. Let's note that as we move closer to the global minimum, the gradient will get smaller and we'll update the weights in finer steps.

To better understand how GD works, we'll use linear regression as an example. Let's recall that linear regression is equivalent to a single NN unit with an identity activation function, $f(x) = x$:

- The linear regression is represented by the function $f_\theta(\mathbf{x}) = \sum_{j=1}^{m} x_j \theta_j$, where m is the dimension of the input vector (equal to the number of weights)

- Then, we have the MSE cost function – $J(\theta) = \frac{1}{2n} \sum_{i=1}^{n} \left(\left(\sum_{j=1}^{m} x_j^{(i)} \theta_j \right) - t^{(i)} \right)^2$

- Then, we compute the partial derivative $\partial J(\theta)/\partial \theta_j$ with respect to a single network weight θ_j using the chain rule and the sum rule:

$$\frac{\partial J(\theta)}{\partial \theta_j} = \frac{\partial \frac{1}{2n} \sum_{i=1}^{n} \left(\left(\sum_{j=1}^{m} x_j^{(i)} \theta_j \right) - t^{(i)} \right)^2}{\partial \theta_j}$$

$$= \frac{1}{2n} \sum_{i=1}^{n} \frac{\partial \left(\left(\sum_{j=1}^{m} x_j^{(i)} \theta_j \right) - t^{(i)} \right)^2}{\partial \theta_j}$$

$$= \frac{1}{n} \sum_{i=1}^{n} \left(\left(\sum_{j=1}^{m} x_j^{(i)} \theta_j \right) - t^{(i)} \right) x_j^{(i)}$$

$$= \frac{1}{n} \sum_{i=1}^{n} \left(f_\theta(\mathbf{x}^{(i)}) - t^{(i)} \right) x_j^{(i)}$$

- Now that we have the gradient, $\partial J(\theta)/\partial \theta_j$, we can update the weight θ_j using the learning rate η

So far, we've discussed a GD that works with NNs with multiple weights. However, for the sake of simplicity, the preceding diagram illustrates the relationship between the cost function and a single-weight NN. Let's remedy this by introducing a more complex cost function for an NN with two weights, θ_1 and θ_2:

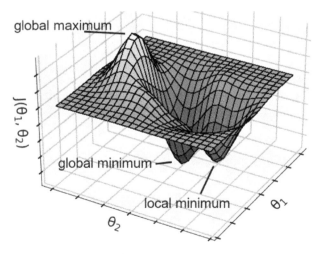

Figure 2.21 – The cost function J with respect to two weights

The function has a local and a global minimum. Nothing prevents GD from converging to the local minimum, instead of the global one, thus finding a sub-optimal approximation to the target function. We can try to mitigate this by increasing the learning rate η. The idea is that even if GD converges toward the local minimum, the larger η will help us *jump* over the saddle and converge toward the global maximum. The risk is that the opposite could happen – if GD correctly converges toward the global maximum, the larger learning rate could make it *jump* over to the local minimum instead.

A more elegant way to prevent this issue is to use **momentum**. This extends vanilla GD by adjusting the current weight update with the values of the previous weight updates – that is, if the weight update at step t-1 was big, it will also increase the weight update of step t. We can explain momentum with an analogy. Think of the loss function surface as the surface of a hill. Now, imagine that we are holding a ball at the top of the hill (maximum). If we drop the ball, thanks to the Earth's gravity, it will start rolling toward the bottom of the hill (minimum). The more distance it travels, the more its speed will increase. In other words, it will gain momentum (hence the name of the optimization).

Now, let's look at how to implement momentum in the weight update rule $\theta_j \leftarrow \theta_j - \eta \partial J(\theta)/\partial \theta_j$. We'll assume that we are at step t of the training process:

1. First, we'll calculate the current weight update value v_t by also including the velocity of the previous update v_{t-1}: $v_t \leftarrow \mu v_{t-1} - \eta \partial J(\theta)/\partial \theta_j$. Here, μ is a hyperparameter in the [0:1] range called the momentum rate. v_t is initialized as 0 during the first iteration.

2. Then, we perform the actual weight update – $\theta_j \leftarrow \theta_j + v_t$.

Finding the best values of the learning rate η and the momentum rate μ is an empirical task. They can depend on the NN architecture, the type and size of a dataset, and other factors. In addition, we might have to adjust them during training. Since the NN weights are randomly initialized, we usually start with a larger η so that GD can quickly advance while the initial value of the cost function (error) is large. Once the decrease in the cost function starts plateauing, we can decrease the learning rate. In this way, GD can find minimums that would have been *jumped over* with the larger learning rate.

Alternatively, we can use an adaptive learning rate algorithm such as **Adam** (see *Adam: A Method for Stochastic Optimization* at `https://arxiv.org/abs/1412.6980`). It calculates individual and adaptive learning rates for every weight, based on previous weight updates (momentum).

The GD we just described is called **batch gradient descent** because it accumulates the error across *all* training samples and then performs a single weight update. This is fine for small datasets but could become impractical for large ones, as the training would take a long time with such sporadic updates. In practice, we would use two modifications:

*   **Stochastic (or online) gradient descent** (**SGD**): Updates the weights after every training sample.

*   **Mini-batch gradient descent**: Accumulates the error over batches of *k* samples (called **mini-batches**) and performs one weight update after each mini-batch. It is a hybrid between online and batch GD. In practice, we'll almost always use mini-batch GD over the other modifications.

The next step in our learning journey is to understand how to apply GD to train networks with more than one layer.

Backpropagation

In this section, we'll discuss how to combine GD with the BP algorithm to update the weights of networks with more than one layer. As we demonstrated in the *GD* section, this means finding the derivative of the cost function $J(\theta)$ with respect to each network weight. We already took a step in this direction with the help of the chain rule:

$$\frac{\partial J(\theta)}{\partial \theta_j} = \frac{\partial J(f(\theta))}{\partial \theta_j} = \frac{\partial J(f(\theta))}{\partial f(\theta)}\frac{\partial f(\theta)}{\partial \theta_j}$$

Here, $f(\theta)$ is the network output and θ_j is the j-th network weight. In this section, we'll push the envelope further, and we'll learn how to derive the NN function itself for all the network weights (hint – the chain rule). We'll do this by propagating the error gradient backward through the network (hence the name). Let's start with a few assumptions:

- For the sake of simplicity, we'll work with a sequential feedforward NN. *Sequential* means that each layer takes input from the preceding layer and sends its output to the following layer.

- We'll define w_{ij} as the weight between the i-th unit of layer l and the j-th unit of the subsequent layer $l+1$. In a multi-layer network, l and $l+1$ can be any two consecutive layers, including input, hidden, and output layers.

- We'll denote the output of the i-th unit of layer l with $y_i^{(l)}$ and the output of the j-th unit of layer $l+1$ with $y_j^{(l+1)}$.

- We'll denote the input to the activation function (that is, the weighted sum of the inputs before activation) of unit j of layer l, with $a_j^{(l)}$.

The following diagram shows all the notations we introduced:

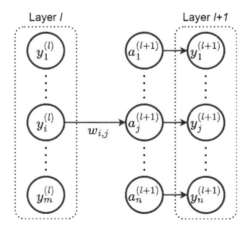

Figure 2.22 – Layer l represents the input, layer l+1 represents the output, and w connects the y activation in layer l to the inputs of the j-th unit of layer l+1

Armed with this useful knowledge, let's get down to business:

1. First, we'll assume that l and $l+1$ are the second-to-last and the last (output) network layers, respectively. Knowing this, the derivative of J with respect to w_{ij} is as follows:

$$\frac{\partial J}{\partial w_{ij}} = \frac{\partial J}{\partial y_j^{(l+1)}} \frac{\partial y_j^{(l+1)}}{\partial a_j^{(l+1)}} \frac{\partial a_j^{(l+1)}}{\partial w_{ij}}$$

2. Let's focus on $\partial a_j^{(l+1)}/\partial w_{ij}$. Here, we compute the partial derivative of the weighted sum of the output of layer l with respect to one of the weights, w_{ij}. As we discussed in the *Differential calculus* section, in partial derivatives, we'll consider all the function parameters except w_{ij} constants. When we derive $a^{(l+1)}$, they all become 0, and we're only left with $\partial\left(y_i^{(l)} w_{ij}\right)/\partial w_{ij} = y_i^{(l)}$. Therefore, we get the following:

$$\frac{\partial a_j^{(l+1)}}{\partial w_{ij}} = y_i^{(l)}$$

3. The formula from point 1 holds for any two consecutive hidden layers, l and $l+1$, of the network. We know that $\partial\left(y_i^{(l)} w_{ij}\right)/\partial w_{ij} = y_i^{(l)}$, and we also know that $\partial y_j^{(l-1)}/\partial a_j^{(l-1)}$ is the derivative of the activation function, which we can calculate (see the *Activation functions* section). All we need to do is calculate the derivative $\partial J/\partial y_j^{(l+1)}$ (recall that, here, $l+1$ is a hidden layer). Let's note that this is the derivative of the error with respect to the activation function in layer $l+1$. We can now calculate all the derivatives, starting from the last layer and moving backward, because the following apply:

   - We can calculate this derivative for the last layer

   - We have a formula that allows us to calculate the derivative for one layer, assuming that we can calculate the derivative for the next

4. With these points in mind, we get the following equation by applying the chain rule:

$$\frac{\partial J}{\partial y_i^{(l)}} = \sum_j \frac{\partial J}{\partial y_j^{(l+1)}} \frac{\partial y_j^{(l+1)}}{\partial y_i^{(l)}} = \sum_j \frac{\partial J}{\partial y_j^{(l+1)}} \frac{\partial y_j^{(l+1)}}{\partial a_j^{(l+1)}} \frac{\partial a_j^{(l+1)}}{\partial y_i^{(l)}}$$

The sum over j reflects the fact that, in the feedforward part of the network, the output $y_i^{(l)}$ is fed to all the units in layer $l+1$. Therefore, they all contribute to $y_i^{(l)}$ when the error is propagated backward.

> **Note**
>
> In the *Layers as operations* section, we discussed that in the forward pass, we can simultaneously compute all outputs of layer $l+1$ as a matrix-matrix multiplication $\mathbf{Y}^{(l)}\mathbf{W}^{(l,l+1)}$. Here, $\mathbf{Y}^{(l)}$ is the layer output of layer l, and $\mathbf{W}^{(l,l+1)}$ is the weight matrix between layers l and $l+1$. In the forward pass, one column of $\mathbf{W}^{(l,l+1)}$ represents the weights from all units of the input layer l to a single unit of the output layer $l+1$. We can also represent the backward pass as a matrix-matrix multiplication by using the transpose weight matrix $\mathbf{W}^{(l+1,l)} = [\mathbf{W}^{(l,l+1)}]^{\top}$. A column of the transposed matrix represents the weights of all units of l, which contributed to a particular unit of $l+1$ during the forward phase.

Once again, we can calculate $\partial y_j^{(l+1)} / \partial a_j^{(l+1)}$. Using the same logic that we followed in *step 3*, we can compute that $\partial a_j^{(l+1)} / \partial y_i^{(l)} = w_{ij}$. Therefore, once we know $\partial J / \partial y_j^{(l+1)}$, we can calculate $\partial J / \partial y_i^{(l)}$. Since we can calculate $\partial J / \partial y_j^{(l+1)}$ for the last layer, we can move backward and calculate $\partial J / \partial y_i^{(l)}$ for any layer and, therefore, $\partial J / \partial w_{ij}$ for any layer.

5. To summarize, let's say we have a sequence of layers where the following applies:

$$y_i \rightarrow y_j \rightarrow y_k$$

Here, we have the following fundamental equations:

$$\frac{\partial J}{\partial w_{ij}} = \frac{\partial J}{\partial y_j^{(l+1)}} \frac{\partial y_j^{(l+1)}}{\partial a_j^{(l+1)}} \frac{\partial a_j^{(l+1)}}{\partial w_{ij}}$$

$$\frac{\partial J}{\partial y_i^{(l)}} = \sum_j \frac{\partial J}{\partial y_j^{(l+1)}} \frac{\partial y_j^{(l+1)}}{\partial y_i^{(l)}} = \sum_j \frac{\partial J}{\partial y_j^{(l+1)}} \frac{\partial y_j^{(l+1)}}{\partial a_j^{(l+1)}} \frac{\partial a_j^{(l+1)}}{\partial y_i^{(l)}}$$

By using these two equations, we can calculate the derivatives for the cost with respect to each layer.

6. If we set $\delta_j^{(l+1)} = \frac{\partial J}{\partial y_j^{(l+1)}} \frac{\partial y_j^{(l+1)}}{\partial a_j^{(l+1)}}$, then $\delta^{(l+1)}$ represents the variation in cost with respect to the activation value, and we can think of $\delta^{(l+1)}$ as the error at unit $y^{(l+1)}$. We can rewrite these equations as follows:

$$\frac{\partial J}{\partial y_i^{(l)}} = \sum_j \frac{\partial J}{\partial y_j^{(l+1)}} \frac{\partial y_j^{(l+1)}}{\partial y_i^{(l)}} = \sum_j \frac{\partial J}{\partial y_j^{(l+1)}} \frac{\partial y_j^{(l+1)}}{\partial a_j^{(l+1)}} \frac{\partial a_j^{(l+1)}}{\partial y_i^{(l)}} = \sum_j \delta_j^{(l+1)} w_{ij}$$

Following this, we can write the following equation:

$$\delta_i^{(l)} = \left(\sum_j \delta_j^{(l+1)} w_{ij} \right) \frac{\partial y_i^{(l)}}{\partial a_i^{(l)}}$$

These two equations provide us with an alternative view of BP, since there is a variation in cost with respect to the activation value. They provide us with a way to calculate the variation for any layer *l* once we know the variation for the following layer, *l+1*.

7. We can combine these equations to show the following:

$$\frac{\partial J}{\partial w_{ij}} = \delta_j^{(l+1)} \frac{\partial a_j^{(l+1)}}{\partial w_{ij}} = \delta_j^{(l+1)} y_i^{(l)}$$

8. The updated rule for the weights of each layer is given by the following equation:

$$w_{ij} \leftarrow w_{ij} - \eta \, \delta_j^{(l+1)} y_i^{(l)}$$

Now that we're familiar with GD and BP, let's implement them in Python.

A code example of an NN for the XOR function

In this section, we'll create a simple network with one hidden layer, which solves the XOR function. Let's recall that XOR is a linearly inseparable problem, hence the need for a hidden layer. The source code will allow you to easily modify the number of layers and the number of units per layer, so you can try a number of different scenarios. We won't use any ML libraries. Instead, we'll implement them from scratch, with only the help of numpy. We'll also use `matplotlib` to visualize the results:

1. Let's start by importing these libraries:

```
import matplotlib.pyplot as plt
import numpy as np

from matplotlib.colors import ListedColormap
```

2. Then, we will define the activation function and its derivative. In this example, we will use `tanh(x)`:

```
def tanh(x):
    return (1.0 - np.exp(-2 * x)) / (1.0 + np.exp(-2 * x))


def tanh_derivative(x):
    return (1 + tanh(x)) * (1 - tanh(x))
```

3. Then, we will start the definition of the `NeuralNetwork` class and its constructor (note that all its methods and properties have to be properly indented):

```
class NeuralNetwork:
    # net_arch consists of a list of integers, indicating
    # the number of units in each layer
    def __init__(self, net_arch):
        self.activation_func = tanh
        self.activation_derivative = tanh_derivative
        self.layers = len(net_arch)
        self.steps_per_epoch = 1000
        self.net_arch = net_arch

        # initialize the weights with random values in the range
    (-1,1)
        self.weights = []
        for layer in range(len(net_arch) - 1):
            w = 2 * np.random.rand(net_arch[layer] + 1, net_
    arch[layer + 1]) - 1
            self.weights.append(w)
```

Here, `net_arch` is a one-dimensional array containing the number of units for each layer. For example [2, 4, 1] means an input layer with two units, a hidden layer with four units, and an output layer with one unit. Since we are studying the XOR function, the input layer will have two units, and the output layer will only have one unit. However, the number of hidden units can vary.

To conclude the constructor, we will initialize the network weights with random values in the range (-1, 1).

4. Now, we need to define the fit function, which will train our network:

```
def fit(self, data, labels, learning_rate=0.1, epochs=10):
```

5. We will start the implementation by concatenating `bias` to the training `data` in a new variable, `input_data` (the source code is indented within the method definition):

```
bias = np.ones((1, data.shape[0]))
input_data = np.concatenate((bias.T, data), axis=1)
```

6. Then, we'll run the training for a number of `epochs`:

```
for k in range(epochs * self.steps_per_epoch):
```

7. Within the loop, we'll visualize the epoch number and the prediction output of the NN at the start of each epoch:

```
print('epochs: {}'.format(k / self.steps_per_epoch))
for s in data:
    print(s, nn.predict(s))
```

8. Within the loop, we select a random sample from the training set and propagate it forward through the hidden network layers:

```
sample = np.random.randint(data.shape[0])
y = [input_data[sample]]

for i in range(len(self.weights) - 1):
    activation = np.dot(y[i], self.weights[i])
    activation_f = self.activation_func(activation)
    # add the bias for the next layer
    activation_f = np.concatenate((np.ones(1),
        np.array(activation_f)))
    y.append(activation_f)
```

9. Outside the loop, we will calculate the last layer output and the error:

```
# last layer
activation = np.dot(y[-1], self.weights[-1])
activation_f = self.activation_func(activation)
y.append(activation_f)

# error for the output layer
error = y[-1] - labels[sample]
delta_vec = [error * self.activation_derivative(y[-1])]
```

10. Then, we will propagate the error back (backward pass):

```
# we need to begin from the back from the next to last layer
for i in range(self.layers - 2, 0, -1):
    error = delta_vec[-1].dot(self.weights[i][1:].T)
    error = error * self.activation_derivative(y[i][1:])
    delta_vec.append(error)

# reverse
# [level3(output)->level2(hidden)] => [level2(hidden)-
>level3(output)]
delta_vec.reverse()
```

11. Finally, we will update the weights based on the errors we just computed. We will multiply its output delta and input activation to get the gradient of the weight. Then, we will update the weight using the learning rate:

```
for i in range(len(self.weights)):
    layer = y[i].reshape(1, nn.net_arch[i] + 1)

    delta = delta_vec[i].reshape(1, nn.net_arch[i + 1])
    self.weights[i] -= learning_rate * layer.T.dot(delta)
```

This concludes the implementation of the `fit` method.

12. We'll now write a `predict` function to check the results, which returns the network output for the given input:

```
def predict(self, x):
    val = np.concatenate((np.ones(1).T, np.array(x)))
    for i in range(0, len(self.weights)):
        val = self.activation_func(
            np.dot(val, self.weights[i]))
        al = np.concatenate((np.ones(1).T,
            np.array(val)))

    return val[1]
```

13. Finally, we'll write the `plot_decision_regions` method, which plots the hyperplane separating the classes (in our case, represented as lines), based on the input variables. We will create a two-dimensional grid with one axis for each input variable. We will plot the NN predictions for all input value combinations across the whole grid. We will take the network output larger than 0.5 to be `true` and `false` otherwise (we'll see the plots at the end of the section):

```
def plot_decision_regions(self, X, y, points=200):
    markers = ('o', '^')
```

```python
colors = ('red', 'blue')
cmap = ListedColormap(colors)

x1_min, x1_max = X[:, 0].min() - 1, X[:, 0].max() + 1
x2_min, x2_max = X[:, 1].min() - 1, X[:, 1].max() + 1

resolution = max(x1_max - x1_min, x2_max - x2_min) /
float(points)

xx1, xx2 = np.meshgrid(np.arange(x1_min,
    x1_max, resolution),
    np.arange(x2_min, x2_max, resolution))
input = np.array([xx1.ravel(), xx2.ravel()]).T
Z = np.empty(0)
for i in range(input.shape[0]):
    val = nn.predict(np.array(input[i]))
    if val < 0.5:
        val = 0
    if val >= 0.5:
        val = 1
    Z = np.append(Z, val)

Z = Z.reshape(xx1.shape)

plt.pcolormesh(xx1, xx2, Z, cmap=cmap)
plt.xlim(xx1.min(), xx1.max())
plt.ylim(xx2.min(), xx2.max())
# plot all samples

classes = ["False", "True"]

for idx, cl in enumerate(np.unique(y)):
    plt.scatter(x=X[y == cl, 0],
        y=X[y == cl, 1],
        alpha=1.0,
        c=colors[idx],
        edgecolors='black',
        marker=markers[idx],
        s=80,
        label=classes[idx])

plt.xlabel('x1')
plt.ylabel('x2')
plt.legend(loc='upper left')
plt.show()
```

This concludes the implementation of the `NeuralNetwork` class.

14. Finally, we can run the program with the following code:

```
np.random.seed(0)

# Initialize the NeuralNetwork with 2 input, 2 hidden, and 1
output units
nn = NeuralNetwork([2, 2, 1])

X = np.array([[0, 0],
              [0, 1],
              [1, 0],
              [1, 1]])

y = np.array([0, 1, 1, 0])

nn.fit(X, y, epochs=10)

print("Final prediction")
for s in X:
    print(s, nn.predict(s))

nn.plot_decision_regions(X, y)
```

We will build the default network, `nn = NeuralNetwork([2,2,1])`. The first and last values (2 and 1) represent the input and output layers and cannot be modified, but we can add different numbers of hidden layers with different numbers of units. For example, (`[2,4,3,1]`) will represent a three-layer NN, with four units in the first hidden layer and three units in the second hidden layer.

We will use `numpy.random.seed(0)` to ensure that the weight initialization is consistent across runs, so we can compare their results. This is a common practice when training NNs.

Now, we will define the training XOR data and labels in x and y respectively. We will run the training for 10 epochs. Finally, we will plot the result.

In the following diagrams, you can see how the `nn.plot_decision_regions` function method plots the hyperplanes, which separate the classes. The circles represent the network output for the (`true, true`) and (`false, false`) inputs, while the triangles represent the (`true, false`) and (`false, true`) inputs for the XOR function. To the left, we can see the hyperplane of an NN with two hidden units, and to the right, we can see an NN with four hidden units:

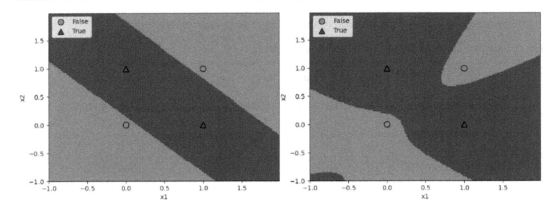

Figure 2.23 – Left: the hyperplane learned by an NN with two hidden units,
and right: the hyperplane of an NN with four hidden units

Networks with different architectures can produce different separating regions. In the preceding figure, we can see that while the network finds the right solution, the curves separating the regions will be different, depending on the chosen architecture.

We are now ready to start looking more closely at what deep neural nets are and their applications.

Summary

In this chapter, we introduced NNs in detail, and we mentioned their success vis-à-vis other competing algorithms. NNs are comprised of interconnected units, where the weights of the connections characterize the strength of the communication between different units. We discussed different network architectures, how an NN can have many layers, and why inner (hidden) layers are important. We explained how information flows from the input to the output by passing from one layer to the next, based on weights and the activation function. Finally, we showed how to train NNs – that is, how to adjust their weights using GD and BP.

In the following chapter, we'll continue discussing deep NNs. We'll explain in particular the meaning of *deep* in deep learning, and that it not only refers to the number of hidden layers in a network but to the quality of the learning of the network. For this purpose, we'll show how NNs learn to recognize features and compile them as representations of larger objects. We'll also describe a few important deep learning libraries and, finally, provide a concrete example of applying NNs to handwritten digit recognition.

3

Deep Learning Fundamentals

In this chapter, we will introduce **deep learning** (**DL**) and **deep neural networks** (**DNNs**) – that is, **neural networks** (**NNs**) with multiple hidden layers. You might be wondering what the point of using more than one hidden layer is, given the universal approximation theorem. This is in no way a naive question, and for a long time, NNs were used in that way.

Without going into too much detail, one reason is that approximating a complex function might require a huge number of units in the hidden layer, making it impractical to use. There is also another, more important, reason for using deep networks, which is not directly related to the number of hidden layers, but to the level of learning. A deep network does not simply learn to predict output Y given input, X; it also understands the basic features of the input. It's able to learn abstractions of features of input samples, understand the basic characteristics of the samples, and make predictions based on those characteristics. This level of abstraction is missing in other basic **machine learning** (**ML**) algorithms and shallow NNs.

In this chapter, we're going to cover the following main topics:

- Introduction to DL
- Fundamental DL concepts
- Deep neural networks
- Training deep networks
- Applications of DL
- Introducing popular DL libraries

Technical requirements

We'll implement the example in this chapter using Python, PyTorch, and Keras as part of **TensorFlow** (**TF**). If you don't have an environment set up with these tools, fret not – the example is available as a Jupyter notebook on Google Colab. You can find the code examples in this book's GitHub repository: https://github.com/PacktPublishing/Python-Deep-Learning-Third-Edition/tree/main/Chapter03.

Introduction to DL

In 2012, Alex Krizhevsky, Ilya Sutskever, and Geoffrey Hinton published a milestone paper titled *ImageNet Classification with Deep Convolutional Neural Networks* (https://papers.nips.cc/paper/4824-imagenet-classification-with-deep-convolutional-neural-networks.pdf). The paper describes their use of NNs to win the ImageNet competition of the same year, which we mentioned in *Chapter 2*. At the end of their paper, they noted that the network's performance degrades even if a single layer is removed. Their experiments demonstrated that removing any of the middle layers resulted in an about 2% top-1 accuracy loss of the model. They concluded that network depth is important for the performance of the network. The basic question is: what makes the network's depth so important?

A typical English saying is a picture is worth a thousand words. Let's use this approach to understand what DL is. We'll use images from the highly cited paper *Convolutional Deep Belief Networks for Scalable Unsupervised Learning of Hierarchical Representations* (https://ai.stanford.edu/~ang/papers/icml09-ConvolutionalDeepBeliefNetworks.pdf). Here, the authors trained an NN with pictures of different categories of either objects or animals. The following figure shows how the different layers of the network learn different characteristics of the input data. In the first layer, the network learns to detect some small basic features, such as lines and edges, which are common for all images in all categories:

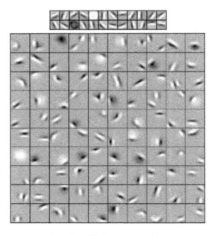

Figure 3.1 – The first layer weights (top) and the second layer weights (bottom) after training

But the next layers, which we can see in the following figure, combine those lines and edges to compose more complex features that are specific to each category. In the first row of the bottom-left image, we can see how the network can detect different features of human faces, such as eyes, noses, and mouths. In the case of cars, these would be wheels, doors, and so on, as seen in the second image from the left in the following figure. These features are **abstract** – that is, the network has learned the generic shape of a feature (such as a mouth or a nose) and can detect this feature in the input data, despite the variations it might have:

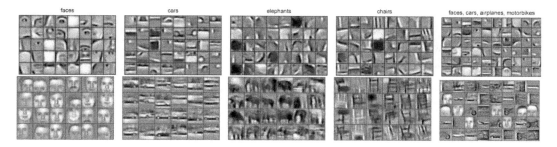

Figure 3.2 – Columns 1 to 4 represent the second-layer (top) and third-layer (bottom) weights learned for a specific object category (class). Column 5 represents the weights learned for a mixture of four object categories (faces, cars, airplanes, and motorbikes)

In the second row of the preceding figure, we can see how, in the deeper layers, the network combines these features in even more complex ones, such as faces and whole cars. One strength of DNNs is that they can learn these high-level abstract representations by themselves, deducting them from the training data.

Next, let's discuss these properties of DNNs in more detail.

Fundamental DL concepts

In 1801, Joseph Marie Charles invented the **Jacquard loom**. Charles was not a scientist, but simply a merchant. The Jacquard loom used a set of punched cards, where each card represented a pattern to be reproduced on the loom. At the same time, each card was an abstract representation of that pattern. Punched cards have been used, for example, in the tabulating machine invented by Herman Hollerith in 1890, or in the first computers as a means to input code. In the tabulating machine, the cards were simply abstractions of samples to be fed into the machine to calculate statistics on a population. But in the Jacquard loom, their use was subtler, and each card represented the abstraction of a pattern that could be combined with others to create more complex patterns. The punched card is an abstract representation of a feature of reality, the final weaved design.

In a way, the Jacquard loom sowed the seeds of what DL is today, the definition of reality through the representations of its features. A DNN does not simply recognize what makes a cat a cat, or a squirrel a squirrel, but it understands what features are present in a cat and a squirrel, respectively. It learns to design a cat or a squirrel using those features. If we were to design a weaving pattern in the shape of a cat using a Jacquard loom, we would need to use punched cards that have whiskers on the nose, such as those of a cat, and an elegant and slender body. Conversely, if we were to design a squirrel, we would need to use a punched card that makes a furry tail. A deep network that learns basic representations of its output can make classifications using the assumptions it has made. For example, if there is no furry tail, it will probably not be a squirrel, but rather a cat. In this way, the amount of information the network learns is much more complete and robust, and the most exciting part is that DNNs learn to do this automatically.

Feature learning

To illustrate how DL works, let's consider the task of recognizing a simple geometric figure, such as a cube, as seen in the following diagram:

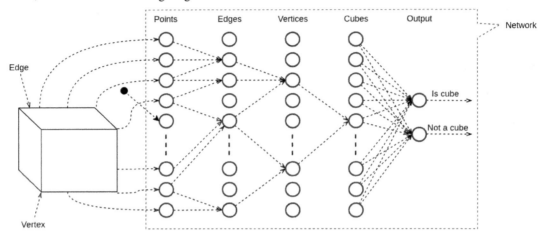

Figure 3.3 – An abstraction of an NN representing a cube. Different
layers encode features with different levels of abstraction

The cube is composed of edges (or lines), which intersect in vertices. Let's say that each possible point in the three-dimensional space is associated with a unit (forget for a moment that this will require an infinite number of units). All the points/units are in the first (input) layer of a multilayer feedforward network. An input point/unit is active if the corresponding point lies on a line. The points/units that lie on a common line (edge) have strong positive connections to a single common edge/unit in the next layer. Conversely, they have negative connections to all other units in the next layer. The only exceptions are the units that lie on the vertices. Each such unit lies simultaneously on three edges and is connected to its three corresponding units in the subsequent layer.

Now, we have two hidden layers, with different levels of abstraction – the first for points and the second for edges. However, this is not enough to encode a whole cube in the network. Let's try this with another layer for vertices. Here, each three active edges/units of the second layer, which form a vertex, have a significant positive connection to a single common vertex/unit of the third layer. Since an edge of the cube forms two vertices, each edge/unit will have positive connections to two vertices/units and negative connections to all others. Finally, we'll introduce the last hidden layer (the cube). The four vertices/units forming the cube will have positive connections to a single cube/unit from the cube/layer.

This cube representation example is oversimplified, but we can draw several conclusions from it. One of them is that DNNs lend themselves well to hierarchically organized data. For example, an image consists of pixels, which form lines, edges, regions, and so on. This is also true for speech, where the building blocks are called **phonemes**, as well as text, where we have characters, words, and sentences.

In the preceding example, we dedicated layers to specific cube features deliberately, but in practice, we wouldn't do that. Instead, a deep network will "discover" features automatically during training. These features might not be immediately obvious and, in general, wouldn't be interpretable by humans. Also, we wouldn't know the level of the features encoded in the different layers of the network. Our example is more akin to classic ML algorithms, where the user has to use their own experience to select what they think are the best features. This process is called **feature engineering**, and it can be labor-intensive and time-consuming. Allowing a network to automatically discover features is not only easier, but those features are highly abstract, which makes them less sensitive to noise. For example, human vision can recognize objects of different shapes, sizes, in different lighting conditions, and even when their view is partly obscured. We can recognize people with different haircuts and facial features, even when they wear a hat or a scarf that covers their mouth. Similarly, the abstract features the network learns will help it recognize faces better, even in more challenging conditions.

In the next section, we'll discuss some of the reasons DL has become so popular.

The reasons for DL's popularity

If you've followed ML for some time, you may have noticed that many DL algorithms are not new. **Multilayer perceptrons** (**MLPs**) have been around for nearly 50 years. Backpropagation was discovered a couple of times but finally gained recognition in 1986. Yann LeCun, a famous computer scientist, perfected his work on convolutional networks in the 1990s. In 1997, Sepp Hochreiter and Jürgen Schmidhuber invented long short-term memory, a type of recurrent NN still in use today. In this section, we'll try to understand why we have AI summer now, and why we only had AI winters (https://en.wikipedia.org/wiki/AI_winter) before.

The first reason is that today, we have a lot more data than in the past. The rise of the internet and software in different industries has generated a lot of computer-accessible data. We also have more benchmark datasets, such as ImageNet. With this comes the desire to extract value from that data by analyzing it. And, as we'll see later, DL algorithms work better when they are trained with a lot of data.

The second reason is the increased computing power. This is most visible in the drastically increased processing capacity of **graphical processing units** (**GPUs**). NNs are organized in such a way as to take advantage of this parallel architecture. Let's see why. As we learned in *Chapter 2*, units from a network layer are not connected to units from the same layer. We also learned that we could represent many layer operations as matrix multiplications. Matrix multiplication is embarrassingly parallel (trust me, this is a term – you can Google it!). The computation of each output cell is not related to the computation of any other output cell. Therefore, we can compute all of the outputs in parallel. Not coincidentally, GPUs are well suited for highly parallel operations like this. On the one hand, a GPU has a high number of computational cores compared to a **central processing unit** (**CPU**). Even though a CPU core is faster than a GPU one, we can still compute a lot more output cells in parallel. But what's even more important is that GPUs are optimized for memory bandwidth, while CPUs are optimized for latency. This means that a CPU can fetch small chunks of memory very quickly but will be slow when it comes to fetching large chunks. The GPU does the opposite. For matrix multiplication in a deep network with a lot of wide layers, bandwidth becomes the bottleneck, not latency. In addition, the L1 cache of the GPU is much faster than the L1 cache for the CPU and is also larger. The L1 cache represents the memory of the information that the program is likely to use next, and storing this data can speed up the process. Much of the memory gets reused in DNNs, which is why L1 cache memory is important.

In the next section, *Deep neural networks*, we'll give a more precise definition of the key NN architectures that will be thoroughly introduced in the coming chapters.

Deep neural networks

We could define DL as a class of ML techniques, where information is processed in hierarchical layers to understand representations and features from data in increasing levels of complexity. In practice, all DL algorithms are NNs, which share some common basic properties. They all consist of a graph of interconnected operations, which operate with input/output tensors. Where they differ is network architecture (or the way units are organized in the network), and sometimes in the way they are trained. With that in mind, let's look at the main classes of NNs. The following list is not exhaustive, but it represents most NN types in use today:

- **Multilayer perceptron** (**MLP**): An NN with feedforward propagation, fully connected layers, and at least one hidden layer. We introduced MLPs in *Chapter 2*.

- **Convolutional neural network** (**CNN**): A CNN is a feedforward NN with several types of special layers. For example, convolutional layers apply a filter to the input image (or sound) by sliding that filter all across the incoming signal, to produce an *n*-dimensional activation map. There is some evidence that units in CNNs are organized similarly to how biological cells are organized in the visual cortex of the brain. We've mentioned CNNs several times so far, and that's not a coincidence – today, they outperform all other ML algorithms on many computer vision and NLP tasks. We'll discuss CNNs in *Chapter 4*.

- **Recurrent neural network** (**RNN**): This type of NN has an internal state (or memory), which is based on all, or part of, the input data that's already been fed to the network. The output of a recurrent network is a combination of its internal state (memory of inputs) and the latest input sample. At the same time, the internal state changes to incorporate newly input data. Because of these properties, recurrent networks are good candidates for tasks that work on sequential data, such as text or time series data. We'll discuss recurrent networks in *Chapter 6*.

- **Transformer**: Like RNNs, the transformer is suited to work with sequential data. It uses a mechanism called **attention**, which allows it *direct simultaneous access* to all elements of the input sequence. This is unlike an RNN, which processes the sequence elements one by one and updates its internal state after each element. As we'll see in *Chapter 7*, the attention mechanism has several major advantages over the classic RNNs. Because of this, in recent years, transformers have superseded RNNs in many tasks.

- **Autoencoders**: As we mentioned in *Chapter 1*, autoencoders are a class of unsupervised learning algorithms, in which the output shape is the same as the input, which allows the network to better learn basic representations.

Now that we've outlined the major types of DNNs, let's discuss how to train them.

Training deep neural networks

Historically, the scientific community has understood that deeper networks have greater representational power compared to shallow ones. However, there were various challenges in training networks with more than a few hidden layers. We now know that we can successfully train DNNs using a combination of gradient descent and backpropagation, just as we discussed in *Chapter 2*. In this section, we'll see how to improve them so that we can solve some of the problems that exist uniquely for DNNs and not shallow NNs.

The first edition of this book included networks such as **Restricted Boltzmann Machines** (**RBMs**) and **Deep Belief Networks** (**DBNs**). They were popularized by Geoffrey Hinton, a Canadian scientist, and one of the most prominent DL researchers. Back in 1986, he was also one of the inventors of backpropagation. RBMs are a special type of generative NN, where the units are organized into two layers, namely visible and hidden. Unlike feedforward networks, the data in an RBM can flow in both directions – from visible to hidden units, and vice versa. In 2002, Prof. Hinton introduced **contrastive divergence**, which is an unsupervised algorithm for training RBMs. In 2006, he introduced **deep belief networks** (**DBNs**), which are DNNs that are formed by stacking multiple RBMs. Thanks to their novel training algorithm, it was possible to create a DBN with more hidden layers than had previously been possible. But even with contrastive divergence, training a DBN is not easy. It is a two-step process:

1. First, we have to train each RBM with contrastive divergence, and gradually stack them on top of each other. This phase is called **pre-training**.

2. In effect, pre-training serves as a sophisticated weight initialization algorithm for the next phase, called **fine-tuning**. With fine-tuning, we transform the DBN into a regular MLP and continue training it using supervised backpropagation and gradient descent, in the same way as we saw in *Chapter 2*.

Thanks to some algorithmic advances, it's now possible to train deep networks using plain old backpropagation, thus effectively eliminating the pre-training phase. These advances rendered DBNs and RBMs obsolete. They are, without a doubt, interesting from a research perspective, but they are rarely used in practice anymore and we'll omit them from this edition.

Next, let's discuss the algorithmic advances that made training of NNs with backpropagation possible.

Improved activation functions

But why is training deep networks so hard? One of the main challenges that pre-training solved is the so-called **vanishing gradients** problem. To understand it, we'll assume that we'll use backpropagation to train a regular MLP with multiple hidden layers and logistic sigmoid activation at each layer. Let's focus on the sigmoid function (the same applies to tanh). As a reminder, it is computed as $\sigma(x) = 1/(1 + e^{-x})$:

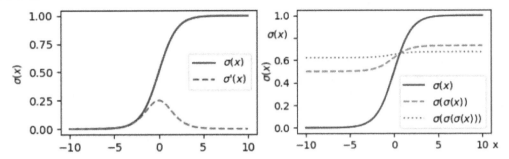

Figure 3.4 – Logistic sigmoid (uninterrupted) and its derivative (interrupted)
(left); consecutive sigmoid activations, which "squash" the data (right)

The vanishing gradients manifest themselves in the following ways:

- In the forward phase, the outputs of the first sigmoid layer are represented by the blue uninterrupted line (both left and right images in the preceding figure) and fall in the range (0, 1). The dotted lines on the right image represent the sigmoid activations of each of the consecutive layers after the first. Even after three layers, we can see that the activation is "squashed" in a narrow range and converges to around 0.66, regardless of the input value. For example, if the input value of the first layer is 2, then $\sigma(2) = 0.881$, $\sigma(\sigma(2)) = 0.71$, $\sigma(\sigma(\sigma(2))) = 0.67$, and so on. This peculiarity of the sigmoid function acts as an eraser of any information coming from the preceding layers.

- We now know that to train an NN, we need to compute the derivative of the activation function (along with all the other derivatives) for the backward phase. The derivative of the sigmoid function is represented by the green interrupted line on the left image in the preceding figure. We can see that it has a significant value in a very narrow interval, centered around 0, and converges toward 0 in all other cases. In networks with many layers, the derivative would likely converge to 0 when propagated to the first layers of the network. Effectively, this means we cannot propagate the error to these layers and we cannot update their weights in a meaningful way.

Thankfully, the **ReLU** activation we introduced in *Chapter 2* can solve both of these problems with a single stroke. To recap, the following figure shows the ReLU graph and its derivative:

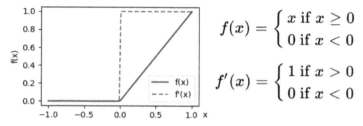

$$f(x) = \begin{cases} x \text{ if } x \geq 0 \\ 0 \text{ if } x < 0 \end{cases}$$

$$f'(x) = \begin{cases} 1 \text{ if } x > 0 \\ 0 \text{ if } x < 0 \end{cases}$$

Figure 3.5 – ReLU activation (uninterrupted) and its derivative (interrupted)

ReLU has the following desirable properties:

- It is **idempotent**. If we pass a value through an arbitrary number of ReLU activations, it will not change; for example, *ReLU(2) = 2*, *ReLU(ReLU(2)) = 2*, and so on. This is not the case for a sigmoid. The idempotence of ReLU makes it theoretically possible to create networks with more layers compared to the sigmoid.

- We can also see that its derivative is either 0 or 1, regardless of the backpropagated value. In this way, we can avoid vanishing gradients in the backward pass as well. Strictly speaking, the derivative ReLU at value 0 is undefined, which makes the ReLU only semi-differentiable (more information about this can be found at `https://en.wikipedia.org/wiki/Semi-differentiability`). But in practice, it works well enough.

- It creates sparse activations. Let's assume that the weights of the network are initialized randomly through normal distribution. Here, there is a 0.5 chance that the input for each ReLU unit is < 0. Therefore, the output of about half of all activations will also be 0. The sparse activations have several advantages, which we can roughly summarize as Occam's razor in the context of NNs – it's better to achieve the same result with a simpler data representation than a complex one.

- It's faster to compute in both the forward and backward passes.

Despite these ReLU advantages, during training, the network weights can be updated in such a way that some of the ReLU units in a layer will always receive inputs smaller than 0, which, in turn, will cause them to permanently output 0 as well. This phenomenon is known as **dying ReLUs**. To solve this, several ReLU modifications have been proposed. The following is a non-exhaustive list:

- **Leaky ReLU**: When the input is larger than 0, leaky ReLU repeats its input in the same way as the regular ReLU does. However, when $x < 0$, the leaky ReLU outputs x multiplied by some constant, α $(0 < α < 1)$, instead of 0. The following diagram shows the leaky ReLU formula, its derivative, and their graphs for $α = 0.2$:

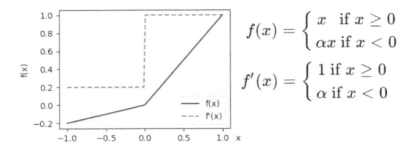

Figure 3.6 – The leaky ReLU activation function

- **Parametric ReLU** (**PReLU**; see *Delving Deep into Rectifiers: Surpassing Human-Level Performance on ImageNet Classification*, `https://arxiv.org/abs/1502.01852`): This activation is the same as the leaky ReLU, but α is tunable and is adjusted during training.

- **Exponential linear unit** (**ELU**; see *Fast and Accurate Deep Network Learning by Exponential Linear Units (ELUs)*, `https://arxiv.org/abs/1511.07289`): When the input is larger than 0, ELU repeats its input in the same way as ReLU does. However, when $x < 0$, the ELU output becomes $f(x) = \alpha(e^x - 1)$, where α is a tunable parameter. The following diagram shows the ELU formula, its derivative, and their graphs for $\alpha = 0.2$:

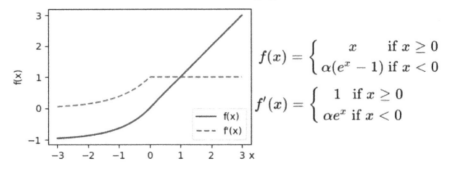

Figure 3.7 – The ELU activation function

- **Scaled exponential linear unit** (**SELU**; see *Self-Normalizing Neural Networks*, `https://arxiv.org/abs/1706.02515`): This activation is like ELU, except that the output (both smaller and larger than 0) is scaled with an additional training parameter, λ. The SELU is part of a larger concept called **self-normalizing NNs** (**SNNs**), which is described in the source paper.

- **Sigmoid Linear Unit** (**SiLU**), **Gaussian Error Linear Unit** (**GELU**; see *Gaussian Error Linear Units (GELUs)*, `https://arxiv.org/abs/1606.08415`), and **Swish** (see *Searching for Activation Functions*, `https://arxiv.org/abs/1710.05941`): This is a collection of three similar (but not the same) functions that closely resemble ReLU but are differentiable at the 0 point. For the sake of simplicity, we'll only show the SiLU graph (σ is the sigmoid function):

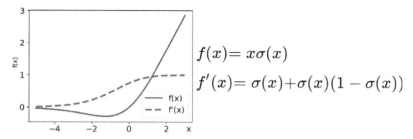

Figure 3.8 – The SiLU activation function

Finally, we have softmax, which is the activation function of the output layer in classification problems. Let's assume that the output of the final network layer is a vector, $\mathbf{z} = (z_1, z_2 ... z_n)$. Each of the n elements represents one of n classes, to which the input sample might belong. To determine the network prediction, we'll take the index, i, of the highest value, z_i, and assign the input sample to the class it represents. However, we can also interpret the network output as a probability distribution of a discrete random variable – that is, each value, z_i, represents the probability that the input sample belongs to that particular class. To help us with this, we'll use the softmax activation:

$$f(z_i) = \frac{exp(z_i)}{\sum_{j=1}^{n} exp(z_j)}$$

It has the following properties:

- The denominator in the formula acts as a normalizer. This is important for the probability interpretation we just introduced:

  - Every value, $f(z_i)$, is constrained within the [0, 1] range, which allows us to treat it as a probability

  - The total sum of values of $f(z_i)$ is equal to 1: $\sum_j f(z_j) = 1$, which also aligns with the probability interpretation

- A bonus (in fact, obligatory) is that the function is differentiable.

- The softmax activation has one more subtle property. Before we normalize the data, we transform each vector component exponentially with e^z. Let's imagine that two of the vector components are $z_1 = 1$ and $z_2 = 2$. Here, we would have exp(1) = 2.7 and exp(2) = 7.39. As we can see, the ratios between the components before and after the transformation are very different – 0.5 and 0.36. In effect, the softmax function increases the probability of higher scores compared to lower ones.

In practice, **softmax** is often used in combination with the **cross-entropy loss** function. It compares the difference between the estimated class probabilities and the actual class distribution (the difference is known as cross-entropy). We can define the cross-entropy loss for a single training sample as follows:

$$H(p, q) = -\sum_{j=1}^{n} p_j(x) log(q_j(x))$$

Here, $q_j(x)$ is the estimated probability of the output belonging to class j (out of n total classes) and p_j (x) is the actual probability. The actual distribution, $P(X)$, is usually a one-hot-encoded vector, where the real class has a probability of 1, and all others have a probability of 0. In this case, the cross-entropy loss will only capture the error on the target class and will discard all other errors.

Now that we've learned how to prevent vanishing gradients and we're able to interpret the NN output as a probability distribution, we'll focus on the next challenge in front of DNNs – overfitting.

DNN regularization

So far, we've learned that an NN can approximate any function. But with great power comes great responsibility. The NN may learn to approximate the noise of the target function rather than its useful components. For example, imagine that we are training an NN to classify whether an image contains a car or not, but for some reason, the training set contains mostly red cars. It may turn out that the NN will associate the color red with the car, rather than its shape. Now, if the network sees a green car in inference mode, it may not recognize it as such because the color doesn't match. As we discussed in *Chapter 1*, this problem is referred to as overfitting and it is central to ML (and even more so in deep networks). In this section, we'll discuss several ways to prevent it. Such techniques are collectively known as regularization.

In the context of NNs, these regularization techniques usually impose some artificial limitations or obstacles on the training process to prevent the network from approximating the target function too closely. They try to guide the network to learn generic rather than specific approximation of the target function in the hope that this representation will generalize well on previously unseen examples of the test dataset. Let's start with regularization techniques that apply to the input data before we feed it to the NN:

- **Min-max normalization**: $x = \frac{x - x_{min}}{x_{max} - x_{min}}$. Here, x is a single element of the input vector, x_{min} is the smallest element of the training dataset, and x_{max} is the largest element. This operation scales all the inputs in the [0, 1] range. For example, a grayscale image will have a min color value of 0 and a max color value of 255. Then, a pixel with an intensity of 125 would have a scaled value of $x = (125 - 0)/(255 - 0) = 0.49$. Min-max is fast and easy to implement. One problem with this normalization is that data outliers could have an outsized impact on the result over the whole dataset. For example, if a single erroneous element has a very large value, it will enter the formula as x_{max} and it will drive all normalized dataset values toward 0.

- **Standard score** (or **z-score**): $z = \frac{x - \mu}{\sigma}$. It handles data outliers better than min-max. To understand how, let's focus on the formula:

  - $\mu = 1/N \sum_{i=1}^{N} x_i$ is the mean value of all elements of the dataset, where x_i is a single element of the input vector and N is the total size of the dataset.

- $\sigma = \sqrt{1/N \sum_{i=1}^{N}(x_i - \mu)^2}$ is the **standard deviation** of all dataset elements. It measures how far apart the dataset values are from the mean value. There is also **variance,** $\sigma^2 = 1/N \sum_{i=1}^{N}(x_i - \mu)^2$, which removes the square root from the standard deviation. The variance is theoretically correct but is less intuitive than standard deviation, which is measured in the same units as the original data, x.

Alternatively, we can compute μ and σ per sample if it's not practical to compute them over the entire dataset. The standard score maintains the dataset's mean value close to 0 and its standard deviation close to 1.

- **Data augmentation**: This is where we artificially increase the size of the training set by applying random modifications to the training samples before feeding them to the network. In the case of images, these would be rotation, skew, scaling, and so on.

The next class of regularization techniques are applied within the DNN structure itself:

- **Dropout**: Here, we randomly and periodically remove some of the units of a layer (along with their input and output connections) from the network. During a training mini-batch, each unit has a probability, p, of being stochastically dropped. This is to ensure that no unit ends up relying too much on other units and "learns" something useful for the NN instead. Dropout is only applied during the training phase and all the units in the network fully participate during the inference phase. In the following figure, we can see a dropout for fully connected layers:

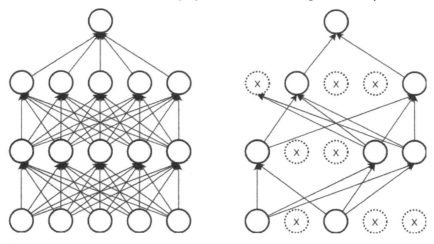

Figure 3.9 – An example of dropout on fully connected layers

- **Batch normalization** (**BN**; see *Batch Normalization: Accelerating Deep Network Training by Reducing Internal Covariate Shift*, https://arxiv.org/abs/1502.03167): This is a way to apply data processing, not unlike the standard score, for the hidden layers of the network. It normalizes the outputs of the hidden layer for each mini-batch (hence the name) in a way that maintains its mean activation value close to 0 (**re-centering**) and its standard deviation close to 1 (**re-scaling**). The intuition is that as information is propagated through the layers, these values can deviate from the desired values. Let's say that the mini-batch is represented by an $m \times n$ matrix, $\mathbf{X}$. Each row of $\mathbf{X}$, $\mathbf{x}_i$, represents a single input vector (this vector is an output of a preceding layer). x_{ij} is the j-th element of the i-th vector. We can compute BN for each matrix element in the following way:

- $\mu_{\mathbf{X}} \leftarrow \frac{1}{m \times n} \sum_{i=1}^{m} \sum_{j=1}^{n} x_{ij}$: This is the mini-batch mean. We compute a single μ value over all cells of the mini-batch matrix.

- $\sigma_{\mathbf{X}}^2 \leftarrow \frac{1}{m \times n} \sum_{i=1}^{m} \sum_{j=1}^{n} (x_{ij} - \mu_{\mathbf{X}})^2$: This is the mini-batch variance. We compute a single σ^2 value over all cells of the mini-batch matrix.

- $\widehat{x_{ij}} \leftarrow \frac{x_{ij} - \mu_{\mathbf{X}}}{\sqrt{\sigma_{\mathbf{X}}^2 + \varepsilon}}$: We normalize each cell of the matrix. ε is a constant that's added for numerical stability, so the denominator cannot become 0.

- $y_{ij} \leftarrow \gamma \widehat{x_{ij}} + \beta \equiv BN_{\gamma,\beta}(x_{ij})$: This formula represents the scale and shift of the original data. γ and β are learnable parameters and we compute them over each location, ij (γ_{ij} and β_{ij}), over all cells of the mini-batch matrix.

- **Layer normalization** (**LN**; see *Layer Normalization*, `https://arxiv.org/abs/1607.06450`): LN is similar to BN, but with one key difference: the mean and variance are computed separately over each mini-batch sample. This is unlike BN, where these values are computed across the whole mini-batch. As with BN, the mini-batch is an $m \times n$ matrix, $\mathbf{X}$, and each row vector, $\mathbf{x}_i$, is the output of a preceding layer, and x_{ij} is the j-th element of the i-th vector. Then, we have the following for the i-th input vector:

  - $\mu_{\mathbf{x}_i} \leftarrow \frac{1}{n} \sum_{j=1}^{n} x_{ij}$
  - $\sigma_{\mathbf{x}_i}^2 \leftarrow \frac{1}{n} \sum_{j=1}^{n} (x_{ij} - \mu_{\mathbf{x}_i})^2$
  - $\widehat{x_{ij}} \leftarrow \frac{x_{ij} - \mu_{\mathbf{x}_i}}{\sqrt{\sigma_{\mathbf{x}_i}^2 + \varepsilon}}$
  - $y_{ij} \leftarrow \gamma \widehat{x_{ij}} + \beta \equiv LN_{\gamma,\beta}(x_{ij})$

- **Root mean square layer normalization** (**RMSNorm**; see `https://arxiv.org/abs/1910.07467`): The authors of RMSNorm argue that the main benefit of LN comes just from the re-scaling, rather than the combination of re-centering and re-scaling. Therefore, RMSNorm is a simplified and faster version of LN, which only applies re-scaling using the root mean square statistic. We'll use the same notation as with LN. So, we can define RMSNorm as follows:

  - $RMS(\mathbf{x}_i) = \sqrt{\frac{1}{n} \sum_{j=1}^{n} x_{ij}^2}$.

  - $x_{ij} \leftarrow \frac{x_{ij}}{RMS(\mathbf{x}_i)} \gamma_{ij}$: Here, γ_{ij} is the gain parameter used to re-scale the standardized summed inputs (it is set to 1 at the beginning). It is equivalent to the γ parameter in BN.

The following figure illustrates the difference between BN and LN. On the left, we compute single μ and σ values across the whole mini-batch. To the right, we can see μ_i and σ_i for each row:

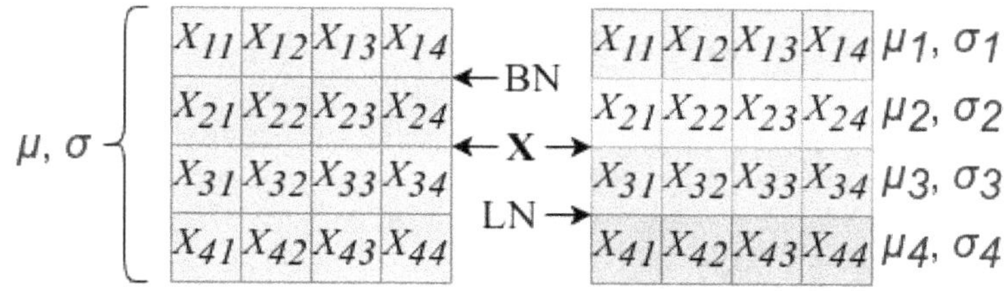
μ, σ
X11 X12 X13 X14
X21 X22 X23 X24
X31 X32 X33 X34
X41 X42 X43 X44
BN
X
LN
X11 X12 X13 X14 μ1, σ1
X21 X22 X23 X24 μ2, σ2
X31 X32 X33 X34 μ3, σ3
X41 X42 X43 X44 μ4, σ4

Figure 3.10 – BN and LN computation of μ and σ

The final type of regularization we'll introduce is **L2 regularization**. This technique adds a special regularization term to the cost function. To understand it, let's take the MSE cost. We can add L2 regularization to it in the following way (the underscored part of the formula):

$$J(\theta) = \frac{1}{2n}\left[\sum_{i=1}^{n}\left(f_\theta(\mathbf{x}^{(i)}) - t^{(i)}\right)^2 + \lambda\underline{\sum_{j=1}^{m}\theta_j^2}\right]$$

Here, θ_j is one of m total network weights and λ is the weight decay coefficient. The rationale is that if the network weights, θ_j, are large, then the cost function will also increase. In effect, weight decay penalizes large weights (hence the name). This prevents the network from relying too heavily on a few features associated with these weights. There is less chance of overfitting when the network is forced to work with multiple features. In practical terms, when we compute the derivative of the weight decay cost function (the preceding formula) concerning each weight and then propagate it to the weights themselves, the weight update rule changes from the following:

$$\theta_j \leftarrow \theta_j - \eta\left(\frac{\partial J(\theta)}{\partial \theta_j}\right) \text{ to } \theta_j \leftarrow \theta_j - \eta\left(\frac{\partial J(\theta)}{\partial \theta_j} + \lambda\theta_j\right)$$

With this discussion of DNN regularization, we've covered our theoretical base. Next, let's see what the real-world applications of DNNs are.

Applications of DL

ML in general, and DL in particular, is producing more and more astonishing results in terms of the quality of predictions, feature detection, and classification. Many of these recent results have made the news. Such is the pace of progress that some experts are worrying that machines will soon be more intelligent than humans. But I hope that any such fears you might have will be alleviated after you have read this book. For better or worse, we're still far from machines having human-level intelligence.

In *Chapter 2*, we mentioned how DL algorithms have occupied the leaderboard of the ImageNet competition. They are successful enough to make the jump from academia to industry.

Let's talk about some real-world use cases of DL:

- Nowadays, new cars have a suite of safety and convenience features that aim to make the driving experience safer and less stressful. One such feature is automated emergency braking if the car sees an obstacle. Another one is lane-keeping assist, which allows the vehicle to stay in its current lane without the driver needing to make corrections with the steering wheel. To recognize lane markings, other vehicles, pedestrians, and cyclists, these systems use a forward-facing camera. One of the most prominent suppliers of such systems, Mobileye (https://www.mobileye.com/), has produced custom chips that use CNNs to detect these objects on the road ahead. To give you an idea of the importance of this sector, in 2017, Intel acquired Mobileye for $15.3 billion. This is not an outlier, and Tesla's famous Autopilot system also relies on CNNs to achieve the same results. The former director of AI at Tesla, Andrej Karpathy (https://karpathy.ai/), is a well-known researcher in the field of DL. We can speculate that future autonomous vehicles will also use deep networks for computer vision.

- Both Google's **Vision API** (https://cloud.google.com/vision/) and Amazon's **Rekognition** (https://aws.amazon.com/rekognition/) services use DL models to provide various computer vision capabilities. These include recognizing and detecting objects and scenes in images, text recognition, face recognition, content moderation, and so on.

- If these APIs are not enough, you can run your own models in the cloud. For example, you can use Amazon's AWS DL AMIs (short for **Amazon Machine Images**; see https://aws.amazon.com/machine-learning/amis/), which are virtual machines that come configured with some of the most popular DL libraries. Google offers a similar service with their Cloud AI (https://cloud.google.com/products/ai/), but they've gone one step further. They created **tensor processing units** (**TPUs**; see https://cloud.google.com/tpu/) – microprocessors that are optimized for fast NN operations such as matrix multiplication and activation functions.

- DL has a lot of potential for medical applications. However, strict regulatory requirements, as well as patient data confidentiality, have slowed down its adoption. Nevertheless, we'll identify several areas in which DL could have a high impact:

  - Medical imaging is an umbrella term for various non-invasive methods of creating visual representations of the inside of the body. Some of these include **magnetic resonance images** (**MRIs**), ultrasound, **computed axial tomography** (**CAT**) scans, X-rays, and histology images. Typically, such an image is analyzed by a medical professional to determine the patient's condition.

  - Computer-aided diagnosis, and computer vision, in particular, can help specialists by detecting and highlighting important features of images. For example, to determine the degree of malignancy of colon cancer, a pathologist would have to analyze the morphology of the glands using histology imaging. This is a challenging task because morphology can vary greatly. A DNN could segment the glands from the image automatically, leaving the pathologist to verify the results. This would reduce the time needed for analysis, making it cheaper and more accessible.

- Another medical area that could benefit from DL is the analysis of medical history records. When a doctor diagnoses a patient's condition and prescribes treatment, they consult the patient's medical history first. A DL algorithm could extract the most relevant and important information from those records, even if they are handwritten. In this way, the doctor's job would be made easier, and the risk of errors would also be reduced.

- One area where DNNs have already had an impact is in protein folding. Proteins are large, complex molecules, whose function depends on their 3D shape. The building blocks of proteins are amino acids, and their sequence determines the shape of the protein. The protein folding problem seeks to understand the relationship between the initial amino acid sequence and the final 3D shape of the protein. DeepMind's **AlphaFold 2** model (believed to be based on transformers; see `https://www.deepmind.com/blog/alphafold-reveals-the-structure-of-the-protein-universe`) has managed to predict 200 million protein structures, which represents almost all known cataloged proteins.

- Google's Neural Machine Translation API (`https://arxiv.org/abs/1609.08144`) uses – you guessed it – DNNs for machine translation.

- Siri (`https://machinelearning.apple.com/2017/10/01/hey-siri.html`), Google Assistant, and Amazon Alexa (`https://aws.amazon.com/deep-learning/`) rely on deep networks for speech recognition.

- **AlphaGo** is an AI machine based on DL that made the news in 2016 by beating the world Go champion, Lee Sedol. AlphaGo had already made the news, in January 2016, when it beat the European champion, Fan Hui. At the time, however, it seemed unlikely that it could go on to beat the world champion. Fast-forward a couple of months and AlphaGo was able to achieve this remarkable feat by sweeping its opponent in a 4-1 victory series. This was an important milestone because Go has many more possible game variations than other games, such as chess, and it's impossible to consider every possible move in advance. Also, unlike chess, in Go, it's very difficult to even judge the current position or value of a single stone on the board. In 2017, DeepMind released an updated version of AlphaGo called **AlphaZero** (`https://arxiv.org/abs/1712.01815`), and in 2019, they released a further update called **MuZero** (`https://arxiv.org/abs/1911.08265`).

- Tools such as GitHub Copilot (`https://github.com/features/copilot`) and ChatGPT (`https://chat.openai.com/`) utilize generative DNN models to transform natural language requests into source code snippets, functions, and entire programs. We already mentioned Stable Diffusion (`https://stability.ai/blog/stable-diffusion-public-release`) and DALL-E (`https://openai.com/dall-e-2/`), which can generate realistic images based on text description.

With this short list, we aimed to cover the main areas in which DL is applied, such as computer vision, NLP, speech recognition, and **reinforcement learning** (**RL**). This list is not exhaustive, however, as there are many other uses for DL algorithms. Still, I hope this has been enough to spark your interest. Next, we'll formally introduce two of the most popular DL libraries – PyTorch and Keras.

Introducing popular DL libraries

We already implemented a simple example with PyTorch in *Chapter 1*. In this section, we'll introduce this library, and Keras, more systemically. Let's start with the common features of most DNN libraries:

- All libraries use Python.

- The basic unit for data storage is the **tensor**. Mathematically, the definition of a tensor is more complex, but in the context of DL libraries, they are multi-dimensional (with an arbitrary number of axes) arrays of base values.

- NNs are represented as a **computational graph** of operations. The nodes of the graph represent the operations (weighted sum, activation function, and so on). The edges represent the flow of data, which is how the output of one operation serves as an input for the next one. The inputs and outputs of the operations (including the network inputs and outputs) are tensors.

- All libraries include **automatic differentiation**. This means that all you need to do is define the network architecture and activation functions, and the library will automatically figure out all of the derivatives required for training with backpropagation.

- So far, we've referred to GPUs in general, but in reality, the vast majority of DL projects work exclusively with NVIDIA GPUs. This is because of the better software support NVIDIA provides. These libraries are no exception – to implement GPU operations, they rely on the CUDA Toolkit (`https://developer.nvidia.com/cuda-toolkit`) in combination with the cuDNN library (`https://developer.nvidia.com/cudnn`). cuDNN is an extension of CUDA, built specifically for DL applications. As mentioned in the *Applications of DL* section, you can also run your DL experiments in the cloud.

PyTorch is an independent library, while Keras is built on top of TF and acts as a user-friendly TF interface. We'll continue by implementing a simple classification example using both PyTorch and Keras.

Classifying digits with Keras

Keras exists either as a standalone library with TF as the backend or as a sub-component of TF itself. You can use it in both flavors. To use Keras as part of TF, we need only to install TF itself. Once we've done this, we can use the library with the following import:

```
import tensorflow.keras
```

The standalone Keras supports different backends besides TF, such as Theano. In this case, we can install Keras itself and then use it with the following import:

```
import keras
```

The large majority of Keras's use is with the TF backend. The author of Keras recommends using the library as a TF component (the first option) and we'll do so in the rest of this book.

In this section, we'll use Keras via TF to classify the images of the MNIST dataset. It's comprised of 70,000 examples of digits that have been handwritten by different people. The first 60,000 are typically used for training and the remaining 10,000 for testing:

Figure 3.11 – A sample of digits taken from the MNIST dataset

We'll build a simple MLP with one hidden layer. Let's start:

1. One of the advantages of Keras is that it can import this dataset for you without you needing to explicitly download it from the web (it will download it for you):

```
import tensorflow as tf
(X_train, Y_train), (X_validation, Y_validation) = \
          tf.keras.datasets.mnist.load_data()
```

Here, (X_train, Y_train) is the training images and labels, and (X_validation, Y_validation) is the test images and labels.

2. We need to modify the data so that we can feed it to the NN. X_train contains 60,000 28×28 pixel images, and X_validation contains 10,000. To feed them to the network as inputs, we want to reshape each sample as a 784-pixel-long array, rather than a 28×28 two-dimensional matrix. We'll also normalize them in the [0:1] range. We can accomplish this with these two lines:

```
X_train = X_train.reshape(60000, 784) / 255
X_validation = X_validation.reshape(10000, 784) / 255
```

3. The labels indicate the value of the digit depicted in the images. We want to convert this into a 10-entry **one-hot-encoded** vector comprised of 0s and just one 1 in the entry corresponding to the digit. For example, 4 is mapped to [0, 0, 0, 0, 1, 0, 0, 0, 0, 0]. Conversely, our network will have 10 output units:

```
classes = 10
Y_train = tf.keras.utils.to_categorical(Y_train,
          classes)
Y_validation = tf.keras.utils.to_categorical(
          Y_validation, classes)
```

4. Define the NN. In this case, we'll use the `Sequential` model, where each layer serves as an input to the next. In Keras, `Dense` means a fully connected layer. We'll use a network with one hidden layer with 100 units, BN, ReLU activation, and softmax output:

```python
from tensorflow.keras.models import Sequential
from tensorflow.keras.layers import Dense, BatchNormalization,
Activation
input_size = 784
hidden_units = 100
model = Sequential([
    Dense(
        hidden_units, input_dim=input_size),
    BatchNormalization(),
    Activation('relu'),
    Dense(classes),
    Activation('softmax')
])
```

5. Now, we can define our gradient descent parameters. We'll use the Adam optimizer and categorical cross-entropy loss (this is cross entropy, optimized for softmax outputs):

```python
model.compile(
    loss='categorical_crossentropy',
    metrics=['accuracy'],
    optimizer='adam')
```

6. Next, run the training for 100 epochs and a batch size of 100. In Keras, we can do this with the `fit` method, which iterates over the dataset internally. Keras will default to GPU training, but if a GPU is not available, it will fall back to the CPU:

```python
model.fit(X_train, Y_train, batch_size=100, epochs=20,
          verbose=1)
```

7. All that's left to do is add code to evaluate the network's accuracy on the test data:

```python
score = model.evaluate(X_validation, Y_validation,
          verbose=1)
print('Validation accuracy:', score[1])
```

And that's it. The validation accuracy will be about 97.7%, which is not a great result, but this example runs in less than 30 seconds on a CPU. We can make some simple improvements, such as a larger number of hidden units, or a higher number of epochs. We'll leave those experiments to you so that you can familiarize yourself with the code.

8. To see what the network has learned, we can visualize the weights of the hidden layer. The following code allows us to obtain them:

```
weights = model.layers[0].get_weights()
```

9. Reshape the weights for each unit back to a 28×28 two-dimensional array and then display them:

```
import matplotlib.pyplot as plt
import matplotlib.cm as cm
import numpy
fig = plt.figure()
w = weights[0].T
for unit in range(hidden_units):
    ax = fig.add_subplot(10, 10, unit + 1)
    ax.axis("off")
    ax.imshow(numpy.reshape(w[unit], (28, 28)),
    cmap=cm.Greys_r)
plt.show()
```

We can see the result in the following figure:

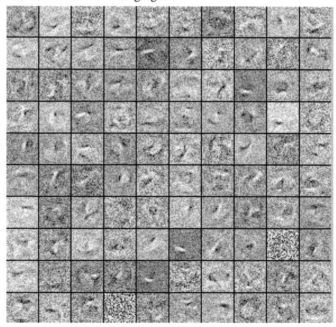

Figure 3.12 – A composite figure of what was learned by all the hidden units

Now, let us see the example for PyTorch.

Classifying digits with PyTorch

In this section, we'll implement the same example that we did in the *Classifying digits with Keras* section but this time with PyTorch. Let's start:

1. First, we'll select the device we're using (CPU or GPU). We'll try with the GPU first and fall back to the CPU if the GPU is not available:

    ```
    import torch
    device = torch.device("cuda:0" if torch.cuda.is_available() else
    "cpu")
    ```

2. Like Keras, PyTorch supports MNIST out of the box. Here's how we can instantiate the train and validation sets:

    ```
    from torchvision import datasets
    from torchvision.transforms import ToTensor, Lambda, Compose
    train_data = datasets.MNIST(
        root='data',
        train=True,
        transform=Compose(
            [ToTensor(),
            Lambda(lambda x: torch.flatten(x))]),
            download=True,
        )
    validation_data = datasets.MNIST(
        root='data',
        train=False,
        transform=Compose(
            [ToTensor(),
            Lambda(lambda x: torch.flatten(x))]),
        )
    ```

    The dataset is automatically downloaded and split into training and validation parts. The `ToTensor()` transformation converts the images from numpy arrays into PyTorch tensors and normalizes them in the [0:1] range (as opposed to [0:255] originally). The `torch.flatten` transform flattens the two-dimensional 28×28 images to a one-dimensional 784 tensor so that we can feed it to the NN.

3. Next, we'll wrap the datasets in `DataLoader` instances:

    ```
    from torch.utils.data import DataLoader
    train_loader = DataLoader(
        dataset=train_data,
        batch_size=100,
        shuffle=True)
    ```

```
validation_loader = DataLoader(
    dataset=validation_data,
    batch_size=100,
    shuffle=True)
```

The data `DataLoader` instance takes care of creating mini-batches and shuffles the data randomly. They are also iterators, which supply mini-batches one at a time.

4. Then, we'll define the NN `model`. We'll use the same MLP with a single hidden layer, as in the Keras example:

```
torch.manual_seed(1234)
hidden_units = 100
classes = 10
model = torch.nn.Sequential(
    torch.nn.Linear(28 * 28, hidden_units),
    torch.nn.BatchNorm1d(hidden_units),
    torch.nn.ReLU(),
    torch.nn.Linear(hidden_units, classes),
)
```

This definition is like the one in Keras. One difference is that the `Linear` (fully connected) layers require both input and output dimensions since they cannot extract the output dimension of the preceding layer. The activations are defined as separate operations.

5. Next, let's define the cross-entropy loss and the Adam optimizer:

```
cost_func = torch.nn.CrossEntropyLoss()
optimizer = torch.optim.Adam(model.parameters())
```

6. Now, we can define the `train_model` function, which, as its name suggests, takes care of training the model. It takes our predefined `model`, `cost_function`, `optimizer`, and `data_loader` and runs the training for a single epoch:

```
def train_model(model, cost_function, optimizer, data_loader):
    # send the model to the GPU
    model.to(device)

    # set model to training mode
    model.train()

    current_loss = 0.0
    current_acc = 0

    # iterate over the training data
    for i, (inputs, labels) in enumerate(data_loader):
        # send the input/labels to the GPU
```

```
    inputs = inputs.to(device)
    labels = labels.to(device)
    # zero the parameter gradients
    optimizer.zero_grad()

    with torch.set_grad_enabled(True):
        # forward
        outputs = model(inputs)
        _, predictions = torch.max(outputs, 1)
        loss = cost_function(outputs, labels)

        # backward
        loss.backward()
        optimizer.step()

    # statistics
    current_loss += loss.item() * inputs.size(0)
    current_acc += torch.sum(predictions == labels.data)
total_loss = current_loss / len(data_loader.dataset)
total_acc = current_acc.double() / len(data_loader.dataset)

print('Train Loss: {:.4f}; Accuracy: /
    {:.4f}'.format(total_loss, total_acc))
```

Unlike Keras and its `fit` function, we have to implement the PyTorch training ourselves. `train_model` iterates over all mini-batches provided by `train_loader`. For each mini-batch, `optimizer.zero_grad()` resets the gradients from the previous iteration. Then, we initiate the forward and backward passes, and finally the weight updates.

7. We'll also define the `test_model` function, which will run the model in inference mode to check its results:

```
def test_model(model, cost_function, data_loader):
    # send the model to the GPU
    model.to(device)

    # set model in evaluation mode
    model.eval()

    current_loss = 0.0
    current_acc = 0
    # iterate over    the validation data
    for i, (inputs, labels) in enumerate(data_loader):
        # send the input/labels to the GPU
        inputs = inputs.to(device)
```

```
        labels = labels.to(device)

        # forward
        with torch.set_grad_enabled(False):
            outputs = model(inputs)
            _, predictions = torch.max(outputs, 1)
            loss = cost_function(outputs, labels)

        # statistics
        current_loss += loss.item() * inputs.size(0)
        current_acc += torch.sum(predictions == labels.data)

    total_loss = current_loss / len(data_loader.dataset)
    total_acc = current_acc.double() / len(data_loader.dataset)
    print('Test Loss: {:.4f}; Accuracy: /
        {:.4f}'.format(total_loss, total_acc))
    return total_loss, total_acc
```

BN and dropout layers are not used in evaluation (only in training), so `model.eval()` turns them off. We iterate over the validation set, initiate a forward pass, and aggregate the validation loss and accuracy.

8. Let's run the training for 20 epochs:

```
for epoch in range(20):
    train_model(model, cost_func, optimizer,
        train_loader)
    test_model(model, cost_func, validation_loader)
```

This model achieves 97.6% accuracy.

Summary

In this chapter, we explained what DL is and how it's related to DNNs. We discussed the different types of DNNs and how to train them, and we paid special attention to various regularization techniques that help with the training process. We also mentioned many real-world applications of DL and tried to analyze the reasons for its efficiency. Finally, we introduced two of the most popular DL libraries, namely PyTorch and Keras. We also implemented identical MNIST classification examples with both libraries.

In the next chapter, we'll discuss how to solve classification tasks over more complex image datasets with the help of convolutional networks – one of the most popular and effective deep network models. We'll talk about their structure, building blocks, and what makes them uniquely suited to computer vision tasks. To spark your interest, let's recall that convolutional networks have consistently won the popular ImageNet challenge since 2012, delivering top-five accuracy from 74.2% to 99%.

Part 2: Deep Neural Networks for Computer Vision

In this part, we'll introduce **convolutional neural networks** (**CNNs**) – a type of neural network suitable for computer vision applications. Building on top of the first three chapters, we'll discuss the rationale behind CNNs, their building blocks, and their architecture. We'll also outline the most popular CNN models in use today. Finally, we'll focus on the advanced applications of CNNs – object detection, image segmentation, and image generation.

This part has the following chapters:

- *Chapter 4, Computer Vision with Convolutional Networks*
- *Chapter 5, Advanced Computer Vision Applications*

4

Computer Vision with Convolutional Networks

In *Chapter 2* and *Chapter 3*, we set high expectations for **deep learning** (**DL**) and computer vision. First, we mentioned the ImageNet competition, and then we talked about some of its exciting real-world applications, such as semi-autonomous cars. In this chapter, and the next two chapters, we'll deliver on those expectations.

Vision is arguably the most important human sense. We rely on it for almost any action we take. But image recognition has (and in some ways still is), for the longest time, been one of the most difficult problems in computer science. Historically, it's been very difficult to explain to a machine what features make up a specified object, and how to detect them. But, as we've seen, in DL, a **neural network** (**NN**) can learn those features by itself.

In this chapter, we will cover the following topics:

- Intuition and justification for **convolutional neural networks** (**CNNs**)
- Convolutional layers
- Pooling layers
- The structure of a convolutional network
- Classifying images with PyTorch and Keras
- Advanced types of convolutions
- Advanced CNN models

Technical requirements

We'll implement the example in this chapter using Python, PyTorch, and Keras. If you don't have an environment set up with these tools, fret not – the example is available as a Jupyter Notebook on Google Colab. You can find the code examples in this book's GitHub repository: `https://github.com/PacktPublishing/Python-Deep-Learning-Third-Edition/tree/main/Chapter04`.

Intuition and justification for CNNs

The information we extract from sensory inputs is often determined by their context. With images, we can assume that nearby pixels are closely related, and their collective information is more relevant when taken as a unit. Conversely, we can assume that individual pixels don't convey information related to each other. For example, to recognize letters or digits, we need to analyze the dependency of pixels close by because they determine the shape of the element. In this way, we could figure out the difference between, say, a 0 or a 1. The pixels in an image are organized in a two-dimensional grid, and if the image isn't grayscale, we'll have a third dimension for the color channels.

Alternatively, a **magnetic resonance image** (**MRI**) also uses three-dimensional space. You might recall that, until now, if we wanted to feed an image to an NN, we had to reshape it from a two-dimensional array into a one-dimensional array. CNNs are built to address this issue: how to make information about units that are closer more relevant than information coming from units that are further apart. In visual problems, this translates into making units process information coming from pixels that are near to one another. With CNNs, we'll be able to feed one-, two-, or three-dimensional inputs and the network will produce an output of the same dimensionality. As we'll see later, this will give us several advantages.

You may recall that at the end of the previous chapter, we successfully classified the MNIST images (with around 98% accuracy) using an NN of **fully connected** (**FC**) layers. Historically, MNIST classification has been an important benchmark for measuring the performance of new computer vision algorithms. But nowadays, it's a toy dataset that we use for educational purposes, as in this book. Another such dataset is CIFAR-10 (*Canadian Institute For Advanced Research*, `https://www.cs.toronto.edu/~kriz/cifar.html`). It consists of 60,000 32×32 **red, green, blue** (**RGB**) images, divided into 10 classes of objects, namely `airplane`, `automobile`, `bird`, `cat`, `deer`, `dog`, `frog`, `horse`, `ship`, and `truck`. Had we tried to classify CIFAR-10 with an FC NN with one or more hidden layers, its validation accuracy would have been just around 50% (trust me, we did just that in the previous edition of this book). Compared to the MNIST result of nearly 98% accuracy, this is a dramatic difference, even though CIFAR-10 is also a toy problem. Therefore, FC NNs are of little practical use for computer vision problems. To understand why, let's analyze the first hidden layer of our hypothetical CIFAR-10 network, which has 1,000 units. The input size of the image is $32 * 32 * 3 = 3,072$. Therefore, the first hidden layer had a total of $2,072 * 1,000 = 2,072,000$ weights. That's no small number! Not only is it easy to overfit such a large network, but it's also memory inefficient. Even more important, each input unit (or pixel) is connected to every unit in the hidden layer. Because of this, the network cannot take advantage of the spatial proximity of the pixels since it doesn't have a way of knowing which pixels are close to each other. In contrast, CNNs have properties that provide an effective solution to these problems:

- They connect units that only correspond to neighboring pixels of the image. In this way, the units are "forced" to only take input from other units that are spatially close. This also reduces the number of weights since not all units are interconnected.

- CNNs use parameter sharing. In other words, a limited number of weights are shared among all units in a layer. This further reduces the number of weights and helps fight overfitting. It might sound confusing, but it will become clear in the next section.

> **Note**
>
> In this chapter, we'll discuss CNNs in the context of computer vision, because computer vision is their most common application. However, CNNs are successfully applied in areas such as speech recognition and **natural language processing** (**NLP**). Many of the explanations we'll describe here are also valid for those areas – that is, the principles of CNNs are the same regardless of the field of use.

To understand CNNs, we'll first discuss their basic building blocks. Once we've done this, we'll show you how to assemble them in a full-fledged NN. Then, we'll demonstrate that such a network is good enough to classify the CIFAR-10 with high accuracy. Finally, we'll discuss advanced CNN models, which can be applied to real-world computer vision tasks.

Convolutional layers

The convolutional layer is the most important building block of a CNN. It consists of a set of **filters** (also known as **kernels** or **feature detectors**), where each filter is applied across all areas of the input data. A filter is defined by a **set of learnable weights**.

To add some meaning to this laconic definition, we'll start with the following figure:

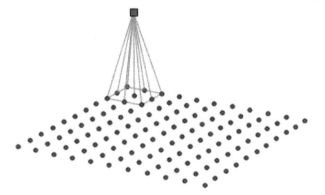

Figure 4.1 – Convolution operation start

The preceding figure shows a two-dimensional input layer of a CNN. For the sake of simplicity, we'll assume that this is the input layer, but it can be any layer of the network. We'll also assume that the input is a grayscale image, and each input unit represents the color intensity of a pixel. This image is represented by a two-dimensional tensor.

We'll start the convolution by applying a 3×3 filter of weights (again, a two-dimensional tensor) in the top-left corner of the image. Each input unit is associated with a single weight of the filter. It has nine weights, because of the nine input units, but, in general, the size is arbitrary (2×2, 4×4, 5×5, and so on). The convolution operation is defined as the following weighted sum:

$$y_{row,col} = \sum_{i=1}^{F_h} \sum_{j=1}^{F_w} x_{row+i-1,col+j-1} \times w_{i,j} + b$$

Here, *row* and *col* represent the input layer position, where we apply the filter (*row=1* and *col=1* in the preceding figure); F_h and F_w are the height and width of the filter size (3×3); *i* and *j* are the filter indices of each filter weight, w_{ij}; *b* is the bias weight. The group of units, $x_{row+i-1,col+j-1}$, which participates in the input, is called the **receptive field**.

We can see that in a convolutional layer, the unit activation value is defined in the same way as the activation value of the unit we defined in *Chapter 2* – that is, a weighted sum of its inputs. But here, the unit takes input only from a limited number of input units in its immediate surroundings (the receptive field). This is opposed to an FC layer, where the input comes from all input units. The difference matters because the purpose of the filter is to highlight a specific feature in the input, for example, an edge or a line in an image. In the context of the NN, the filter output represents the activation value of a unit in the next layer. The unit will be active if the feature is present at this spatial location. In hierarchically structured data, such as images, neighboring pixels form meaningful shapes and objects such as an edge or a line. However, a pixel at one end of the image is unlikely to have a relationship with a pixel at another end. Because of this, using an FC layer to connect all of the input pixels with each output unit is like asking the network to find a needle in a haystack. It has no way of knowing whether an input pixel is relevant (in the immediate surroundings) to the output unit or not (the other end of the image). Therefore, the limited receptive field of the convolutional layer is better suited to highlight meaningful features in the input data.

We've calculated the activation of a single unit, but what about the others? It's simple! For each new unit, we'll slide the filter across the input image, and we'll compute its output (the weighted sum) with each new set of input units. The following diagram shows how to compute the activations of the next two positions (one pixel to the right):

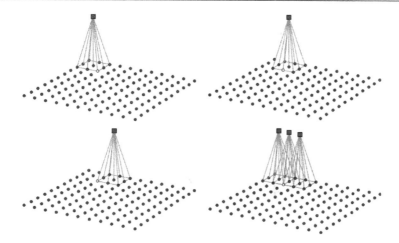

Figure 4.2 – The first three steps of a convolution operation

By "slide," we mean that the weights of the filter don't change across the image. In effect, we'll use the same nine filter weights and the single bias weight to compute the activations of all output units, each time with a different set of inputs. We call this **parameter sharing**, and we do it for two reasons:

- By reducing the number of weights, we reduce the memory footprint and prevent overfitting.

- The filter highlights a specific visual feature in the image. We can assume that this feature is useful, regardless of its position on the image. Since we apply the same filter throughout the image, the convolution is translation invariant; that is, it can detect the same feature, regardless of its location on the image. However, the convolution is neither rotation-invariant (it is not guaranteed to detect a feature if it's rotated) nor scale-invariant (it is not guaranteed to detect the same artifact in different scales).

To compute all output activations, we'll repeat the sliding process until we've covered the whole input. The spatially arranged input and output units are called **depth slices** (**feature maps** or **channels**), implying that there is more than one slice. The slices, like the image, are represented by tensors. A slice tensor can serve as an input to other layers in the network. Finally, just as with regular layers, we can use an activation function, such as the **rectified linear unit** (**ReLU**), after each unit.

> **Note**
>
> It's interesting to note that each input unit is part of the input of multiple output units. For example, as we slide the filter, the green unit in the preceding diagram will form the input of nine output units.

We can illustrate what we've learned so far with a simple example. The following diagram illustrates a 2D convolution with a 2×2 filter applied over a single 3×3 slice:

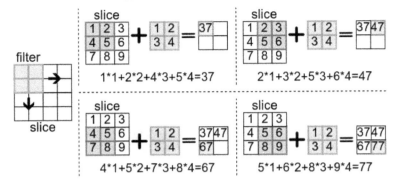

Figure 4.3 – A 2D convolution with a 2×2 filter applied over a single 3×3 slice for a 2×2 output slice

This example also shows us that the input and output feature maps have different dimensions. Let's say we have an input layer with a size of (`width_i, height_i`) and a filter with dimensions, (`filter_w, filter_h`). After applying the convolution, the dimensions of the output layer are `width_o = width_i - filter_w + 1` and `height_o = height_i - filter_h + 1`. In this example, we have `width_o = height_o = 3 - 2 + 1 = 2`.

In the next section, we'll illustrate convolutions with a simple coding example.

A coding example of the convolution operation

We've now described how convolutional layers work, but we'll gain better intuition with a visual example. Let's implement a convolution operation by applying a couple of filters across an image. For the sake of clarity, we'll implement the sliding of the filters across the image manually and we won't use any DL libraries. We'll only include the relevant parts and not the full program, but you can find the full example in this book's GitHub repository. Let's start:

1. Import numpy:

    ```
    import numpy as np
    ```

2. Define the `conv` function, which applies the convolution across the image. `conv` takes two parameters, both two-dimensional numpy arrays: `image`, for the pixel intensities of the grayscale image itself, and the hardcoded `im_filter`, for the filter:

    I.   First, we'll compute the output image size, which depends on the input `image` and `im_filter` sizes. We'll use it to instantiate the output image, `im_c`.

II.   Then, we'll iterate over all pixels of `image`, applying `im_filter` at each location. This operation requires four nested loops: the first two for the `image` dimensions and the second two for iterating over the two-dimensional filter.

III.   We'll check if any value is out of the [0, 255] interval and fix it, if necessary.

This is shown in the following example:

```
def conv(image, im_filter):
    # input dimensions
    height = image.shape[0]
    width = image.shape[1]
    # output image with reduced dimensions
    im_c = np.zeros((height - len(im_filter) + 1,
        width - len(im_filter) + 1))
    # iterate over all rows and columns
    for row in range(len(im_c)):
        for col in range(len(im_c[0])):
            # apply the filter
            for i in range(len(im_filter)):
                for j in range(len(im_filter[0])):
                    im_c[row, col] += image[row + i, /
                        col + j] * im_filter[i][j]
    # fix out-of-bounds values
    im_c[im_c > 255] = 255
    im_c[im_c < 0] = 0
    return im_c
```

3.   Apply different filters across the image. To better illustrate our point, we'll use a 10×10 blur filter, as well as Sobel edge detectors, as shown in the following example (`image_grayscale` is the two-dimensional numpy array, which represents the pixel intensities of a grayscale image):

```
# blur filter
blur = np.full([10, 10], 1. / 100)
conv(image_grayscale, blur)
# sobel filters
sobel_x = [[-1, -2, -1],
           [0, 0, 0],
           [1, 2, 1]]
conv(image_grayscale, sobel_x)
sobel_y = [[-1, 0, 1],
           [-2, 0, 2],
           [-1, 0, 1]]
conv(image_grayscale, sobel_y)
```

The full program will produce the following output:

Figure 4.4 – The first image is the grayscale input. The second image is the result of a 10×10 blur filter. The third and fourth images use detectors and vertical Sobel edge detectors

In this example, we used filters with hardcoded weights to visualize how the convolution operation works in NNs. In reality, the weights of the filter will be set during the network's training. All we'll need to do is define the network architecture, such as the number of convolutional layers, the depth of the output volume, and the size of the filters. The network will figure out the features highlighted by each filter during training.

Note

As we saw in this example, we had to implement four nested loops to implement the convolution. However, with some clever transformations, the convolution operation can be implemented with matrix-matrix multiplication. In this way, it can take full advantage of GPU parallelization.

In the next few sections, we'll discuss some of the finer details of the convolutional layers.

Cross-channel and depthwise convolutions

So far, we have described the one-to-one slice relation, where we apply a single filter over a single input slice to produce a single output slice. But this arrangement is limiting for the following reasons:

- A single input slice works well for a grayscale image, but it doesn't work for color images with multiple channels or any other multi-dimensional input

- A single filter can detect a single feature in the slice, but we are interested in detecting many different features

How do we solve these limitations? It's simple:

- For the input, we'll split the image into color channels. In the case of an RGB image, that would be three. We can think of each color channel as a depth slice, where the values are the pixel intensities for the given color (R, G, or B), as shown in the following example:

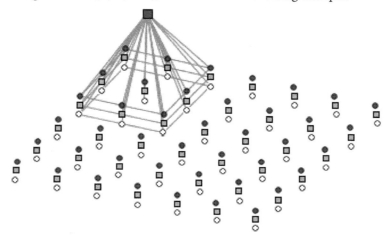

Figure 4.5 – An example of an input slice with a depth of 3

The combination of input slices is called **input volume** with a **depth** of 3. An RGB image is represented by a 3D tensor of three 2D slices (one slice per color channel).

- The CNN convolution can have multiple filters, highlighting different features, which results in multiple output feature maps (one for each filter), combined in an **output volume**.

Let's say we have C_{in} input (uppercase C) and C_{out} output slices. Depending on the relationship of the input and output slice, we get cross-channel and depthwise convolutions, as illustrated in the following diagram:

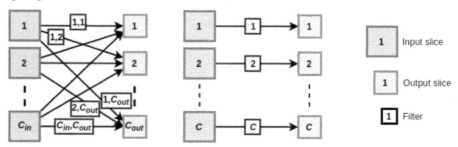

Figure 4.6 – Cross-channel convolution (left); depthwise convolution (right)

Let's discuss their properties:

- **Cross-channel convolutions**: One output slice receives input from all input slices ($C_{in} - to - one$ relationship). With multiple output slices, the relationship becomes $C_{in} - to - C_{out}$. In other words, each input slice contributes to the output of each output slice. Each pair of input/output slices uses a separate filter slice that's unique to that pair. Let's denote the index of the input slice with c_{in} (lowercase c); the index of the output slice with c_{out}; the dimensions of the filter with F_h and F_w. Then, the cross-channel 2D convolution of a single output cell in one of the output slices is defined as the following weighted sum:

$$y_{c_{out},row,col} = \sum_{c_{in}=1}^{C_{in}} \sum_{i=1}^{F_h} \sum_{j=1}^{F_w} x_{c_{in},row+i-1,col+j-1} \times w_{c_{in},c_{out},i,j} + b_{c_{out}}$$

Note that we have a unique bias, $b_{c_{out}}$, for each output slice.

We can also compute the total number of weights, W, in a cross-channel 2D convolution with the following equation:

$$W = (C_{in} \times F_h \times F_w + 1) \times C_{out}$$

Here, $+1$ represents the bias weight for each filter. Let's say we have three input slices and want to apply four 5×5 filters to them. If we did this, the convolution filter would have a total of (3 * 5 * 5 + 1) * 4 = 304 weights, four output slices (an output volume with a depth of 4), and one bias per slice. The filter for each output slice will have three 5×5 filter patches for each of the three input slices and one bias for a total of 3 * 5 * 5 + 1 = 76 weights.

- **Depthwise convolutions**: One output slice receives input from a single input slice. It's a kind of reversal of the previous case. In its simplest form, we apply a filter over a single input slice to produce a single output slice. In this case, the input and output volumes have the same depth – that is, C. We can also specify a channel multiplier (an integer, M), where we apply M filters over a single output slice to produce M output slices per input slice. In this case, the total number of output slices is $C_{in} \times M$. The depthwise 2D convolution is defined as the following weighted sum:

$$y_{c,m,row,col} = \sum_{m=1}^{M} \sum_{i=1}^{F_h} \sum_{j=1}^{F_w} x_{c,row+i-1,col+j-1} \times w_{c,m,i,j} + b_{c,m}$$

We can compute the number of weights, W, in a 2D depthwise convolution with the following formula:

$$W = C \times (M \times F_h \times F_w + M)$$

Here, $+M$ represents the biases of each output slice.

Next, we'll discuss some more properties of the convolution operation.

Stride and padding in convolutional layers

So far, we've assumed that sliding of the filter happens one pixel at a time, but that's not always the case. We can slide the filter over multiple positions. This parameter of the convolutional layers is called **stride**. Usually, the stride is the same across all dimensions of the input. In the following diagram, we can see a convolutional layer with *stride = 2* (also called **stride convolution**):

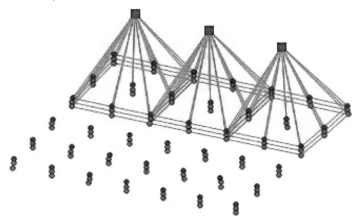

Figure 4.7 – With *stride = 2*, the filter is translated by two pixels at a time

The main effect of the larger stride is an increase in the receptive field of the output units at the expense of the size of the output slice itself. To understand this, let's recall that in the previous section, we introduced a simple formula for the output size, which included the sizes of the input and the kernel. Now, we'll extend it to also include the stride: `width_o = (width_i - filter_w) / stride_w + 1` and `height_o = 1 + (height_i - filter_h) / stride_h`. For example, the output size of a square slice generated by a 28×28 input image, convolved with a 3×3 filter with *stride = 1*, would be 1 + 28 - 3 = 26. But with *stride = 2*, we get 1 + (28 - 3) / 2 = 13. Therefore, if we use *stride = 2*, the size of the output slice will be roughly four times smaller than the input. In other words, one output unit will "cover" an area, which is four times larger compared to the input units. The units in the following layers will gradually capture input from larger regions from the input image. This is important because it would allow them to detect larger and more complex features of the input.

The convolution operations we have discussed so far have produced smaller output than the input (even with *stride = 1*). But, in practice, it's often desirable to control the size of the output. We can solve this by **padding** the edges of the input slice with rows and columns of zeros before the convolution operation. The most common way to use padding is to produce output with the same dimensions as the input. In the following diagram, we can see a convolutional layer with *padding = 1* and *stride = 1*:

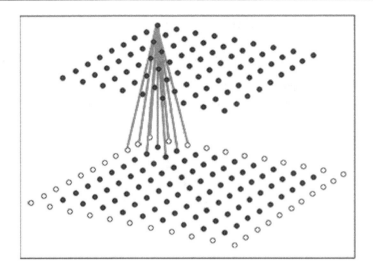

Figure 4.8 – A convolutional layer with padding = 1

The white units represent the padding. The input and the output slices have the same dimensions (dark units). The newly padded zeros will participate in the convolution operation with the slice, but they won't affect the result. The reason is that, even though the padded areas are connected with weights to the following layer, we'll always multiply those weights by the padded value, which is 0. At the same time, sliding the filter across the padded input slice will produce an output slice with the same dimensions as the unpadded input.

Now that we know about stride and padding, we can introduce the full formula for the size of the output slice:

```
height_o = 1 + (height_i + 2*padding_h - filter_h) / stride
width_o = 1 + (width_i + 2*padding_w - filter_w) / stride
```

We now have a basic knowledge of convolutions, and we can continue to the next building block of CNNs – the pooling layer. Once we know all about pooling layers, we'll introduce our first full CNN, and we'll implement a simple task to solidify our knowledge. Then, we'll focus on more advanced CNN topics.

Pooling layers

In the previous section, we explained how to increase the receptive field of the units by using *stride > 1*. But we can also do this with the help of pooling layers. A pooling layer splits the input slice into a grid, where each grid cell represents a receptive field of several units (just as a convolutional layer does). Then, a pooling operation is applied over each cell of the grid. Pooling layers don't change the volume depth because the pooling operation is performed independently on each slice. They are defined by two parameters: stride and receptive field size, just like convolutional layers (pooling layers usually don't use padding).

In this section, we'll discuss three types of pooling layers – max pooling, average pooling, and **global average pooling** (**GAP**). These three types of pooling are displayed in the following diagram:

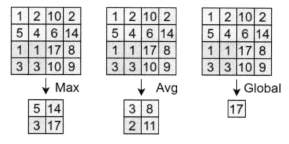

Figure 4.9 – Max, average, and global average pooling

Max pooling is the most common way of pooling. The max pooling operation takes the unit with the highest activation value in each local receptive field (grid cell) and propagates only that value forward. In the preceding figure (left), we can see an example of max pooling with a receptive field of 2×2 and *stride = 2*. This operation discards 3/4 of the input units. Pooling layers don't have any weights. In the backward pass of max pooling, the gradient is routed only to the unit with the highest activation during the forward pass. The other units in the receptive field backpropagate zeros.

Average pooling is another type of pooling, where the output of each receptive field is the mean value of all activations within the field. In the preceding figure (middle), we can see an example of average pooling with a receptive field of 2×2 and *stride = 2*.

GAP is similar to average pooling, but a single pooling region covers the whole input slice. We can think of GAP as an extreme type of dimensionality reduction because it outputs a single value that represents the average of the whole slice. This type of pooling is usually applied at the end of the convolutional portion of a CNN. In the preceding figure (right), we can see an example of a GAP operation. Stride and receptive field size don't apply to the GAP operation.

In practice, only two combinations of stride and receptive field size are used. The first is a 2×2 receptive field with *stride = 2*, and the second is a 3×3 receptive field with *stride = 2* (overlapping). If we use a larger value for either parameter, the network loses too much information. Alternatively, if the stride is 1, the size of the layer wouldn't be smaller, and nor will the receptive field increase.

Based on these parameters, we can compute the output size of a pooling layer:

```
height_o = 1 + (height_i - filter_h) / stride
width_o = 1 + (width_i - filter_w) / stride
```

Pooling layers are still very much used, but often, we can achieve similar or better results by simply using convolutional layers with larger strides. (See, for example, *J. Springerberg, A. Dosovitskiy, T. Brox, and M. Riedmiller, Striving for Simplicity: The All Convolutional Net, (2015),* https://arxiv.org/abs/1412.6806.)

We now have sufficient knowledge to introduce our first full CNN.

The structure of a convolutional network

The following figure shows the structure of a basic classification CNN:

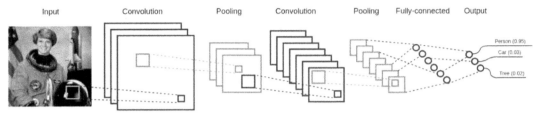

Figure 4.10 – A basic convolutional network with convolutional, FC, and pooling layers

Most CNNs share basic properties. Here are some of them:

- We would typically alternate one or more convolutional layers with one pooling layer (or a stride convolution). In this way, the convolutional layers can detect features at every level of the receptive field size. The aggregated receptive field size of deeper layers is larger than the ones at the beginning of the network. This allows them to capture more complex features from larger input regions. Let's illustrate this with an example. Imagine that the network uses 3×3 convolutions with *stride = 1* and 2×2 pooling with *stride = 2*:

  - The units of the first convolutional layer will receive input from 3×3 pixels of the image.

  - A group of 2×2 output units of the first layer will have a combined receptive field size of 4×4 (because of the stride).

  - After the first pooling operation, this group will be combined in a single unit of the pooling layer.

  - The second convolution operation takes input from 3×3 pooling units. Therefore, it will receive input from a square with side 3×4 = 12 (or a total of 12×12 = 144) pixels from the input image.

- We use the convolutional layers to extract features from the input. The features detected by the deepest layers are highly abstract, but they are also not readable by humans. To solve this problem, we usually add one or more FC layers after the last convolutional/pooling layer. In this example, the last FC layer (output) will use softmax to estimate the class probabilities of the input. You can think of the FC layers as translators between the network's language (which we don't understand) and ours.

- The deeper convolutional layers usually have more filters (hence higher volume depth), compared to the initial ones. A feature detector at the beginning of the network works on a small receptive field. It can only detect a limited number of features, such as edges or lines, shared among all classes. On the other hand, a deeper layer would detect more complex and numerous features. For example, if we have multiple classes such as cars, trees, or people, each will have its own set of features, such as tires, doors, leaves and faces, and so on. This would require more feature detectors.

Now that we know the structure of a CNN, let's implement one with PyTorch and Keras.

Classifying images with PyTorch and Keras

In this section, we'll try to classify the images of the CIFAR-10 dataset with both PyTorch and Keras. It consists of 60,000 32x32 RGB images, divided into 10 classes of objects. To understand these examples, we'll first focus on two prerequisites that we haven't covered until now: how images are represented in DL libraries and data augmentation training techniques.

Convolutional layers in deep learning libraries

PyTorch, Keras, and **TensorFlow** (**TF**) have out-of-the-gate support for 1D, 2D, and 3D convolutions. The inputs and outputs of the convolution operation are tensors. A 1D convolution with multiple input/output slices would have 3D input and output tensors. Their axes can be in either SCW or SWC order, where we have the following:

- S: The index of the sample in the mini-batch
- C: The index of the depth slice in the volume
- W: The content of the slice

In the same way, a 2D convolution will be represented by $SCHW$ or $SHWC$ ordered tensors, where H and W are the height and width of the slices. A 3D convolution will have $SCDHW$ or $SDHWC$ order, where D stands for the depth of the slice.

Data augmentation

One of the most efficient regularization techniques is data augmentation. If the training data is too small, the network might start overfitting. Data augmentation helps counter this by artificially increasing the size of the training set. In the CIFAR-10 examples, we'll train a CNN over multiple epochs. The network will "see" every sample of the dataset once per epoch. To prevent this, we can apply random augmentations to the images, before feeding them to train the CNN. The labels will stay the same. Some of the most popular image augmentations are as follows:

- Rotation
- Horizontal and vertical flip

- Zoom in/out

- Crop

- Skew

- Contrast and brightness adjustment

The emboldened augmentations are shown in the following example:

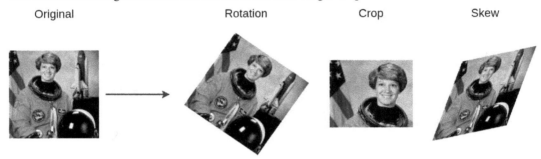

Figure 4.11 – Examples of different image augmentations

With that, we're ready to proceed with the examples.

Classifying images with PyTorch

We'll start with PyTorch first:

1. Select the device, preferably a GPU. This NN is larger than the MNIST ones and the CPU training would be very slow:

   ```
   import torch
   from torchsummary import summary

   device = torch.device("cuda:0" if torch.cuda.is_available() else
   "cpu")
   ```

2. Load the training dataset (followed by the validation):

   ```
   import torchvision.transforms as transforms
   from torchvision import datasets
   from torch.utils.data import DataLoader

   # Training dataset
   train_transform = transforms.Compose([
       transforms.RandomHorizontalFlip(),
   ```

```
        transforms.RandomVerticalFlip(),
        transforms.ToTensor(),
        transforms.Normalize(
            [0.485, 0.456, 0.406],
            [0.229, 0.224, 0.225])
    ])

    train_data = datasets.CIFAR10(
        root='data',
        train=True,
        download=True,
        transform=train_transform)

    batch_size = 50
    train_loader = DataLoader(
        dataset=train_data,
        batch_size=batch_size,
        shuffle=True,
        num_workers=2)
```

`train_transform` is of particular interest. It performs random horizontal and vertical flips, and it normalizes the dataset with `transforms.Normalize` using z-score normalization. The hardcoded numerical values represent the manually computed channel-wise mean and `std` values for the CIFAR-10 dataset. `train_loader` takes care of providing training minibatches.

3. Load the validation dataset. Note that we normalize the validation set with the mean and `std` values of the training dataset:

```
    validation_transform = transforms.Compose([
        transforms.ToTensor(),
        transforms.Normalize(
            [0.485, 0.456, 0.406],
            [0.229, 0.224, 0.225])
    ])

    validation_data = datasets.CIFAR10(
        root='data',
        train=False,
        download=True,
        transform=validation_transform)

    validation_loader = DataLoader(
        dataset=validation_data,
        batch_size=100,
        shuffle=True)
```

4. Define our CNN using the `Sequential` class. It has the following properties:

- Three blocks of two convolutional layers (3×3 filters) and one max pooling layer.

- Batch normalization after each convolutional layer.

- The first two blocks apply `padding=1` to the convolutions, so they don't decrease the size of the feature maps.

- **Gaussian Error Linear Unit (GELU)** activation functions.

- The feature maps after the last pooling are flattened to a tensor of size 512 and serve as input to a `Linear` (FC) layer with 10 outputs (one of each class). The final activation is softmax.

Let's see the definition:

```
from torch.nn import Sequential, Conv2d, BatchNorm2d, GELU,
MaxPool2d, Dropout2d, Linear, Flatten

model = Sequential(
    Conv2d(in_channels=3, out_channels=32,
        kernel_size=3, padding=1),
    BatchNorm2d(32),
    GELU(),
    Conv2d(in_channels=32, out_channels=32,
        kernel_size=3, padding=1),
    BatchNorm2d(32),
    GELU(),
    MaxPool2d(kernel_size=2, stride=2),
    Dropout2d(0.2),

    Conv2d(in_channels=32, out_channels=64,
        kernel_size=3, padding=1),
    BatchNorm2d(64),
    GELU(),
    Conv2d(in_channels=64, out_channels=64,
        kernel_size=3, padding=1),
    BatchNorm2d(64),
    GELU(),
    MaxPool2d(kernel_size=2, stride=2),
    Dropout2d(p=0.3),

    Conv2d(in_channels=64, out_channels=128,
        kernel_size=3),
    BatchNorm2d(128),
    GELU(),
    Conv2d(in_channels=128, out_channels=128,
```

```
            kernel_size=3),
    BatchNorm2d(128),
    GELU(),
    MaxPool2d(kernel_size=2, stride=2),
    Dropout2d(p=0.5),
    Flatten(),

    Linear(512, 10),
)
```

5. Run the training and validation. We'll use the same `train_model` and `test_model` functions that we implemented in the MNIST PyTorch example in *Chapter 3*. Because of this, we won't implement them here, but the full source code is available in this chapter's GitHub repository (including a Jupyter Notebook). We can expect the following results: 51% accuracy in 1 epoch, 70% accuracy in 5 epochs, and around 82% accuracy in 75 epochs.

This concludes our PyTorch example.

Classifying images with Keras

Our second example is the same task, but this time implemented with Keras:

1. Start by downloading the dataset. We'll also convert the numerical labels into one-hot-encoded tensors:

```
import tensorflow as tf

(X_train, Y_train), (X_validation, Y_validation) = \
    tf.keras.datasets.cifar10.load_data()
Y_train = tf.keras.utils.to_categorical(Y_train, 10)
Y_validation = \
    tf.keras.utils.to_categorical(Y_validation, 10)
```

2. Create an instance of `ImageDataGenerator`, which applies z-normalization over each channel of the training set images. It also provides data augmentation (random horizontal and vertical flips) during training. Also, note that we apply the mean and standard variation of the training over the test set for the best performance:

```
from tensorflow.keras.preprocessing.image import
ImageDataGenerator

data_generator = ImageDataGenerator(
    featurewise_center=True,
    featurewise_std_normalization=True,
    horizontal_flip=True,
```

```
            vertical_flip=True)

    # Apply z-normalization on the training set
    data_generator.fit(X_train)

    # Standardize the validation set
    X_validation = \
        data_generator.standardize( \
        X_validation.astype('float32'))
```

3.   Then, we can define our CNN using the Sequential class. We'll use the same architecture
     we defined in the *Classifying images with PyTorch* section. The following is the Keras definition
     of that model:

```
from tensorflow.keras.models import Sequential
from tensorflow.keras.layers import Conv2D, Dense, MaxPooling2D,
Dropout, BatchNormalization, Activation, Flatten

model = Sequential(layers=[
    Conv2D(32, (3, 3),
        padding='same',
        input_shape=X_train.shape[1:]),
    BatchNormalization(),
    Activation('gelu'),
    Conv2D(32, (3, 3), padding='same'),
    BatchNormalization(),
    Activation('gelu'),
    MaxPooling2D(pool_size=(2, 2)),
    Dropout(0.2),
    Conv2D(64, (3, 3), padding='same'),
    BatchNormalization(),
    Activation('gelu'),
    Conv2D(64, (3, 3), padding='same'),
    BatchNormalization(),
    Activation('gelu'),
    MaxPooling2D(pool_size=(2, 2)),
    Dropout(0.3),
    Conv2D(128, (3, 3)),
    BatchNormalization(),
    Activation('gelu'),
    Conv2D(128, (3, 3)),
    BatchNormalization(),
    Activation('gelu'),
    MaxPooling2D(pool_size=(2, 2)),
```

```
        Dropout(0.5),
        Flatten(),
        Dense(10, activation='softmax')
    ])
```

4. Define the training parameters (we'll also print the model summary for clarity):

```
model.compile(loss='categorical_crossentropy',
    optimizer='adam', metrics=['accuracy'])
print(model.summary())
```

5. Run the training for 50 epochs:

```
batch_size = 50

model.fit(
    x=data_generator.flow(x=X_train,
        y=Y_train,
        batch_size=batch_size),
    steps_per_epoch=len(X_train) // batch_size,
    epochs=100,
    verbose=1,
    validation_data=(X_validation, Y_validation),
    workers=4)
```

Depending on the number of epochs, this model will produce the following results: 50% accuracy in 1 epoch, 72% accuracy in 5 epochs, and around 85% accuracy in 45 epochs. Our Keras example has slightly higher accuracy compared to the one in PyTorch, although they should be identical. Maybe we've got a bug somewhere. We might never know, but we can learn a lesson, nevertheless: ML models aren't easy to debug because they can fail with slightly degraded performance, instead of outright error. Finding the exact reason for this performance penalty can be hard.

Now that we've implemented our first full CNN twice, we'll focus on some more advanced types of convolutions.

Advanced types of convolutions

So far, we've discussed the "classic" convolutional operation. In this section, we'll introduce several new variations and their properties.

1D, 2D, and 3D convolutions

In this chapter, we've used **2D convolutions** because computer vision with two-dimensional images is the most common CNN application. But we can also have 1D and 3D convolutions, where the units are arranged in one-dimensional or three-dimensional space, respectively. In all cases, the filter has the same number of dimensions as the input, and the weights are shared across the input. For example, we would use 1D convolution with time series data because the values are arranged across a single time axis. In the following diagram, on the left, we can see an example of 1D convolution:

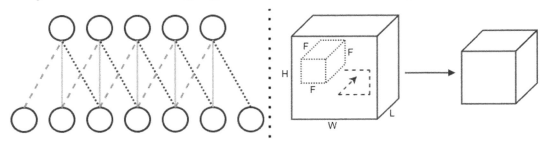

Figure 4.12 –1D convolution (left); 3D convolution (right)

The weights with the same dashed lines share the same value. The output of the 1D convolution is also 1D. If the input is 3D, such as a 3D MRI, we could use 3D convolution, which will also produce 3D output. In this way, we'll maintain the spatial arrangement of the input data. We can see an example of 3D convolution in the preceding diagram, on the right. The input has dimensions of H/W/L, and the filter has a single size, F, for all dimensions. The output is also 3D.

1×1 convolutions

A 1×1 (pointwise) convolution is a special case of convolution where each dimension of the convolution filter is of size 1 (1×1 in 2D convolutions and 1×1×1 in 3D). At first, this doesn't make sense – a 1×1 filter doesn't increase the receptive field size of the output units. The result of such a convolution would be pointwise scaling. But it can be useful in another way – we can use them to change the depth between the input and output volumes. To understand this, let's recall that, in general, we have an input volume with a depth of C_{in} slices and C_{out} filters for C_{out} output slices. Each output slice is generated by applying a unique filter over all the input slices. If we use a 1×1 filter and $C_{in} \neq C_{out}$, we'll have output slices of the same size, but with different volume depths. At the same time, we won't change the receptive field size between the input and output. The most common use case is to reduce the output volume, or $C_{in} > C_{out}$ (dimension reduction), nicknamed the **"bottleneck" layer**.

Depthwise separable convolutions

An output slice in a cross-channel convolution receives input from all of the input slices using a single filter. The filter tries to learn features in a 3D space, where two of the dimensions are spatial (the height and width of the slice) and the third is the channel. Therefore, the filter maps both spatial and cross-channel correlations.

Depthwise separable convolutions (**DSCs**, *Xception: Deep Learning with Depthwise Separable Convolutions*, https://arxiv.org/abs/1610.02357) can completely decouple cross-channel and spatial correlations. A DSC combines two operations: a depthwise convolution and a 1×1 convolution. In a depthwise convolution, a single input slice produces a single output slice, so it only maps spatial (and not cross-channel) correlations. With 1×1 convolutions, we have the opposite. The following diagram represents the DSC:

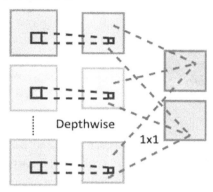

Figure 4.13 – A depth-wise separable convolution

The DSC is usually implemented without non-linearity after the first (depthwise) operation.

> **Note**
>
> Let's compare the standard and depthwise separable convolutions. Imagine that we have 32 input and output channels and a filter with a size of 3×3. In a standard convolution, one output slice is the result of applying one filter for each of the 32 input slices for a total of 32 * 3 * 3 = 288 weights (excluding bias). In a comparable depthwise convolution, the filter has only 3 * 3 = 9 weights and the filter for the 1×1 convolution has 32 * 1 * 1 = 32 weights. The total number of weights is 32 + 9 = 41. Therefore, the depthwise separable convolution is faster and more memory-efficient compared to the standard one.

Dilated convolutions

The regular convolution applies an $n×n$ filter over an $n×n$ receptive field. With dilated convolutions, we apply the same filter sparsely over a receptive field of size $(n * l - 1) × (n * l - 1)$, where l is the **dilation factor**. We still multiply each filter weight by one input slice cell, but these cells are at a distance of l away from each other. The regular convolution is a special case of dilated convolution with $l = 1$. This is best illustrated with the following diagram:

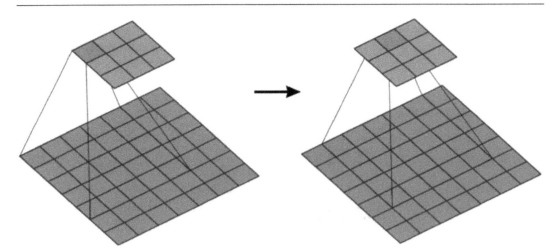

Figure 4.14 – A dilated convolution with a dilation factor of I=2. Here, the first two
steps of the operation are displayed. The bottom layer is the input while the top
layer is the output. Source: https://github.com/vdumoulin/conv_arithmetic

Dilated convolutions can increase the receptive field's size exponentially without losing resolution or coverage. We can also increase the receptive field with stride convolutions or pooling but at the cost of resolution and/or coverage. To understand this, let's imagine that we have a stride convolution with stride $s > 1$. In this case, the output slice is s times smaller than the input (loss of resolution). If we increase $s > F$ further (F is the size of either the pooling or convolutional kernel), we get a loss of coverage because some of the areas of the input slice will not participate in the output at all. Additionally, dilated convolutions don't increase the computation and memory costs because the filter uses the same number of weights as the regular convolution.

Transposed convolutions

In the convolutional operations we've discussed so far, the output dimensions are either equal or smaller than the input dimensions. In contrast, transposed convolutions (first proposed in *Deconvolutional Networks by Matthew D. Zeiler, Dilip Krishnan, Graham W. Taylor, and Rob Fergus*: https://www.matthewzeiler.com/mattzeiler/deconvolutionalnetworks.pdf) allow us to upsample the input data (their output is larger than the input). This operation is also known as **deconvolution, fractionally strided convolution**, or **sub-pixel convolution**. These names can sometimes lead to confusion. To clarify things, note that the transposed convolution is, in fact, a regular convolution with a slightly modified input slice or convolutional filter.

For the longer explanation, we'll start with a 1D regular convolution over a single input and output slice:

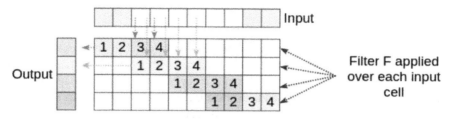

Figure 4.15 – 1D regular convolution

It uses a filter with *size = 4*, *stride = 2*, and *padding = 2* (denoted with gray in the preceding diagram). The input is a vector of size 6 and the output is a vector of size 4. The filter, a vector, $\mathbf{f} = [1,2,3,4]$, is always the same, but it's denoted with different colors for each position we apply it to. The respective output cells are denoted with the same color. The arrows show which input cells contribute to one output cell.

> **Note**
>
> The example that is being discussed in this section is inspired by the paper *Is the deconvolution layer the same as a convolutional layer?* (https://arxiv.org/abs/1609.07009).

Next, we'll discuss the same example (1D, single input and output slices, and a filter with *size = 4*, *padding = 2*, and *stride = 2*), but for transposed convolution. The following diagram shows two ways we can implement it:

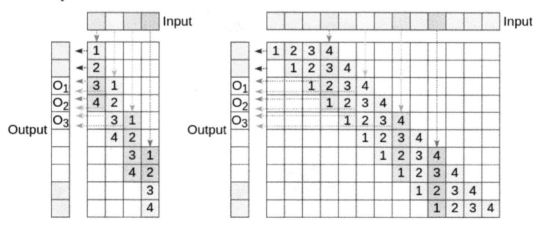

Figure 4.16 – A convolution with stride = 2, applied with the transposed filter f. The 2 pixels at the beginning and the end of the output are cropped (left); a convolution with stride 0.5, applied over input data, padded with subpixels. The input is filled with 0-valued pixels (gray) (right)

Let's discuss them in detail:

- In the first case, we have a regular convolution with *stride* = 2 and a filter represented as a transposed row matrix (equivalent to a column matrix) with size 4: $\mathbf{f}^\top = [1,2,3,4]^\top$ (shown in the preceding diagram, left). Note that the stride is applied over the output layer as opposed to the regular convolution, where we stride over the input. By setting the stride larger than 1, we can increase the output size, compared to the input. Here, the size of the input slice is I, the size of the filter is F, the stride is S, and the input padding is P. Due to this, the size, O, of the output slice of a transposed convolution is given by the following formula: $O = S(I - 1) + F - 2P$. In this scenario, an input of size 4 produces an output of size 2 * (4 - 1) + 4 - 2 * 2 = 6. We also crop the two cells at the beginning and end of the output vector because they only gather input from a single input cell.

- In the second case, the input is filled with imaginary 0-valued subpixels between the existing ones (shown in the preceding diagram, right). This is where the name subpixel convolution comes from. Think of it as padding but within the image itself and not only along the borders. Once the input has been transformed in this way, a regular convolution is applied.

Let's compare the two output cells, o_1 and o_3, in both scenarios. As shown in the preceding diagram, in either case, o_1 receives input from the first and the second input cells and o_3 receives input from the second and third cells. The only difference between these two cases is the index of the weight, which participates in the computation. However, the weights are learned during training, and, because of this, the index is not important. Therefore, the two operations are equivalent.

Next, let's take a look at a 2D transposed convolution from a subpixel point of view. As with the 1D case, we insert 0-valued pixels and padding in the input slice to achieve upsampling (the input is at the bottom):

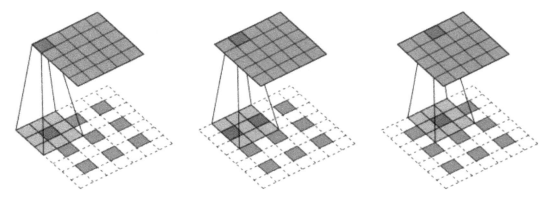

Figure 4.17 – The first three steps of a 2D transpose convolution with padding = 1 and stride = 2.
Source: https://github.com/vdumoulin/conv_arithmetic, https://arxiv.org/abs/1603.07285

The backpropagation operation of a regular convolution is a transposed convolution.

This concludes our extended introduction to the various types of convolutions. In the next section, we'll learn how to build some advanced CNN architectures with the advanced convolutions we've learned about so far.

Advanced CNN models

In this section, we'll discuss some complex CNN models. They are available in both PyTorch and Keras, with pre-trained weights on the ImageNet dataset. You can import and use them directly, instead of building them from scratch. Still, it's worth discussing their central ideas as an alternative to using them as black boxes.

Most of these models share a few architectural principles:

- They start with an "entry" phase, which uses a combination of stride convolutions and/or pooling to reduce the input image size at least two to eight times, before propagating it to the rest of the network. This makes a CNN more computationally- and memory-efficient because the deeper layers work with smaller slices.

- The main network body comes after the entry phase. It is composed of multiple repeated composite modules. Each of these modules utilizes padded convolutions in such a way that its input and output slices are the same size. This makes it possible to stack as many modules as necessary to reach the desired depth. The deeper modules utilize a higher number of filters (output slices) per convolution, compared to the earlier ones.

- The downsampling in the main body is handled by special modules with stride convolutions and/or pooling operations.

- The convolutional phase usually ends with GAP over all slices.

- The output of the GAP operation can serve as input for various tasks. For example, we can add an FC layer for classification.

We can see a prototypical CNN built with these principles in the following figure:

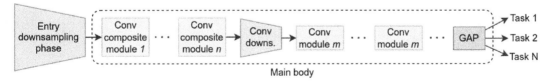

Figure 4.18 – A prototypical CNN

With that, let's dig deeper into deep CNNs (get it?).

Introducing residual networks

Residual networks (**ResNets**, *Deep Residual Learning for Image Recognition*, `https://arxiv.org/abs/1512.03385`) were released in 2015 when they won all five categories of the ImageNet challenge that year. In *Chapter 2*, we discussed that the layers of an NN are not restricted to sequential order but form a directed graph instead. This is the first architecture we'll learn about that takes advantage of this flexibility. This is also the first network architecture that has successfully trained a network with a depth of more than 100 layers.

Thanks to better weight initializations, new activation functions, as well as normalization layers, it's now possible to train deep networks. However, the authors of the paper conducted some experiments and observed that a network with 56 layers had higher training and testing errors compared to a network with 20 layers. They argue that this should not be the case. In theory, we can take a shallow network and stack identity layers (these are layers whose output just repeats the input) on top of it to produce a deeper network that behaves in the same way as the shallow one. Yet, their experiments have been unable to match the performance of the shallow network.

To solve this problem, they proposed a network constructed of residual blocks. A residual block consists of two or three sequential convolutional layers and a separate parallel **identity** (repeater) shortcut connection, which connects the input of the first layer and the output of the last one. We can see three types of residual blocks in the following figure:

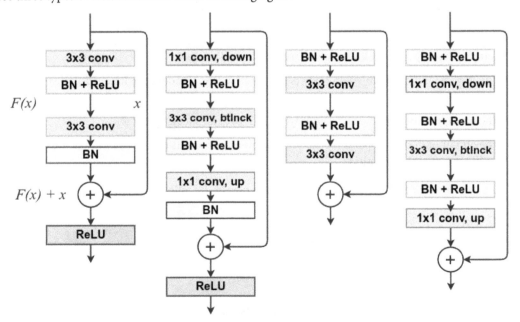

Figure 4.19 – From left to right – original residual block; original bottleneck residual block; pre-activation residual block; pre-activation bottleneck residual block

Each block has two parallel paths. The left-hand path is similar to the other networks we've seen and consists of sequential convolutional layers and batch normalization. The right path contains the identity shortcut connection (also known as the **skip connection**). The two paths are merged via an element-wise sum – that is, the left and right tensors have the same shape, and an element of the first tensor is added to the element in the same position in the second tensor. The output is a single tensor with the same shape as the input. In effect, we propagate the features learned by the block forward, but also the original unmodified signal. In this way, we can get closer to the original scenario, as described by the authors. The network can decide to skip some of the convolutional layers thanks to the skip connections, in effect reducing its depth. The residual blocks use padding in such a way that the input and the output of the block have the same dimensions. Thanks to this, we can stack any number of blocks for a network with an arbitrary depth.

Now, let's see how the blocks in the diagram differ:

- The first block contains two 3×3 convolutional layers. This is the original residual block, but if the layers are wide, stacking multiple blocks becomes computationally expensive.

- The second block is equivalent to the first, but it uses a **bottleneck layer**. First, we use a 1×1 convolution to downsample the input volume depth (we discussed this in the *1×1 convolutions* section). Then, we apply a 3×3 (bottleneck) convolution to the reduced input. Finally, we expand the output back to the desired depth with another 1×1 upsampling convolution. This layer is less computationally expensive than the first.

- The third block is the latest revision of the idea, published in 2016 by the same authors (*Identity Mappings in Deep Residual Networks*, `https://arxiv.org/abs/1603.05027`). It uses pre-activations, and the batch normalization and the activation function come before the convolutional layer. This may seem strange at first, but thanks to this design, the skip connection path can run uninterrupted throughout the network. This is contrary to the other residual blocks, where at least one activation function is on the path of the skip connection. A combination of stacked residual blocks still has the layers in the right order.

- The fourth block is the bottleneck version of the third layer. It follows the same principle as the bottleneck residual layer v1.

In the following table, we can see the family of networks proposed by the authors of the paper:

output size	18-layer	34-layer	50-layer	101-layer	152-layer
112x112	7x7 conv, stride 2				
56x56	3x3 max pool, stride 2				
56x56	3x3, 64 3x3, 64 x2	3x3, 64 3x3, 64 x3	1x1, 64 3x3, 64 1x1, 256 x3	1x1, 64 3x3, 64 1x1, 256 x3	1x1, 64 3x3, 64 1x1, 256 x3
28x28	3x3, 128 3x3, 128 x2	3x3, 128 3x3, 128 x4	1x1, 128 3x3, 128 1x1, 512 x4	1x1, 128 3x3, 128 1x1, 512 x4	1x1, 128 3x3, 128 1x1, 512 x8
14x14	3x3, 256 3x3, 256 x2	3x3, 256 3x3, 256 x6	1x1, 256 3x3, 256 1x1, 1024 x6	1x1, 256 3x3, 256 1x1, 1024 x23	1x1, 256 3x3, 256 1x1, 1024 x36
7x7	3x3, 512 3x3, 512 x2	3x3, 512 3x3, 512 x3	1x1, 512 3x3, 512 1x1, 2048 x3	1x1, 512 3x3, 512 1x1, 2048 x3	1x1, 512 3x3, 512 1x1, 2048 x3
1x1	average pool, 1000-d fc, softmax				

Figure 4.20 – The family of the most popular residual networks. The residual blocks are represented by rounded rectangles. Inspired by https://arxiv. org/abs/1512.03385

Some of their properties are as follows:

- They start with a 7×7 convolutional layer with *stride = 2*, followed by 3×3 max pooling. This phase serves as a downsampling step, so the rest of the network can work with a much smaller slice of 56×56, compared to 224×224 of the input.

- Downsampling in the rest of the network is implemented with a modified residual block with *stride = 2*.

- GAP downsamples the output after all residual blocks and before the 1,000-unit FC softmax layer.

- The number of parameters for the various ResNets range from 25.6 million to 60.4 million and their depth ranges from 18 to 152 layers.

The ResNet family of networks is popular not only because of their accuracy but also because of their relative simplicity and the versatility of the residual blocks. As we mentioned previously, the input and output shape of the residual block can be the same due to the padding. We can stack residual blocks in different configurations to solve various problems with wide-ranging training set sizes and input dimensions.

Inception networks

Inception networks (*Going Deeper with Convolutions*, https://arxiv.org/abs/1409.4842) were introduced in 2014 when they won the ImageNet challenge of that year (there seems to be a pattern here). Since then, the authors have released multiple improvements (versions) of the architecture.

Fun fact

The name *inception* comes in part from the *We need to go deeper* internet meme, related to the movie Inception.

The idea behind inception networks started from the basic premise that the objects in an image have different scales. A distant object might take up a small region of the image, but the same object, once nearer, might take up a large part of the image. This presents a difficulty for standard CNNs, where the units in the different layers have a fixed receptive field size, as imposed on the input image. A regular network might be a good detector of objects at a certain scale but could miss them otherwise. To solve this problem, the authors of the paper proposed a novel architecture: one composed of inception blocks. An inception block starts with a common input and then splits it into different parallel paths (or towers). Each path contains either convolutional layers with a different-sized filter or a pooling layer. In this way, we apply different receptive fields to the same input data. At the end of the Inception block, the outputs of the different paths are concatenated.

In the next few sections, we'll discuss the different variations of Inception networks.

Inception v1

The following diagram shows the first version of the inception block, which is part of the **GoogLeNet** network architecture (https://arxiv.org/abs/1409.4842):

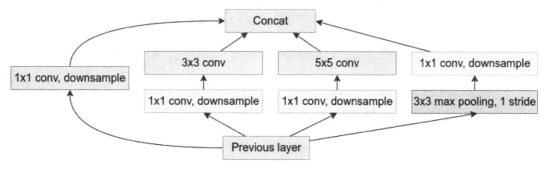

Figure 4.21 – Inception v1 block; inspired by https://arxiv.org/abs/1409.4842

The v1 block has four paths:

- 1×1 convolution, which acts as a kind of repeater to the input
- 1×1 convolution, followed by a 3×3 convolution
- 1×1 convolution, followed by a 5×5 convolution
- 3×3 max pooling with *stride = 1*

The layers in the block use padding in such a way that the input and the output have the same shape (but different depths). The padding is also necessary because each path would produce an output with a different shape, depending on the filter size. This is valid for all versions of inception blocks.

The other major innovation of this inception block is the use of downsampling 1×1 convolutions. They are needed because the output of all paths is concatenated to produce the final output of the block. The result of the concatenation is an output with a quadrupled depth. If another inception block followed the current one, its output depth would quadruple again. To avoid such exponential growth, the block uses 1×1 convolutions to reduce the depth of each path, which, in turn, reduces the output depth of the block. This makes it possible to create deeper networks, without running out of resources.

The full GoogLeNet has the following properties:

- Like ResNets, it starts with a downsampling phase, which utilizes two convolutional and two max pooling layers to reduce the input size from 224×224 to 56×56, before the inception blocks get involved.
- The network has nine inception v1 blocks.
- The convolutional phase ends with global average pooling.
- The network utilizes auxiliary classifiers—that is, it has two additional classification outputs (with the same ground truth labels) at various intermediate layers. During training, the total value of the loss is a weighted sum of the auxiliary losses and the real loss.
- The model has a total of 6.9 million parameters and a depth of 22 layers.

Inception v2 and v3

Inception v2 and v3 were released together and proposed several improved inception blocks over the original v1 (*Rethinking the Inception Architecture for Computer Vision*, https://arxiv.org/abs/1512.00567). We can see the first new inception block, A, in the following diagram:

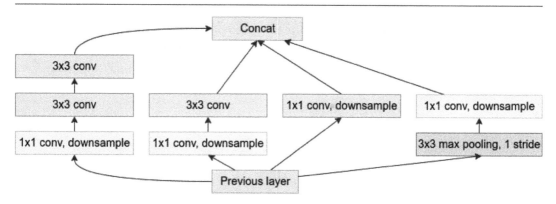

Figure 4.22 – Inception block A, inspired by https://arxiv.org/abs/1512.00567

The first new property of block A is the factorization of the 5×5 convolution in two stacked 3×3 convolutions. This structure has several advantages.

The receptive field of the units of the last stacked layer is equivalent to the receptive field of a single layer with a large convolutional filter. The stacked layers achieve the same receptive field size with fewer parameters, compared to a single layer with a large filter. For example, let's replace a single 5×5 layer with two stacked 3×3 layers. For the sake of simplicity, we'll assume that we have single input and output slices. The total number of weights (excluding biases) of the 5×5 layer is 5 * 5 = 25. On the other hand, the total weights of a single 3×3 layer is 3 * 3 = 9, and simply 2 * (3 * 3) = 18 for two layers, which makes this arrangement 28% more efficient (18/25 = 0.72). The efficiency gain is preserved even with multiple input and output slices for the two layers. The next improvement is the factorization of an $n \times n$ convolution in two stacked asymmetrical $1 \times n$ and $n \times 1$ convolutions. For example, we can split a single 3×3 convolution into two 1×3 and 3×1 convolutions, where the 3×1 convolution is applied over the output of the 1×3 convolution. In the first case, the filter size would be 3 * 3 = 9, while in the second case, we would have a combined size of (3 * 1) + (1 * 3) = 3 + 3 = 6, resulting in 33% efficiency, as seen in the following diagram:

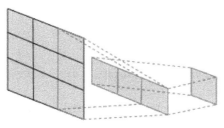

Figure 4.23 – Factorization of a 3×3 convolution in 1×3 and 3×1 convolutions; inspired by https://arxiv.org/abs/1512.00567

The authors introduced two new blocks that utilize factorized convolutions. The first of these blocks (and the second in total), inception block B, is equivalent to inception block A:

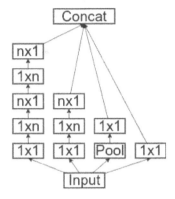

Figure 4.24 – Inception block B. When n=3, it is equivalent to block
A; inspired by https://arxiv.org/abs/1512.00567

The second (third in total) inception block, C, is similar, but the asymmetrical convolutions are parallel, resulting in a higher output depth (more concatenated paths). The hypothesis here is that the more features (different filters) the network has, the faster it learns. On the other hand, the wider layers take more memory and computation time. As a compromise, this block is only used in the deeper part of the network, after the other blocks:

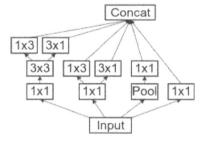

Figure 4.25 – Inception block C; inspired by https://arxiv.org/abs/1512.00567

Another major improvement in this version is the use of batch normalization, which was introduced by the same authors.

These new blocks create two new inception networks: v2 and v3. Inception v3 uses batch normalization and is the more popular of the two. It has the following properties:

- The network starts with a downsampling phase, which utilizes stride convolutions and max pooling to reduce the input size from 299×299 to 35×35 before the inception blocks get involved

- The layers are organized into three inception blocks, A, five inception blocks, B, and two inception blocks, C

- The convolutional phase ends with global average pooling

- It has 23.9 million parameters and a depth of 48 layers

Inception v4 and Inception-ResNet

The latest revisions of inception networks introduce three new streamlined inception blocks (**Inception-v4**, *Inception-v4, Inception-ResNet and the Impact of Residual Connections on Learning*, https:// arxiv.org/abs/1602.07261). More specifically, the new versions introduce 7×7 asymmetric factorized convolutions average pooling instead of max pooling and new Inception-ResNet blocks with residual connections. We can see one such block in the following diagram:

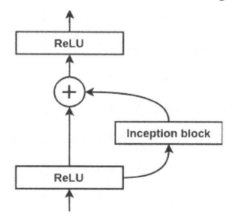

Figure 4.26 – An inception block (any kind) with a residual skip connection

The Inception-ResNet family of models share the following properties:

- The networks start with a downsampling phase, which utilizes stride convolutions and max pooling to reduce the input size from 299×299 to 35×35 before the inception blocks get involved.

- The main body of the model consists of three groups of four residual-inception-A blocks, seven residual-inception-B blocks, three residual inception-B blocks, and special reduction modules between the groups. The different models use slightly different variations of these blocks.

- The convolutional phase ends with global average pooling.

- The models have around 56 million weights.

In this section, we discussed different types of inception networks and the different principles used in the various inception blocks. Next, we'll talk about a newer CNN architecture that takes the inception concept to a new depth (or width, as it should be).

Introducing Xception

All inception blocks we've discussed so far start by splitting the input into several parallel paths. Each path continues with a dimensionality-reduction 1×1 cross-channel convolution, followed by regular cross-channel convolutions. On one hand, the 1×1 connection maps cross-channel correlations, but not spatial ones (because of the 1×1 filter size). On the other hand, the subsequent cross-channel convolutions map both types of correlations. Let's recall that earlier in this chapter, we introduced DSCs, which combine the following two operations:

- **A depthwise convolution**: In a depthwise convolution, a single input slice produces a single output slice, so it only maps spatial (and not cross-channel) correlations

- **A 1×1 cross-channel convolution**: With 1×1 convolutions, we have the opposite – that is, they only map cross-channel correlations

The author of Xception (*Xception: Deep Learning with Depthwise Separable Convolutions*, https://arxiv.org/abs/1610.02357) argues that we can think of DSC as an extreme (hence the name) version of an inception block, where each depthwise input/output slice pair represents one parallel path. We have as many parallel paths as the number of input slices. The following diagram shows a simplified inception block and its transformation to an Xception block:

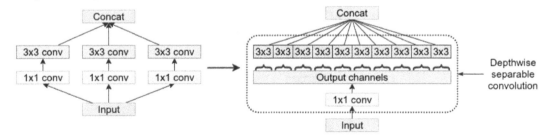

Figure 4.27 – A simplified inception module (left); an Xception block
(right); inspired by https://arxiv.org/abs/1610.02357

The Xception block and the DSC have two differences:

- In Xception, the 1×1 convolution comes first, instead of last as in DSC. However, these operations are meant to be stacked anyway, and we can assume that the order is of no significance.

- The Xception block uses ReLU activations after each convolution, while the DSC doesn't use non-linearity after the cross-channel convolution. According to the author's experiments, networks with absent non-linearity depthwise convolution converged faster and were more accurate.

The full Xception network has the following properties:

- It starts with an entry flow of convolutional and pooling operations, which reduces the input size from 299×299 to 19×19.

- It has 14 Xception modules, all of which have linear residual connections around them, except for the first and last modules.

- All convolutions and DSCs are followed by batch normalization. All DSCs have a depth multiplier of 1 (no depth expansion).

- The convolutional phase ends with global average pooling.

- A total of 23 million parameters and a depth of 36 convolutional layers.

This section concludes the series of inception-based models. In the next section, we'll focus on a novel NN architectural element.

Squeeze-and-Excitation Networks

Squeeze-and-Excitation Networks (**SENet**, *Squeeze-and-Excitation Networks*, https://arxiv.org/abs/1709.01507) introduce a new NN architectural unit, which the authors call – you guessed it – the **Squeeze-and-Excitation** (**SE**) block. Let's recall that the convolutional operation applies multiple filters across the input channels to produce multiple output feature maps (or channels). The authors of SENet observe that each of these channels has "equal weight" when it serves as input to the next layer. However, some channels could be more informative than others. To emphasize their importance, the authors propose the content-aware SE block, which weighs each channel adaptively. We can also think of the SE block as an **attention mechanism**. To understand how it works, let's start with the following figure:

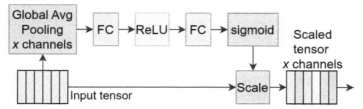

Figure 4.28 – The Squeeze-and-Excitation block

The block introduces a parallel path to the main NN data flow. Let's see its steps:

1. **Squeeze phase**: A GAP operation is applied across the channels. The output of the GAP is a single scalar value for each channel. For example, if the input is an RGB image, the unique GAP operations across each of the R, G, and B channels will produce a one-dimensional tensor with size 3. Think of these scalar values as the distilled state of the channels.

2. **Excitement phase**: The output tensor of the squeeze phase serves as input to a two-layer FC NN. This NN is in the shape of `FC layer -> ReLU -> FC layer -> sigmoid`. It resembles an autoencoder because the first hidden layer reduces the size of the input tensor and the second hidden layer upscales it to the original size (3 in the case of RGB input). The final sigmoid activation ensures that all values of the output are in the (0:1) range.

3. **Scale**: The output values of the excitement NN serve as scaling coefficients of the channels of the original input tensor. All the values of a channel are scaled (or excited) by its corresponding coefficient produced by the excitement phase. In this way, the excitement NN can emphasize the importance of a given channel.

The authors added SE blocks to different existing models, which improved their accuracy. In the following figure, we can see how we can add SE blocks to inception and residual modules:

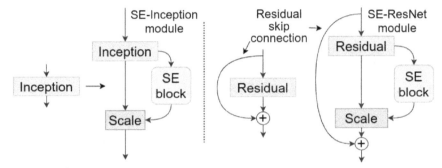

Figure 4.29 – An SE-inception module (left) and an SE-ResNet module (right)

In the next section, we'll see the SE block applied to a model, which prioritizes a small footprint and computational efficiency.

Introducing MobileNet

In this section, we'll discuss a lightweight CNN model called **MobileNet** (*MobileNetV3: Searching for MobileNetV3*, `https://arxiv.org/abs/1905.02244`). We'll focus on the third revision of this idea (MobileNetV1 was introduced in *MobileNets: Efficient Convolutional Neural Networks for Mobile Vision Applications*, `https://arxiv.org/abs/1704.04861` and MobileNetV2 was introduced in *MobileNetV2: Inverted Residuals and Linear Bottlenecks*, `https://arxiv.org/abs/1801.04381`).

MobileNet is aimed at devices with limited memory and computing power, such as mobile phones (the name kind of gives it away). The NN introduces a new **inverted residual block** (or **MBConv**) with a reduced footprint. MBConv uses DSC, **linear bottlenecks**, and **inverted residuals**. V3 also introduces SE blocks. To understand all this, here's the structure of the MBConv block:

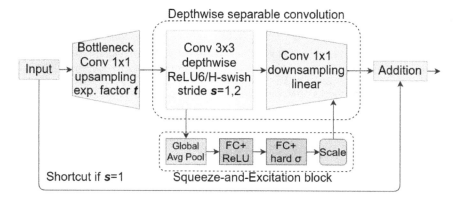

Figure 4.30 – MobileNetV3 building block. The shortcut connection exists only if the stride s=1

Let's discuss its properties:

- **Linear bottlenecks**: We'll assume that our input is an RGB image. As it's propagated through the NN, each layer produces an activation tensor with multiple channels. It has long been assumed that the information encoded in these tensors can be compressed in the so-called "manifold of interest," which is represented by a smaller tensor than the original. One way to force the NN to seek such manifolds is with 1×1 bottleneck convolutions. However, the authors of the paper argue that if this convolution is followed by non-linearity like ReLU, this might lead to a loss of manifold information because of the dying-ReLUs problem. To avoid this, MobileNet uses a 1×1 bottleneck convolution without non-linear activation.

- **Inverted residuals**: In the *Residual networks* section, we introduced the bottleneck residual block, where the data flow in the non-shortcut path is `input -> 1x1 bottleneck conv -> 3x3 conv -> 1x1 unsampling conv`. In other words, it follows a `wide -> narrow -> wide` data representation. On the other hand, the inverted residual block follows a `narrow -> wide -> narrow` representation. Here, the bottleneck convolution expands its input with an **expansion factor**, *t*.

  The authors argue that the bottlenecks contain all the necessary information, while an expansion layer acts merely as an implementation detail that accompanies a non-linear transformation of the tensor. Because of this, they propose having shortcut connections between the bottleneck connections instead.

- **DSC**: We already introduced this operation earlier in this chapter. MobileNet V3 introduces **H-swish** activation in the DSC. H-swish resembles the swish function, which we introduced in *Chapter 2*. The V3 architecture includes alternating ReLU and H-swish activations.

- **SE blocks**: We're already familiar with this block. The difference here is the **hard sigmoid** activation, which approximates the sigmoid but is computationally more efficient. The module is placed after the expanding depthwise convolution, so the attention can be applied to the largest representation. The SE block is a new addition to V3 and was not present in V2.

- **Stride** s: The block implements downsampling with stride convolutions. The shortcut connection exists only when $s=1$.

MobileNetV3 introduces large and small variations of the network with the following properties:

- Both networks start with a stride convolution that downsamples the input from 224×224 to 112×112

- The small and large variations have 11 and 15 MBConv blocks, respectively

- The convolutional phase ends with global average pooling for both networks

- The small and large networks have 3 and 5 million parameters, respectively

In the next section, we'll discuss an improved version of the MBConv block.

EfficientNet

EfficientNet (*EfficientNet: Rethinking Model Scaling for Convolutional Neural Networks*, https://arxiv.org/abs/1905.11946, and *EfficientNetV2: Smaller Models and Faster Training*, https://arxiv.org/abs/2104.00298) introduces the concept of **compound scaling**. It starts with a small baseline model and then simultaneously expands it in three directions: depth (more layers), width (more feature maps per layer), and higher input resolution. The compound scaling produces a series of new models. The EfficientNetV1 baseline model uses the MBConv building block of MobileNetV2. EfficientNetV2 introduces the new **fused-MBConv** block, which replaces the expanding 1×1 bottleneck convolution and the 3×3 depthwise convolution of MBConv, with a single expanding 3×3 cross-channel convolution:

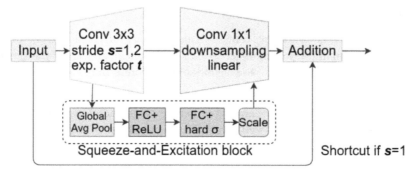

Figure 4.31 – Fused-MBConv block

The new 3×3 convolution handles both the expanding (with a factor of *t*) and the stride (1 or 2).

The authors of EfficientNetV2 observed that a CNN, which uses a combination of fused-MBConv and MBConv blocks, trains faster compared to a CNN with MBConv blocks only. However, the fused-MBConv block is computationally more expensive, compared to the plain MBConv block. Because of this, EfficientNetV2 replaces the blocks gradually, starting from the early stages. This makes sense because the earlier convolutions use a smaller number of filters (and hence slices), so the memory and computational penalty are less pronounced at this stage. Finding the right combination of the two blocks is not trivial, hence the need for compound scaling. This process produced multiple models with the following properties:

- The networks start with a stride convolution that downsamples the input twice
- The early stages of the main body use fused-MBConv blocks, and the later stages use MBConv blocks
- The convolutional phase ends with global average pooling for all networks
- The number of parameters ranges between 5.3 million and 119 million

This concludes our introduction to advanced CNN models. We didn't discuss all the available models, but we focused on some of the most popular ones. I hope that you now have sufficient knowledge to explore new models yourself. In the next section, we'll demonstrate how to use these advanced models in PyTorch and Keras.

Using pre-trained models with PyTorch and Keras

Both PyTorch and Keras have a collection of pre-trained ready-to-use models. All the models we discussed in the *Advanced network models* section are available in this way. The models are usually pre-trained on classifying the ImageNet dataset and can serve as backbones to various computer vision tasks, as we'll see in *Chapter 5*.

We can load a pre-trained model in PyTorch with the following code:

```
from torchvision.models import mobilenet_v3_large, MobileNet_V3_Large_
Weights
# With pretrained weights:
model = mobilenet_v3_large(
        weights=MobileNet_V3_Large_Weights.IMAGENET1K_V1)
model = mobilenet_v3_large(weights="IMAGENET1K_V1")
# Using no weights:
model = mobilenet_v3_large(weights=None)
```

The weights will be automatically downloaded. In addition, we can list all available models and load an arbitrary model using the following code:

```
from torchvision.models import list_models, get_model
# List available models
all_models = list_models()
model = get_model(all_models[0], weights="DEFAULT")
```

Keras supports similar functionality. We can load a pre-trained model with the following code:

```
from keras.applications.mobilenet_v3 import MobileNetV3Large
model = MobileNetV3Large(weights='imagenet')
```

These short but very useful code examples conclude this chapter.

Summary

In this chapter, we introduced CNNs. We talked about their main building blocks – convolutional and pooling layers – and we discussed their architecture and features. We paid special attention to the different types of convolutions. We also demonstrated how to use PyTorch and Keras to implement the CIFAR-10 classification CNN. Finally, we discussed some of the most popular CNN models in use today.

In the next chapter, we'll build upon our new-found computer vision knowledge with some exciting additions. We'll discuss how to train networks faster by transferring knowledge from one problem to another. We'll also go beyond simple classification with object detection, or how to find the object's location on the image. We'll even learn how to segment each pixel of an image.

5

Advanced Computer Vision Applications

In *Chapter 4*, we introduced **convolutional networks** (**CNNs**) for computer vision and some of the most popular and best-performing CNN models. In this chapter, we'll continue with more of the same, but at a more advanced level. Our *modus operandi* so far has been to provide simple classification examples to support your theoretical knowledge of **neural networks** (**NNs**). In the universe of computer vision tasks, classification is fairly straightforward as it assigns a single label to an image. This also makes it possible to manually create large, labeled training datasets. In this chapter, we'll introduce **transfer learning** (**TL**), a technique that will allow us to transfer the knowledge of pre-trained NNs to a new and unrelated task. We'll also see how TL makes it possible to solve two interesting computer vision tasks – object detection and semantic segmentation. We can say that these tasks are more complex compared to classification because the model has to obtain a more comprehensive understanding of the image. It has to be able to detect different objects as well as their positions in the image. At the same time, the task's complexity allows for more creative solutions.

Finally, we'll introduce a new class of algorithms called generative models, which will help us generate new images.

This chapter will cover the following topics:

- **Transfer learning** (**TL**)
- Object detection
- Semantic segmentation
- Image generation with diffusion models

Technical requirements

We'll implement the example in this chapter using Python, PyTorch, Keras, and Ultralytics YOLOv8 (`https://github.com/ultralytics/ultralytics`). If you don't have an environment set up with these tools, fret not – the example is available as a Jupyter notebook on Google Colab. You can find the code examples in this book's GitHub repository: `https://github.com/PacktPublishing/Python-Deep-Learning-Third-Edition/tree/main/Chapter05`.

Transfer learning (TL)

So far, we've trained small models on toy datasets, where the training took no more than an hour. But if we want to work with large datasets, such as ImageNet, we will need a much bigger network that trains for a lot longer. More importantly, large datasets are not always available for the tasks we're interested in. Keep in mind that besides obtaining the images, they have to be labeled, and this could be expensive and time-consuming. So, what does a humble engineer do when they want to solve a real ML problem with limited resources? Enter TL.

TL is the process of applying an existing trained ML model to a new, but related, problem. For example, we can take a network trained on ImageNet and repurpose it to classify grocery store items. Alternatively, we could use a driving simulator game to train an NN to drive a simulated car, and then use the network to drive a real car (but don't try this at home!). TL is a general ML concept that applies to all ML algorithms – we'll also use TL in *Chapter 8*. But in this chapter, we'll talk about TL in CNNs. Here's how it works.

We start with an existing pre-trained net. The most common scenario is to take a network pre-trained with ImageNet, but it could be any dataset. PyTorch, **TensorFlow** (**TF**), and Keras all have popular ImageNet pre-trained neural architectures that we can use. Alternatively, we can train our network with a dataset of our choice.

In *Chapter 4*, we mentioned how the **fully connected** (**FC**) layers at the end of a CNN act as translators between the network's language (the abstract feature representations learned during training) and our language, which is the class of each sample. You can think of TL as a translation to another language. We start with the network's features, which is the output of the last convolutional or pooling layer. Then, we translate them to a different set of classes for the new task. We can do this by removing the last layers of an existing pre-trained network and replacing them with a different set of layers, which represents the classes of the new problem. Here is a diagram of the TL scenario:

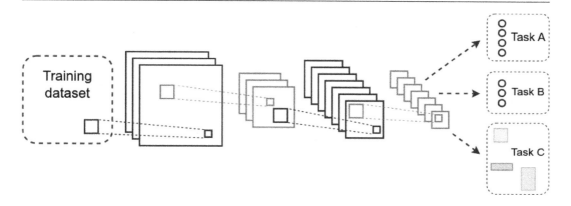

Figure 5.1 – A TL scenario, where we replace the last layer(s) of a
pre-trained network and repurpose it for a new problem

However, we cannot do this mechanically and expect the new network to work because we still have to train the new layer with data related to the new task. We have two options to do this:

- **Use the original part of the network as a feature extractor and only train the new layer(s):** First, we feed the network a training batch of the new data and propagate it forward and backward to see the network's output and error gradients. This part works just like regular training would. But during the weight updates phase, we lock the weights of the original network and only update the weights of the new layers. This is the recommended approach when we have limited training data for the new problem. By locking most of the network weights, we prevent overfitting on the new data.

- **Fine-tune the whole network:** We train the whole network and not just the newly added layers at the end. It is possible to update all the network weights, but we can also lock some of the weights in the first layers. The idea here is that the initial layers detect general features – not related to a specific task – and it makes sense to reuse them. On the other hand, the deeper layers may detect task-specific features and it would be better to update them. We can use this method when we have more training data and don't need to worry about overfitting.

Before we continue, let's note that TL is not limited to classification-to-classification problems. As we'll see later in this chapter, we can use pre-trained CNN as a backbone NN for object detection and semantic segmentation tasks. With that, let's see how to implement TL in practice.

Transfer learning with PyTorch

In this section, we'll apply an advanced ImageNet pre-trained network on the CIFAR-10 images. We'll implement both types of TL. It's preferable to run this example on GPU:

1. To define the training dataset, we have to consider a few things:

    *   Use mini-batch with size 50.

    *   The CIFAR-10 images are 32×32, while the ImageNet network expects 224×224 input. As we are using an ImageNet-based network, we'll upsample the 32×32 CIFAR images to 224×224 using `transforms.Resize`.

    *   Standardize the CIFAR-10 data using the ImageNet mean and standard deviation, because this is what the network expects.

    *   Add minor data augmentation (flip).

    We can do all this with the following code:

    ```
    import torch
    from torch.utils.data import DataLoader
    from torchvision import datasets
    from torchvision import transforms
    batch_size = 50
    # training data
    train_data_transform = transforms.Compose([
        transforms.Resize(224),
        transforms.RandomHorizontalFlip(),
        transforms.RandomVerticalFlip(),
        transforms.ToTensor(),
        transforms.Normalize(
            [0.485, 0.456, 0.406],
            [0.229, 0.224, 0.225])
    ])
    train_set = datasets.CIFAR10(
        root='data',
        train=True,
        download=True,
        transform=train_data_transform)
    train_loader = DataLoader(
        dataset=train_set,
        batch_size=batch_size,
        shuffle=True,
        num_workers=2)
    ```

2. Follow the same steps with the validation data (except for the data augmentation):

```
val_data_transform = transforms.Compose([
    transforms.Resize(224),
    transforms.ToTensor(),
    transforms.Normalize(
        [0.485, 0.456, 0.406],
        [0.229, 0.224, 0.225])
])
val_set = datasets.CIFAR10(
    root='data',
    train=False,
    download=True,
    transform=val_data_transform)
val_order = DataLoader(
    dataset=val_set,
    batch_size=batch_size,
    shuffle=False,
    num_workers=2)
```

3. Choose a device – preferably a GPU with a fallback on CPU:

```
device = torch.device("cuda:0" if torch.cuda.is_available() else
"cpu")
```

4. To train and validate the model, we'll use the `train_model(model, loss_function, optimizer, data_loader)` and `test_model(model, loss_function, data_loader)` functions. We first implemented them in *Chapter 3*, so we will not repeat the implementation here (it is available in the source code example on GitHub).

5. Define the first TL scenario, where we use the pre-trained network as a feature extractor:

    • We'll use a popular network, **MobileNetV3** (we discussed it in *Chapter 4*).

    • PyTorch will automatically download the pre-trained ImageNet weights.

    • Replace the last network layer with a new layer with 10 outputs, one for each CIFAR-10 class.

    • Exclude the existing network layers from the weight updates phase, and only pass the newly added FC layer to the Adam optimizer.

    • Run the training for `epochs` and evaluate the network accuracy after each epoch.

    • Use the `plot_accuracy` accuracy function, which plots the validation accuracy on a `matplotlib` graph. We won't include the full implementation here, but it is available on GitHub.

The following is the `tl_feature_extractor` function, which implements all this:

```python
import torch.nn as nn
import torch.optim as optim
from torchvision.models import MobileNet_V3_Small_Weights,
mobilenet_v3_small
def tl_feature_extractor(epochs=5):
    # load the pre-trained model
    model = mobilenet_v3_small(
        weights=MobileNet_V3_Small_Weights.IMAGENET1K_V1)
    # exclude existing parameters from backward pass
    # for performance
    for param in model.parameters():
        param.requires_grad = False
    # newly constructed layers have requires_grad=True by
default
    num_features = model.classifier[0].in_features
    model.classifier = nn.Linear(num_features, 10)
    # transfer to GPU (if available)
    model = model.to(device)
    loss_function = nn.CrossEntropyLoss()
    # only parameters of the final layer are being optimized
    optimizer = optim.Adam(model.classifier.parameters())
    # train
    test_acc = list()  # collect accuracy for plotting
    for epoch in range(epochs):
        print('Epoch {}/{}'.format(epoch + 1,
            epochs))
        train_model(model, loss_function,
        optimizer, train_loader)
        _, acc = test_model(model, loss_function,
            val_order)
        test_acc.append(acc.cpu())
    plot_accuracy(test_acc)
```

6. Implement the fine-tuning approach with the `tl_fine_tuning` function. This function is similar to `tl_feature_extractor`, but now, we'll train the whole network:

```python
def tl_fine_tuning(epochs=5):
    # load the pre-trained model
    model = mobilenet_v3_small(
        weights=MobileNet_V3_Small_Weights.IMAGENET1K_V1)
    # replace the last layer
    num_features = model.classifier[0].in_features
    model.classifier = nn.Linear(num_features, 10)
```

```
# transfer the model to the GPU
model = model.to(device)
# loss function
loss_function = nn.CrossEntropyLoss()
# We'll optimize all parameters
optimizer = optim.Adam(model.parameters())
# train
test_acc = list()  # collect accuracy for plotting
for epoch in range(epochs):
    print('Epoch {}/{}'.format(epoch + 1,
        epochs))
    train_model(model, loss_function,
    optimizer, train_loader)
    _, acc = test_model(model, loss_function,
        val_order)
    test_acc.append(acc.cpu())
plot_accuracy(test_acc)
```

7. We can run the whole thing in one of two ways:

    I.   Call `tl_fine_tuning(epochs=5)` to use the fine-tuning approach for five epochs.

    II. Call `tl_feature_extractor(epochs=5)` to train the network with the feature extractor approach for five epochs.

With a network as a feature extractor, we'll get about 81% accuracy, while with fine-tuning, we'll get 89%. But if we run the fine-tuning for more epochs, the network will start overfitting. Next, let's see the same example but with Keras.

Transfer learning with Keras

In this section, we'll implement the two TL scenarios again, but this time using Keras and TF. In this way, we can compare the two libraries. Again, we'll use the `MobileNetV3Small` architecture. In addition to Keras, this example also requires the TF Datasets package (`https://www.tensorflow.org/datasets`), a collection of various popular ML datasets. Let's start:

> **Note**
>
> This example is partially based on `https://github.com/tensorflow/docs/blob/master/site/en/tutorials/images/transfer_learning.ipynb`.

1. Define the mini-batch and input image sizes (the image size is determined by the network architecture):

    ```
    IMG_SIZE = 224
    BATCH_SIZE = 50
    ```

2. Load the CIFAR-10 dataset with the help of TF datasets. The `repeat()` method allows us to reuse the dataset for multiple epochs:

```
import tensorflow as tf
import tensorflow_datasets as tfds
data, metadata = tfds.load('cifar10', with_info=True,
    as_supervised=True)
raw_train, raw_test = data['train'].repeat(),
    data['test'].repeat()
```

3. Define the `train_format_sample` and `test_format_sample` functions, which will transform the initial images into suitable CNN inputs. These functions play the same roles that the `transforms.Compose` object plays, which we defined in the *Implementing transfer learning with PyTorch* section. The input is transformed as follows:

    *   The images are resized to 224×224, which is the expected network input size

    *   Each image is standardized by transforming its values so that it's in the (-1; 1) interval

    *   The labels are converted into one-hot encodings

    *   The training images are randomly flipped horizontally and vertically

    Let's look at the actual implementation:

```
def train_format_sample(image, label):
    """Transform data for training"""
    image = tf.cast(image, tf.float32)
    image = tf.image.resize(image, (IMG_SIZE,
        IMG_SIZE))
    image = tf.image.random_flip_left_right(image)
    image = tf.image.random_flip_up_down(image)
    label = tf.one_hot(label,
        metadata.features['label'].num_classes)
    return image, label
def test_format_sample(image, label):
    """Transform data for testing"""
    image = tf.cast(image, tf.float32)
    image = tf.image.resize(image, (IMG_SIZE,
        IMG_SIZE))
    label = tf.one_hot(label,
        metadata.features['label'].num_classes)
    return image, label
```

4. Next is some boilerplate code that assigns these transformers to the train/test datasets and splits them into mini-batches:

```
# assign transformers to raw data
train_data = raw_train.map(train_format_sample)
test_data = raw_test.map(test_format_sample)
# extract batches from the training set
train_batches = train_data.shuffle(1000).batch(BATCH_SIZE)
test_batches = test_data.batch(BATCH_SIZE)
```

5. Define the feature extraction model:

- Use Keras for the pre-trained network and model definition since it is an integral part of TF

- Load the MobileNetV3Small pre-trained net, excluding the final FC layers

- Call base_model.trainable = False, which freezes all the network weights and prevents them from training

- Add a GlobalAveragePooling2D operation, followed by a new and trainable FC trainable layer at the end of the network

The following code implements this:

```
def build_fe_model():
    """"Create feature extraction model from the pre-trained
model ResNet50V2"""
    # create the pre-trained part of the network, excluding FC
layers
    base_model = tf.keras.applications.MobileNetV3Small(
        input_shape=(IMG_SIZE, IMG_SIZE, 3),
        include_top=False,
        classes=10,
        weights='imagenet',
        include_preprocessing=True)
    # exclude all model layers from training
    base_model.trainable = False
    # create new model as a combination of the pre-trained net
    # and one fully connected layer at the top
    return tf.keras.Sequential([
        base_model,
        tf.keras.layers.GlobalAveragePooling2D(),
        tf.keras.layers.Dense(
            metadata.features['label'].num_classes,
            activation='softmax')
    ])
```

6. Define the fine-tuning model. The only difference it has from the feature extraction is that we only freeze some of the bottom pre-trained network layers (as opposed to all of them). The following is the implementation:

```
def build_ft_model():
    """Create fine tuning model from the pre-trained model
MobileNetV3Small"""
    # create the pre-trained part of the network, excluding FC
layers
    base_model = tf.keras.applications.MobileNetV3Small(
        input_shape=(IMG_SIZE, IMG_SIZE, 3),
        include_top=False,
        weights='imagenet',
        include_preprocessing=True
    )
    # Fine tune from this layer onwards
    fine_tune_at = 100
    # Freeze all the layers before the `fine_tune_at` layer
    for layer in base_model.layers[:fine_tune_at]:
        layer.trainable = False
    # create new model as a combination of the pre-trained net
    # and one fully connected layer at the top
    return tf.keras.Sequential([
        base_model,
        tf.keras.layers.GlobalAveragePooling2D(),
        tf.keras.layers.Dense(
            metadata.features['label'].num_classes,
            activation='softmax')
    ])
```

7. Implement the train_model function, which trains and evaluates the models that are created by either the build_fe_model or build_ft_model function. The plot_accuracy function is not implemented here but is available on GitHub:

```
def train_model(model, epochs=5):
    """Train the model. This function is shared for both FE and
FT modes"""
    # configure the model for training
    model.compile(
        optimizer=tf.keras.optimizers.Adam(
            learning_rate=0.0001),
        loss='categorical_crossentropy',
        metrics=['accuracy'])
    # train the model
    history = model.fit(
```

```
        train_batches,
        epochs=epochs,
        steps_per_epoch=metadata.splits['train'].num_examples /
            BATCH_SIZE,
        validation_data=test_batches,
        validation_steps=metadata.splits['test'].num_examples /
            BATCH_SIZE,
        workers=4)
    # plot accuracy
    plot_accuracy(history.history['val_accuracy'])
```

8. We can run either the feature extraction or fine-tuning TL using the following code:

    • `train_model(build_ft_model())`

    • `train_model(build_fe_model())`

With a network as a feature extractor, we'll get about 82% accuracy, while with fine-tuning, we'll get 89% accuracy. The results are similar to the PyTorch example.

Next, let's turn our attention to object detection – a task we can solve with the help of TL.

Object detection

Object detection is the process of finding object instances of a certain class, such as people, cars, and trees, in images or videos. Unlike classification, object detection can detect multiple objects as well as their location in the image.

An object detector would return a list of detected objects with the following information for each object:

- The class of the object (person, car, tree, and so on).

- A probability (or objectness score) in the [0, 1] range, which conveys how confident the detector is that the object exists in that location. This is similar to the output of a regular binary classifier.

- The coordinates of the rectangular region of the image where the object is located. This rectangle is called a **bounding box**.

We can see the typical output of an object-detection algorithm in the following figure. The object type and objectness score are above each bounding box:

Figure 5.2 – The output of an object detector. Source: https://en.wikipedia.
org/wiki/File:2011_FIA_GT1_Silverstone_2.jpg

Next, let's outline the different approaches to solving an object detection task.

Approaches to object detection

In this section, we'll outline three approaches:

- **Classic sliding window**: Here, we'll use a regular classification network (classifier). This approach can work with any type of classification algorithm, but it's relatively slow and error-prone:

  - **Build an image pyramid**: This is a combination of different scales of the same image (see the following figure). For example, each scaled image can be two times smaller than the previous one. In this way, we'll be able to detect objects regardless of their size in the original image.

  - **Slide the classifier across the whole image**: We'll use each location of the image as an input to the classifier, and the result will determine the type of object that is in the location. The bounding box of the location is just the image region that we used as input.

  - **Multiple overlapping bounding boxes for each object**: We'll use some heuristics to combine them into a single prediction.

Here is a figure showing the sliding window approach:

Figure 5.3 – Sliding window plus image pyramid object detection

- **Two-stage detection methods**: These methods are very accurate but relatively slow. As its name suggests, this involves two steps:

  - A special type of CNN, called a **Region Proposal Network** (**RPN**), scans the image and proposes several possible bounding boxes, or **regions of interest** (**RoI**), where objects might be located. However, this network doesn't detect the type of object, but only whether an object is present in the region.

  - The RoI is sent to the second stage for object classification, which determines the actual object in each bounding box.

- **One-stage (or one-shot) detection methods**: Here, a single CNN produces both the object type and the bounding box. These approaches are usually faster but less accurate than the two-stage methods.

In the next section, we'll introduce **YOLO** – an accurate and efficient one-stage detection algorithm.

Object detection with YOLO

YOLO is one of the most popular one-stage detection algorithms. The name is an acronym for the popular motto "You only live once", which reflects the one-stage nature of the algorithm. Since its original release, there have been multiple YOLO versions, with different authors. For the sake of clarity, we'll list all versions here:

- *You Only Look Once: Unified, Real-Time Object Detection* (`https://arxiv.org/abs/1506.02640`), by Joseph Redmon, Santosh Divvala, Ross Girshick, and Ali Farhadi.

- *YOLO9000: Better, Faster, Stronger* (`https://arxiv.org/abs/1612.08242`), by Joseph Redmon and Ali Farhadi.

- *YOLOv3: An Incremental Improvement* (`https://arxiv.org/abs/1804.02767`, `https://github.com/pjreddie/darknet`), by Joseph Redmon and Ali Farhadi.

- *YOLOv4: Optimal Speed and Accuracy of Object Detection* (`https://arxiv.org/abs/2004.10934`, `https://github.com/AlexeyAB/darknet`), by Alexey Bochkovskiy, Chien-Yao Wang, and Hong-Yuan Mark Liao.

- **YOLOv5** and **YOLOv8** (`https://github.com/ultralytics/yolov5`, `https://github.com/ultralytics/ultralytics`), by Ultralitics (`https://ultralytics.com/`). V5 and v8 have no official paper.

- *YOLOv6 v3.0: A Full-Scale Reloading* (`https://arxiv.org/abs/2301.05586`, `https://github.com/meituan/YOLOv6`), by Chuyi Li, Lulu Li, Yifei Geng, Hongliang Jiang, Meng Cheng, Bo Zhang, Zaidan Ke, Xiaoming Xu, and Xiangxiang Chu.

- *YOLOv7: Trainable bag-of-freebies sets new state-of-the-art for real-time object detectors* (`https://arxiv.org/abs/2207.02696`, Mark L`https://github.com/WongKinYiu/yolov7`), by Chien-Yao Wang, Alexey Bochkovskiy, and Hong-Yuan iao.

> **Note**
>
> v3 is the last version, released by the original authors of the algorithm. v4 is a fork of v3 and was endorsed by the main author of v1-v3, Joseph Redmon (`https://twitter.com/pjreddie/status/1253891078182199296`). On the other hand, v5 is an independent implementation, inspired by YOLO. This sparked a controversy regarding the name of v5. You can follow some of the discussion at `https://github.com/AlexeyAB/darknet/issues/5920`, where Alexey Bochkovskiy, the author of v4, has also posted. The authors of v5 have also addressed the controversy here: `https://blog.roboflow.com/yolov4-versus-yolov5/`. Regardless of this discussion, v5 and v8 have proven to work and are popular detection algorithms in their own right.

We'll discuss the YOLO properties shared among all versions and we'll point out some of the differences.

Let's start with the YOLO architecture:

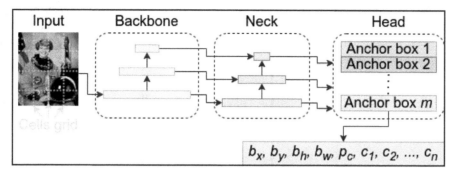

Figure 5.4 – The YOLO architecture

It contains the following components:

- **Backbone**: This is a CNN model that's responsible for extracting features from the input image. These features are then passed to the next components for object detection. Usually, the backbone is an ImageNet pre-trained CNN, similar to the advanced models we discussed in *Chapter 4*.

  The backbone is an example of TL – we take a CNN trained for classification and repurpose it for object detection. The different YOLO versions use different backbones. For example, v3 uses a special fully convolutional CNN called DarkNet-53 with 53 layers. Subsequent YOLO versions introduce various improvements to this architecture, while others use their own unique backbone.

- **Neck**: This is an intermediate part of the model that connects the backbone to the head. It concatenates the output at different stages of the backbone feature maps before sending the combined result to the next component (the head). This is an alternative to the standard approach, where we would just send the output of the last backbone convolution for further processing. To understand the need for the neck, let's recall that our goal is to create a precise bounding box around the edges of the detected object. The object itself might be big or small, relative to the image. However, the receptive field of the deeper layers of the backbone is large because it aggregates the receptive fields of all preceding layers. Hence, the features detected at the deeper layers encompass large parts of the input image. This runs contrary to our goal of fine-grained object detection, regardless of the object's size. To solve this, the neck combines the feature maps at different backbone stages, which makes it possible to detect objects at different scales. However, the feature maps at each backbone stage have different dimensions and cannot be combined directly. The neck applies different techniques, such as upsampling or downsampling, to equalize these dimensions, so that they can be concatenated.

- **Head**: This is the final component of the model, which outputs the detected objects. Each detected object is represented by its bounding box coordinates and its class.

With that, we have gained a bird's-eye view of the YOLO architecture. But it doesn't answer some inconvenient (yet intriguing) questions, such as how the model detects multiple objects on the same image, or what happens if two or more objects overlap and one is only partially visible. To find the answers to these questions, let's introduce the following diagram, which consists of two overlapping objects:

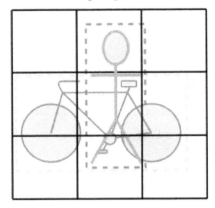

Figure 5.5 – An object detection YOLO example with two overlapping objects and their bounding boxes

These are the steps that YOLO implements to detect them:

1. Split the input image into a grid of $S \times S$ cells (the preceding diagram uses a 3×3 grid):

    - The center of a cell represents the center of a region where an object might be located.

    - The model can detect both objects that span multiple cells and ones that lie entirely within the cell. Each object is associated with a single cell, even if it covers multiple cells. In this case, we'll associate the object with the cell, where the center of its bounding box lies. For example, the two objects in the diagram span multiple cells, but they are both assigned to the central cell because their centers lie in it.

    - A cell can contain multiple objects (*1-to-n* relationship) or no objects at all. We're only interested in the cells with objects.

2. The model outputs multiple possible detected objects for each grid cell. Each detected object is represented by the following array of values: $\left[b_x, b_y, b_h, b_w, p_c, c_1, c_2 \ldots c_n\right]$. Let's discuss them:

    - $\left[b_x, b_y, b_h, b_w\right]$ describes the object bounding box. $\left[b_x, b_y\right]$ are the coordinates of the center of the box concerning the whole image. They are normalized in the [0, 1] range. For example, if the image size is 100×100 and the center of the bounding box is located at [40, 70], then $\left[b_x, b_y\right]$ = [0.4, 0.7]. $\left[b_h, b_w\right]$ represent the normalized bounding box height and width concerning the whole image. If the bounding box's size is 80×50, then $\left[b_h, b_w\right]$ = [0.8, 0.5] for the same 100×100 image. In practice, a YOLO implementation usually includes helper methods, which will allow us to obtain the absolute coordinates of the bounding boxes.

- p_c is an objectness score, which represents the confidence of the model (in the [0, 1] range) that an object is present in the cell. If p_c is closer to 1, then the model is confident that an object is present and vice versa.

- $[c_1, c_2 ... c_n]$ is a one-hot encoding of the class of the detected object. For example, if we have bicycle, flower, person, and fish classes, and the current object is a person, its encoding will be [0, 0, 1, 0].

3. So far, we've demonstrated that the model can detect multiple objects on the same image. Next, let's focus on the trickier case with multiple objects in the same cell. YOLO has an elegant solution to this problem in the form of **anchor boxes** (also known as **priors**). To understand this concept, we'll start with the following diagram, which shows the grid cell (square, uninterrupted line) and two anchor boxes – vertical and horizontal (dashed lines):

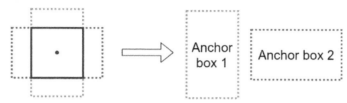

Figure 5.6 – A grid cell (a square, uninterrupted line) with two anchor boxes (dashed lines)

For each cell, we'll have multiple candidate anchor boxes with different scales and aspect ratios. If we have multiple objects in the same cell, we'll associate each object with a single anchor box. If an anchor box doesn't have an associated object, it will have an objectness score of zero ($p_c = 0$). We can detect as many objects as there are anchor boxes per cell. For example, our example 3×3 grid with two anchor boxes per cell can detect a total of 3*3*2 = 18 objects. Because we have a fixed number of cells ($S \times S$) and a fixed number of anchor boxes per cell, the size of the network output doesn't change with the number of detected objects. Instead, we'll output results for all possible anchor boxes, but we'll only consider the ones with an objectness score of $p_c \sim 1$.

4. The YOLO algorithm uses the **Intersection over Union** (**IoU**) technique both during training and inference to improve its performance:

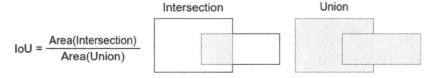

Figure 5.7 – Intersection over Union (IoU)

IoU is the ratio between the area of the intersection and the area of the union of the detected object bounding box and the ground truth (or another object's) bounding box.

During training, we can compute the IoU between the anchor boxes and ground truth bounding boxes. Then, we can assign each ground truth object to its highest overlapping anchor box to generate labeled training data. In addition, we can compute the IoU between the detected bounding box and the ground truth (label) box. The higher value of IoU indicates a better overlap between ground truth and prediction. This can help us evaluate the detector.

During inference, the output of the model includes all possible anchor boxes for each cell, regardless of whether an object is present in them. Many of the boxes will overlap and predict the same object. We can filter the overlapping objects with the help of IoU and **non-maximum suppression** (**NMS**). Here's how it works:

1. Discard all bounding boxes with an objectness score of p_c < 0.6.

2. Pick the box with the highest objectness score, p_c, from the remaining boxes.

3. Discard all boxes with IoU >= 0.5 with the box we selected in the previous step.

Now that we are (hopefully) familiar with YOLO, let's learn how to use it in practice.

Using Ultralytics YOLOv8

In this section, we'll demonstrate how to use the YOLOv8 algorithm, developed by Ultralytics. For this example, you'll need to install the `ultralytics` Python package. Let's start:

1. Import the YOLO module. We'll load a pre-trained YOLOv8 model:

   ```
   from ultralytics import YOLO
   model = YOLO("yolov8n.pt")
   ```

2. Use `model` to detect the objects on a Wikipedia image:

   ```
   results = model.predict('https://raw.githubusercontent.com/
   ivan-vasilev/Python-Deep-Learning-3rd-Edition/main/Chapter05/
   wikipedia-2011_FIA_GT1_Silverstone_2.jpg')
   ```

   `results` is a list, composed of a single instance of the `ultralytics.yolo.engine.results.Results` class. The instance contains the list of detected objects: their bounding boxes, classes, and objectness scores.

3. We can display the results with the help of the `results[0].plot()` method, which overlays the detected object on the input image. The result of this operation is the first image, we introduced at the start of the *Introduction to object detection* section:

   ```
   from PIL import Image
   Image.fromarray(results[0].plot()).show()
   ```

This concludes our introduction to the YOLO family of single-shot object detection models. Next, we'll focus on a popular example of a two-shot detection algorithm.

Object detection with Faster R-CNN

In this section, we'll discuss the **Faster R-CNN** (*Faster R-CNN: Towards Real-Time Object Detection with Region Proposal Networks*, https://arxiv.org/abs/1506.01497) two-stage object detection algorithm. It is an evolution of the earlier two-stage detectors, **Fast R-CNN** (*Fast R-CNN*, https://arxiv.org/abs/1504.08083) and **R-CNN** (*Rich feature hierarchies for accurate object detection and semantic segmentation*, https://arxiv.org/abs/1311.2524).

The general structure of the Faster R-CNN model is outlined in the following diagram:

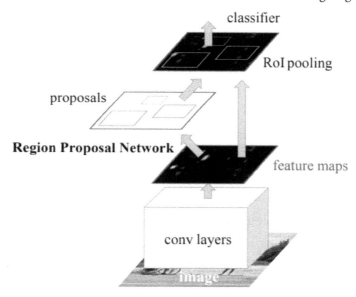

Figure 5.8 – The structure of Faster R-CNN. Source: https://arxiv.org/abs/1506.01497

Let's keep this figure in mind while we explain the algorithm. Like YOLO, Faster R-CNN starts with a backbone classification network trained on ImageNet, which serves as a base for the different modules of the model. Originally, the authors of the paper experimented with classic backbone architectures, such as **VGG-16** (*Very Deep Convolutional Networks for Large-Scale Image Recognition*, https://arxiv.org/abs/1409.1556) and **ZFNet** (*Visualizing and Understanding Convolutional Networks*, https://arxiv.org/abs/1311.2901). Today, the model is available with more contemporary backbones, such as ResNet and MobileNet.

Unlike YOLO, Faster R-CNN doesn't have a neck module and only uses the feature maps of the last backbone convolutional layer as input to the next components of the algorithm. More specifically, the backbone serves as a backbone (get it?) to the two other components of the model (hence two-stage) – the **region proposal network** (**RPN**) and the detection network. Let's discuss the RPN first.

The region proposal network

In the first stage, the RPN takes an image (of any size) as input and outputs a set of rectangular RoI, where an object might be located. The RoI is equivalent to the bounding box in YOLO. The RPN itself is created by taking the first p convolutional layers of the backbone model (see the preceding diagram). Once the input image is propagated to the last shared convolutional layer, the algorithm takes the feature map of that layer and slides another small network over each location of the feature map. The small network outputs whether an object is present at any of the k anchor boxes (the concept of anchor box is the same as in YOLO), as well as the coordinates of its potential bounding box. This is illustrated on the left-hand side image of the following diagram, which shows a single location of the RPN sliding over a single feature map of the last convolutional layer:

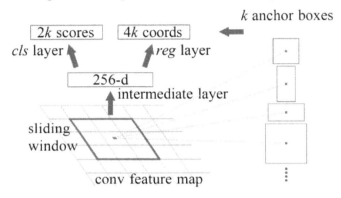

Figure 5.9 – RPN proposals over a single location. Source: https://arxiv.org/abs/1506.01497

The small network takes an $n \times n$ region at the same location across all input feature maps as input ($n = 3$ according to the paper). For example, if the final convolutional layer has 512 feature maps, the small network's input size at one location is 512*3*3 = 4608. The 512 3×3 feature maps are flattened to a 4,608-dimensional vector. It serves as input to a fully connected layer, which maps it to a lower dimensional (usually 512) vector. This vector itself serves as input to the following two parallel fully connected layers:

- A classification layer with $2k$ units organized into k 2-unit binary softmax outputs. Like YOLO, the output of each softmax represents the objectness score (in the [0, 1] range) of whether an object exists in each of the k anchor boxes. During training, an object is assigned to an anchor box based on the IoU formula in the same way as in YOLO.

- A regression layer with $4k$ units organized into k 4-unit RoI arrays. Like YOLO, the first array elements represent the coordinates of the RoI center in the [0:1] range relative to the whole image. The other two elements represent the height and width of the region, relative to the whole image (again, similar to YOLO).

The authors of the paper experimented with three scales and three aspect ratios, resulting in nine possible anchor boxes over each location. The typical $H \times W$ size of the final feature map is around 2,400, which results in 2,400*9 = 21,600 anchor boxes.

RPN as a cross-channel convolution

In theory, we slide the small network over the feature map of the last convolutional layer. However, the small network weights are shared along all locations. Because of this, the sliding can be implemented as a cross-channel convolution. Therefore, the network can produce output for all anchor boxes in a single image pass. This is an improvement over Fast R-CNN, which requires a separate network pass for each anchor box.

The RPN is trained with backpropagation and stochastic gradient descent (what a surprise!). The weights of the shared convolutional layers are initialized with the pre-trained weights of the backbone network and the rest are initialized randomly. The samples of each mini-batch are extracted from a single image. Each mini-batch contains an equal number of positive (objects) and negative (background) anchor boxes. There are two kinds of anchors with positive labels: the anchor/anchors with the highest IoU overlap with a ground truth box and an anchor that has an IoU overlap of higher than 0.7 with any ground truth box. If the IoU ratio of an anchor is lower than 0.3, the box is assigned a negative label. Anchors that are neither positive nor negative do not participate in the training.

As the RPN has two output layers (classification and regression), the training uses the following composite cost function with classification (L_{cls}) and regression (L_{reg}) parts:

$$L(\{p_i\}, \{\mathbf{t}_i\}) = \frac{1}{N_{cls}} \sum_i L_{cls}(p_i, p_i^*) + \lambda \frac{1}{N_{reg}} \sum_i p_i^* L_{reg}(\mathbf{t}_i, \mathbf{t}_i^*)$$

Let's discuss its components:

- i: The index of the anchor in the mini-batch.

- p_i: The classification output, which represents the predicted objectness score of an anchor, i, being an object or background. p_i^* is the target data for the same (0 or 1).

- $\mathbf{t}_i$: The regression output vector with size 4, which represents the RoI parameters.

- $\mathbf{t}_i^*$: The target vector for the same.

- L_{cls}: A cross-entropy loss for the classification layer.

- N_{cls}: A normalization term, equal to the mini-batch size.

- L_{reg}: The regression loss, $L_{reg} = R(\mathbf{t}_i - \mathbf{t}_i^*)$, where R is the mean absolute error (https://en.wikipedia.org/wiki/Mean_absolute_error).

- N_{reg}: A normalization term equal to the total number of anchor locations (around 2400).

- λ: This helps combine the classification and regression components of the cost function. Since $N_{reg} \sim 2400$ and $N_{cls} = 256$, λ is set to 10 to preserve the balance between the two losses.

Now that we've discussed the RPN, let's focus on the detection network.

Detection network

Let's go back to the diagram that was shown at the beginning of the *Object detection with Faster R-CNN* section. Recall that in the first stage, the RPN has already generated the RoI coordinates and their objectness scores. The detection network is a regular classifier, which determines the class of objects in the current RoI. Both the RPN and the detection network share their first convolutional layers, borrowed from the backbone network. In addition, the detection network incorporates the proposed regions from the RPN, along with the feature maps of the last shared layer.

But how do we combine the backbone feature maps and the proposed regions in a unified input format? We can do this with the help of **RoI pooling**, which is the first layer of the second part of the detection network:

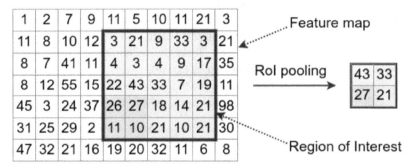

Figure 5.10 – An example of 2×2 RoI pooling with a 10×7 feature map and a 5×5 RoI (bold rectangle)

To understand how RoI pooling works, let's assume that we have a single 10×7 feature map and a single RoI. As we learned in the *Region proposal network* section, a RoI is defined by its center coordinates, width, and height. The RoI pooling first converts these parameters into actual coordinates on the feature map. In this example, the region size is $h \times w = 5 \times 5$. The RoI pooling is further defined by its output height and width, H and W. In this example, $H \times W = 2 \times 2$, but in practice, the values could be larger, such as 7×7. The operation splits the $h \times w$ RoI into a grid of subregions with different sizes (displayed in the figure with different background colors). Once this is done, each subregion is downsampled to a single output cell by taking the maximum value of that region. In other words, RoI pooling can transform inputs with arbitrary sizes into a fixed-size output window. In this way, the transformed data can propagate through the network in a consistent format.

As we mentioned in the *Object detection with Faster R-CNN* section, the RPN and the detection network share their initial layers. However, they start their lives as separate networks. The training alternates between the two in a four-step process:

1.   Train the RPN, which is initialized with the ImageNet weights of the backbone.

2. Train the detection network, using the proposals from the freshly trained RPN from *step 1*. The training also starts with the weights of the ImageNet backbone. At this point, the two networks don't share weights.

3. Use the detection network shared layers to initialize the weights of the RPN. Then, train the RPN again, but freeze the shared layers and fine-tune the RPN-specific layers only. The two networks share their weights now.

4. Train the detection network by freezing the shared layers and fine-tuning the detection-net-specific layers only.

Now that we've introduced Faster R-CNN, let's discuss how to use it in practice with the help of a pre-trained PyTorch model.

Using Faster R-CNN with PyTorch

In this section, we'll use a pre-trained PyTorch Faster R-CNN model with a ResNet50 backbone for object detection. PyTorch has out-of-the-box support for Faster R-CNN, which makes it easy for us to use. This example is implemented with PyTorch. In addition, it uses the `torchvision` and `opencv-python` packages. We will only include the relevant parts of the code, but you can find the full version in this book's GitHub repository. Let's start:

1. Load the pre-trained model with the latest available weights. Ensure this by using the DEFAULT option:

```
from torchvision.models.detection import \
FasterRCNN_ResNet50_FPN_V2_Weights, \
fasterrcnn_resnet50_fpn_v2
model = fasterrcnn_resnet50_fpn_v2(
    weights=FasterRCNN_ResNet50_FPN_V2_Weights.DEFAULT)
```

2. We are going to use the model for inference and not for training, so we'll enable the `eval()` mode:

```
model.eval()
```

3. Use `opencv-python` to read the RGB image located at `image_file_path`. We'll omit the code, which downloads the image from this book's repository if it doesn't already exist locally:

```
import cv2
img = cv2.imread(image_file_path)
```

Here, `img` is a three-dimensional numpy array of integers.

4. Implement the single-step image pre-processing pipeline. It transforms the `img` numpy array into `torch.Tensor`, which will serve as input to the model:

```
import torchvision.transforms as transforms
transform = transforms.ToTensor()
```

5. Run the detection model:

```
nn_input = transform(img)
detected_objects = model([nn_input])
```

Here, `detected_objects` is a dictionary with three items:

- `boxes`: A list of bounding boxes, represented by their top-left and bottom-right pixel coordinates
- `labels`: A list of labels for each detected object
- `scores`: A list of objectness scores for each detected object

6. Use the initial `img` array and `detected_objects` as parameters for the `draw_bboxes` function, which overlays the bounding boxes and their labels on the original input image (the implementation of `draw_bboxes` is available in the full example):

```
draw_bboxes(img, detected_objects)
```

7. Display the result with `opencv-python`:

```
cv2.imshow("Object detection", img)
cv2.waitKey()
```

The output image looks like this:

Figure 5.11 – Object detection with Faster R-CNN

We're now familiar with two of the most popular object detection algorithms. In the next section, we'll focus on the next major computer vision task, called **image segmentation**.

Introducing image segmentation

Image segmentation is the process of assigning a class label (such as person, bicycle, or animal) to each pixel of an image. You can think of it as classification but on a pixel level – instead of classifying the entire image under one label, we'll classify each pixel separately. The output of an image segmentation operation is known as a **segmentation mask**. It is a tensor with the same dimensions as the original input image, but instead of color, each pixel is represented by the class of object, to which it belongs. There are two types of segmentation:

- **Semantic segmentation**: This assigns a class to each pixel but doesn't differentiate between object instances. For example, the middle image in the following figure shows a semantic segmentation mask, where the pixels of each separate vehicle have the same value. Semantic segmentation can tell us that a pixel is part of a vehicle but cannot make a distinction between two vehicles.

- **Instance segmentation**: This assigns a class to each pixel and differentiates between object instances. For example, the image on the right in the following figure shows an instance segmentation mask, where each vehicle is segmented as a separate object.

The following figure shows an example of semantic and instance segmentation:

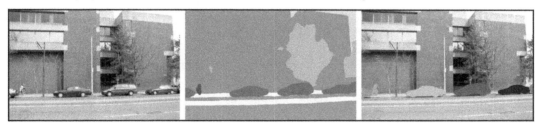

Figure 5.12 – Left: input image; middle: semantic segmentation mask; right:
instance segmentation mask. Source: http://sceneparsing.csail.mit.edu/

To train a segmentation algorithm, we'll need a special type of ground truth data, where the labels of each image are the segmented version of the image.

The easiest way to segment an image is by using the familiar sliding-window technique, which we described in the *Approaches to object detection* section – that is, we'll use a regular classifier, and we'll slide it in either direction with stride 1. After we get the prediction for a location, we'll take the pixel that lies in the middle of the input region, and we'll assign it to the predicted class. Predictably, this approach is very slow because of the large number of pixels in an image (even a 1,024×1,024 image has more than 1 million pixels). Thankfully, there are faster and more accurate algorithms, which we'll discuss in the following sections.

Semantic segmentation with U-Net

The first approach to segmentation we'll discuss is called **U-Net** (*U-Net: Convolutional Networks for Biomedical Image Segmentation*, `https://arxiv.org/abs/1505.04597`). The name comes from the visualization of the network architecture:

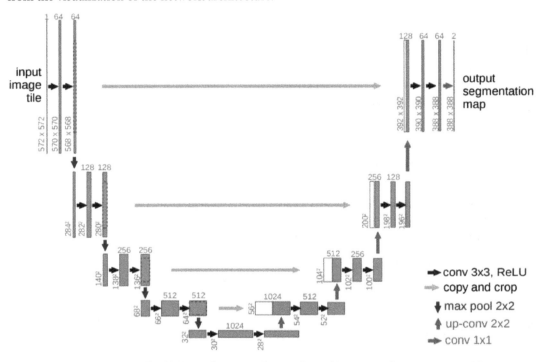

Figure 5.13 – The U-Net architecture. Source: https://arxiv.org/abs/1505.04597

U-Net is a type of **fully convolutional network** (**FCN**), called so because it contains only convolutional layers and doesn't use any fully connected layers at its output. An FCN takes the whole image as input and outputs its segmentation map in a single pass. To better understand this architecture, let's clarify the figure notations first:

- The horizontal dark blue arrows correspond to 3×3 cross-channel convolutions with ReLU activation. The single light blue arrow at the end of the model represents a 1×1 bottleneck convolution to reduce the number of channels.

- All feature maps are denoted with blue boxes. The number of feature maps is on top of the box, and the feature map's size is at the lower-left edge of the box.

- The horizontal gray arrows represent copy and crop operation (more on that later).

- The red vertical arrows represent 2×2 max pooling operations.

- The vertical green arrows represent 2×2 up-convolutions (or transposed convolutions; see *Chapter 4*).

We can separate the U-Net model into two virtual components (in reality, this is just a single network):

- **Encoder**: The first part of the network (the left part of the *U*) is similar to a regular CNN but without the fully connected layers at the end. Its role is to learn highly abstract representations of the input image (nothing new here). The input image itself can be an arbitrary size, so long as the input feature maps of every max pooling operation have even (and not odd) dimensions. Otherwise, the output segmentation mask will be distorted. By default, the input size is 572×572. From there, it continues like a regular CNN with alternating convolutional and max pooling layers. The encoder consists of four identical blocks of two consecutive valid (unpadded) cross-channel 3×3 convolutions with stride 1, optional batch normalization, ReLU activations, and a 2×2 max pooling layer. Each downsampling step doubles the number of feature maps. The final encoder convolution ends with 1,024 28×28 feature maps.

- **Decoder**: The second part of the network (the right part of the *U*) is symmetrical to the encoder. The decoder takes the innermost 28×28 encoder feature maps and simultaneously upsamples and converts them into a 388×388 segmentation map. It contains four identical upsampling blocks:

  - The upsampling works with 2×2 transposed cross-channel convolutions with stride 2.

  - The output of each upsampling step is concatenated with the cropped high-resolution feature maps of the corresponding encoder step (gray horizontal arrows). The cropping is necessary because of the loss of border pixels in every unpadded encoder and decoder convolution.

  - Each transposed convolution is followed by two regular convolutions to smooth the expanded representation.

  - The upsampling steps halve the number of feature maps. The final output uses a 1×1 bottleneck convolution to map the 64-component feature map tensor to the desired number of classes (light blue arrow). The authors of the paper have demonstrated the binary segmentation of medical images of cells.

The network's output is a softmax over each pixel of the segmentation mask – that is, the output contains as many independent softmax operations as the number of pixels. The softmax output for one pixel determines the pixel class. U-Net is trained like a regular classification network. However, the cost function is a combination of the cross-entropy losses of the softmax outputs over all pixels.

We can see that because of the unpadded convolutions of the network, the output segmentation map is smaller than the input image (388 versus 572). However, the output map is not a rescaled version of the input image. Instead, it has a one-to-one scale compared to the input, but only covers the central part of the input tile. This is illustrated in the following diagram:

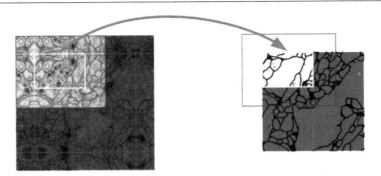

Figure 5.14 – An overlap-tile strategy for segmenting large
images. Source: https://arxiv.org/abs/1505.04597

The unpadded convolutions are necessary so that the network doesn't produce noisy artifacts at the borders of the segmentation map. This makes it possible to segment images with arbitrary large sizes using the so-called overlap-tile strategy. The input image is split into overlapping input tiles, like the one shown on the left of the preceding figure. The segmentation map of the small light area in the image on the right requires the large light area (one tile) on the left image as input.

The next input tile overlaps with the previous one in such a way that their segmentation maps cover adjacent areas of the image. To predict the pixels in the border region of the image, the missing context is extrapolated by mirroring the input image.

We're not going to implement a code example with U-Net, but you can check out `https://github.com/mateuszbuda/brain-segmentation-pytorch` for U-Net brain MRI image segmentation.

Instance segmentation with Mask R-CNN

Mask R-CNN (`https://arxiv.org/abs/1703.06870`) is an extension of Faster R-CNN for instance segmentation. Faster R-CNN has two outputs for each candidate object: bounding box parameters and class labels. In addition to these, Mask R-CNN adds a third output – an FCN that produces a binary segmentation mask for each RoI. The following diagram shows the structure of Mask R-CNN:

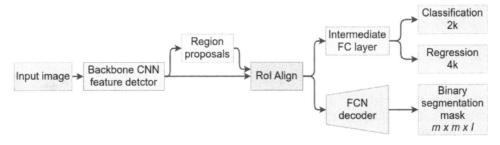

Figure 5.15 – Mask R-CNN structure

The segmentation and detection paths both use the RoI predictions of the RPN but are otherwise independent and *parallel* to each other. The segmentation path produces *I m×m* segmentation masks, one for each of the *I* RoIs. Since the detection path handles the classification of the object, the segmentation mask is *binary* and independent of the object class. The segmented pixels are automatically assigned to the class produced by the detection path. This is opposed to other algorithms, such as U-Net, where the segmentation is combined with classification and an individual softmax is applied at each pixel. At training or inference, only the mask related to the predicted object of the classification path is considered; the rest are discarded.

Mask R-CNN replaces the RoI max pooling operation with a more accurate RoI align layer. The RPN outputs the anchor box center and its height and width as four floating-point numbers. Then, the RoI pooling layer translates them into integer feature map cell coordinates (quantization). Additionally, the division of the RoI to $H×W$ bins (the same size as the RoI pooling regions) also involves quantization. The RoI example from the *Object detection with Faster R-CNN* section shows that the bins have different sizes (3×3, 3×2, 2×3, 2×2). These two quantization levels can introduce misalignment between the RoI and the extracted features. The following diagram shows how RoI alignment solves this problem:

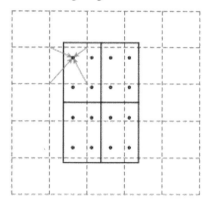

Figure 5.16 – RoI align example. Source: https://arxiv.org/abs/1703.06870

The dashed lines represent the feature map cells. The region with solid lines in the middle is a 2×2 RoI overlaid on the feature map. Note that it doesn't match the cells exactly. Instead, it is located according to the RPN prediction without quantization. In the same way, a cell of the RoI (the black dots) doesn't match one particular cell of the feature map. The **RoI align** operation computes the value of a RoI cell with a bilinear interpolation of its adjacent cells. In this way, RoI align is more accurate than RoI pooling.

At training, a RoI is assigned a positive label if it has IoU with a ground truth box of at least 0.5, and negative otherwise. The mask target is the intersection between a RoI and its associated ground truth mask. Only the positive RoIs participate in the segmentation path training.

Using Mask R-CNN with PyTorch

In this section, we'll use a pre-trained PyTorch Mask R-CNN model with a ResNet50 backbone for instance segmentation. Like Faster R-CNN, PyTorch has out-of-the-box support for Mask R-CNN. The program structure and the requirements are the same as the ones in the *Using Faster R-CNN with PyTorch* section. We will only include the relevant parts of the code, but you can find the full version in this book's GitHub repository. Let's start:

1. Load the pre-trained model with the latest available weights, which you can ensure by using the DEFAULT option:

   ```
   from torchvision.models.detection import \
   maskrcnn_resnet50_fpn_v2, \
   MaskRCNN_ResNet50_FPN_V2_Weights
   model = maskrcnn_resnet50_fpn_v2(
       weights=MaskRCNN_ResNet50_FPN_V2_Weights.DEFAULT)
   ```

2. We are going to use the model for inference and not for training, so we'll enable the eval() mode:

   ```
   model.eval()
   ```

3. Use opencv-python to read the RGB image located at image_file_path. We'll omit the code, which downloads the image from this book's repository if it doesn't already exist locally:

   ```
   import cv2
   img = cv2.imread(image_file_path)
   ```

   Here, img is a three-dimensional numpy array of integers.

4. Implement the single-step image pre-processing pipeline. It transforms the img numpy array into torch.Tensor, which will serve as input to the model:

   ```
   import torchvision.transforms as transforms
   transform = transforms.ToTensor()
   ```

5. Run the detection model:

   ```
   nn_input = transform(image)
   segmented_objects = model([nn_input])
   ```

   Here, segmented_objects is a dictionary with four items: boxes, labels, scores, and masks. The first three are the same as in Faster R-CNN. masks is a tensor with a shape of [number_of_detected_objects, 1, image_height, image_width]. We have one binary segmentation mask that covers the entire image for each detected object. Each such mask has zeroes at all pixels, except the pixels where the object is detected with a value of 1.

6. Use the initial `img` array and `segmented_objects` as parameters for the `draw_segmentation_masks` function. It overlays the bounding boxes, the segmentation masks, and the labels of the detected objects on the original input image (the implementation of `draw_segmentation_masks` is available in the full example):

```
draw_segmentation_masks(image, segmented_objects)
```

7. Display the result with `opencv`:

```
cv2.imshow("Object detection", img)
cv2.waitKey()
```

The output image looks like this:

Figure 5.17 – Instance segmentation with Mask R-CNN

We've now discussed object detection and semantic segmentation. In the next section, we'll discuss how to use CNNs to generate new images, instead of simply processing existing ones.

Image generation with diffusion models

So far, we've used NNs as **discriminative models**. This simply means that, given input data, a discriminative model will **map** it to a certain label (in other words, a classification). A typical example is the classification of MNIST images in one of ten digit classes, where the NN maps input data features (pixel intensities) to the digit label. We can also say this in another way: a discriminative model gives us the probability of y (class), given x (input). In the case of MNIST, this is the probability of the digit when given the pixel intensities of the image. In the next section, we'll introduce NNs as generative models.

Introducing generative models

A **generative model** learns the distribution of data. In a way, it is the opposite of the discriminative model we just described. It predicts the probability of the input sample, given its class, $y - P(X|Y = y)$. For example, a generative model will be able to create an image based on textual description. Most often, y is tensor, rather than scalar. This tensor exists in the so-called **latent space** (or **latent feature space**), and we'll refer to it as the **latent representation** (or **latent space representation**) of the original data, which itself exists in its own **feature space**. We can think of the latent representation as a **compressed** (or simplified) version of the original feature space. The digit-to-class case serves as an extreme example of this paradigm – after all, we're compressing an entire image into a single digit. For the latent representation to work, it will have to capture the most important hidden properties of the original data and discard the noise.

Because of its relative simplicity, we can reasonably expect that we have some knowledge of the structure and properties of the latent space. This is opposed to the feature space, which is complex beyond our comprehension. Therefore, if we know the reverse mapping from the latent space to the feature space, we could generate different feature space representations (that is, images) based on different latent representations. More importantly, we can influence the output image properties by modifying (in a conscious way) the initial latent representation.

To illustrate this, let's imagine that we've managed to create a reverse mapping between latent vectors with $n=3$ elements and full-fledged images of vehicles. Each vector element represents one vehicle property, such as length, height, and width (as shown in the following diagram):

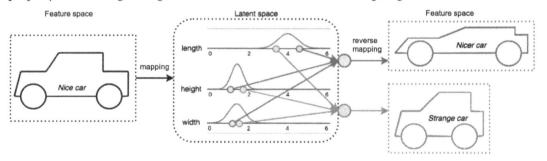

Figure 5.18 – An example of feature space-latent space and latent space-feature space mapping

Say that the average vehicle length is four meters. Instead of a discrete value, we can represent this property as a **normal (Gaussian) distribution** (https://en.wikipedia.org/wiki/Normal_distribution) with a mean of 4, making the latent space continuous (the same applies to the other properties). Then, we can choose to sample new values for each element from the ranges of their distributions. They will form a new latent vector (in this case, a **latent variable**), which we can use as a seed to generate new images. For example, we can create longer and lower vehicles (as illustrated previously).

> **Note**
>
> The second edition of this book included a whole chapter on NN-based generative models, where we discussed two particular architectures: **variational autoencoders** (**VAE**, *Auto-Encoding Variational Bayes*, `https://arxiv.org/abs/1312.6114`) and **generative adversarial networks** (**GAN**, `https://arxiv.org/abs/1406.2661`). At the time, these were the state-of-the-art generative models for images. Since then, they've been surpassed by a new class of algorithms called **diffusion models**. As we have to move with the times, in this edition, we'll omit VAEs and GANs, and we'll focus on diffusion models instead.

Denoising Diffusion Probabilistic Models

Diffusion models are a particular class of generative models, first introduced in 2015 (*Deep Unsupervised Learning using Nonequilibrium Thermodynamics*, `https://arxiv.org/abs/1503.03585`). In this section, we'll focus on **Denoising Diffusion Probabilistic Models** (**DDPM**, `https://arxiv.org/abs/2006.11239`), which form the foundation of some of the most impressive generative tools such as **Stable Diffusion** (`https://github.com/CompVis/stable-diffusion`).

DDPM follows a similar pattern to the generative models we've already discussed: it starts with a latent variable and uses it to generate a full-fledged image. The DDPM training algorithm is split into two parts:

- **Forward diffusion**: This starts with an initial image and then gradually adds random **Gaussian noise** (`https://en.wikipedia.org/wiki/Gaussian_noise`) to it through a series of small steps until the final (latent) representation is pure noise.

- **Reverse diffusion**: This is the opposite of the forward process. It starts with pure noise and gradually tries to restore the original image.

The following diagram illustrates the forward (top) and reverse (bottom) diffusion processes:

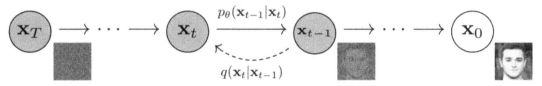

Figure 5.19 – The forward (bottom) and reverse (top) diffusion processes. Source: https://arxiv.org/abs/2006.11239

Let's discuss it in detail:

- x_0: The initial image from the original feature space, represented as a tensor.

- T: The number of steps in the forward and reverse processes. Originally, the authors used $T=1000$. More recently, $T=4000$ has been proposed (*Improved Denoising Diffusion Probabilistic Models*). Each forward or reverse step adds or removes small amounts of noise.

- $\mathbf{x}_T$: The final result of the forward diffusion, which represents pure noise. We can think of $\mathbf{x}_T$ as a peculiar latent representation of $\mathbf{x}_0$. The two tensors have the same dimensions, unlike the example we discussed in the *Introducing generative models* section.

- $\mathbf{x}_t$ (note the lowercase t): The noise-augmented tensor at an intermediate step, t. Again, it has the same dimensions as $\mathbf{x}_0$ and $\mathbf{x}_T$.

- $q(\mathbf{x}_t|\mathbf{x}_{t-1})$: This is the **probability density function** (**PDF**) of the forward diffusion process at an intermediate step, t. PDF sounds scary, but it isn't. It simply means that we add small amounts of Gaussian noise to the already noisy tensor, $\mathbf{x}_{t-1}$, to produce a new, noisier, tensor, $\mathbf{x}_t$ ($\mathbf{x}_t$ is conditioned on $\mathbf{x}_{t-1}$). The forward diffusion doesn't involve ML or NNs and has no learnable parameters. We just add noise and that's it. Still, it represents a mapping from the original feature space to the latent representation space.

  Note that we need to know $\mathbf{x}_{t-1}$ to produce $\mathbf{x}_t$, $\mathbf{x}_{t-2}$ for $\mathbf{x}_{t-1}$ and so on – that is, we need all tensors $\mathbf{x}_0 \ldots \mathbf{x}_{t-1}$ to produce $\mathbf{x}_t$. Thankfully, the authors have proposed an optimization that allows us to derive the value of any $\mathbf{x}_t$ using only the initial tensor, $\mathbf{x}_0$:

  $$\mathbf{x}_t = \sqrt{\overline{a}_t}\,\mathbf{x}_0 + \sqrt{1 - \overline{a}_t}\,\boldsymbol{\epsilon}, \boldsymbol{\epsilon} \sim N(0, I) (1)$$

  Here, $a_t \in [0{:}1]$ is a coefficient, which changes on a pre-defined schedule, but generally increases with t. $\boldsymbol{\epsilon}$ is the Gaussian random noise tensor with the same size as $\mathbf{x}_t$. The square root ensures that the new $\mathbf{x}_t$ will still follow a Gaussian distribution. We can see that $\mathbf{x}_t$ is a mixture of $\mathbf{x}_0$ and $\boldsymbol{\epsilon}$ and a_t determines the balance between the two. If $t \rightarrow 0$, then $\mathbf{x}_0$ will have more weight. The more $t \rightarrow T$, the more the noise, $\boldsymbol{\epsilon}$, will prevail. Because of this optimization, we don't have a real multi-step forward diffusion process. Instead, we generate the desired noisy representation at step t, $\mathbf{x}_t$, in a single operation.

- $p_\theta(\mathbf{x}_{t-1}|\mathbf{x}_t)$: This is the PDF of the reverse diffusion process at an intermediate step, t-1. This is the opposite function of $q(\mathbf{x}_t|\mathbf{x}_{t-1})$. It is a mapping from the latent space to the original feature space – that is, we start from the pure noise tensor, $\mathbf{x}_T$, and we gradually try to remove the noise until we reach the original image in T steps. The reverse diffusion is a lot more challenging compared to simply adding noise to an image, as in the forward phase. This is the primary reason to split the denoising process into multiple steps with small amounts of noise in the first place. Our best chance is to train an NN with the hope that it will learn a reasonable approximation of the actual mapping between the latent and the original feature spaces. Therefore, p_θ is an NN, where the θ index indicates its weights. The authors have proposed a **U-Net** type of network. It takes the noisy tensor, $\mathbf{x}_t$, as input and outputs its approximation of the noise (that is, only the noise and not the image itself) that was added to the original image, $\boldsymbol{\epsilon}_\theta$. The input and output tensors have the same dimensions. DDPM was released later than the original U-Net, so their NN architecture uses some improvements that were introduced in the meantime. These include residual blocks, **group normalization** (an alternative to batch normalization), and **attention** (*Attention Is All You Need*, `https://arxiv.org/abs/1706.03762`), `https://arxiv.org/abs/1803.08494`), and **attention** (*Attention Is All You Need*, `https://arxiv.org/abs/1706.03762`).

Next, let's focus on the DDPM training, which is displayed on the left in the following figure:

Algorithm 1 Training	**Algorithm 2** Sampling
1: **repeat**	1: $\mathbf{x}_T \sim \mathcal{N}(\mathbf{0}, \mathbf{I})$
2: $\mathbf{x}_0 \sim q(\mathbf{x}_0)$	2: **for** $t = T, \ldots, 1$ **do**
3: $t \sim \text{Uniform}(\{1, \ldots, T\})$	3: $\mathbf{z} \sim \mathcal{N}(\mathbf{0}, \mathbf{I})$ if $t > 1$, else $\mathbf{z} = \mathbf{0}$
4: $\epsilon \sim \mathcal{N}(\mathbf{0}, \mathbf{I})$	4: $\mathbf{x}_{t-1} = \frac{1}{\sqrt{\alpha_t}}\left(\mathbf{x}_t - \frac{1-\alpha_t}{\sqrt{1-\bar{\alpha}_t}}\epsilon_\theta(\mathbf{x}_t, t)\right) + \sigma_t \mathbf{z}$
5: Take gradient descent step on $\nabla_\theta \left\| \epsilon - \epsilon_\theta(\sqrt{\bar{\alpha}_t}\mathbf{x}_0 + \sqrt{1 - \bar{\alpha}_t}\epsilon, t) \right\|^2$	5: **end for**
6: **until** converged	6: **return** $\mathbf{x}_0$

Figure 5.20 – DDPM training (left); DDPM sampling (right). Source: https://arxiv.org/abs/2006.11239

A single training episode involves the following steps (line 1):

- Start with a random sample (image), $\mathbf{x}_0$, from the training set (line 2).

- Sample the random noise step, t, in the range $[1{:}T]$ (line 3).

- Sample the random noise tensor, ϵ, from a Gaussian distribution (line 4). Within the NN itself, the step, t, is embedded in the values of ϵ using **sinusoidal position embeddings**. Don't worry if you don't understand the concept of positional embeddings. We'll discuss it in detail in *Chapter 7*, as it was first introduced in that context. All we need to know now is that the step number, t, is implicitly encoded in the elements of ϵ, in a way that allows the model to use this information. The step-adjusted noise is denoted with ϵ, t in the preceding diagram.

- Produce a corrupt image tensor, $\mathbf{x}_t$, conditioned on the initial image, $\mathbf{x}_0$, and based on the sampled noise step, t, and the random noise, ϵ, t. To do this, we'll use formula (1), which we introduced earlier in this section. Thanks to it, this single step constitutes the entire forward diffusion phase (line 5).

- Perform a single gradient descent step and weight update. The training uses the **mean squared error** (MSE). It measures the difference between the sampled noise, ϵ (line 4), and the noise predicted by the model, ϵ_θ (line 5). The loss equation seems deceptively simple. The paper's authors made a long chain of transformations and assumptions to reach this simple result. This is one of the main contributions of the paper.

Once the model has been trained, we can use it to sample new images based on random initial tensors, $\mathbf{x}_T$. We can do this with the following procedure (preceding diagram, right):

1. Sample the initial random latent tensor, $\mathbf{x}_T$, from a Gaussian distribution (line 1).

2. Repeat the next steps T times (line 2):

    *   Sample random noise tensor, $\mathbf{z}$, from a Gaussian distribution (line 3). We do this for all reverse steps, except for the final one.

    *   Use the trained U-Net model to predict the noise, $\boldsymbol{\epsilon}_\theta$, at step t. Subtract this noise from the current sample, $\mathbf{x}_t$, to produce the new, less noisy, $\mathbf{x}_{t-1}$ (line 4). The scheduling coefficient, a_t, also takes part in this formula, as it did in the forward phase. The formula also preserves the mean and the variance of the original distribution.

3. The final denoising step produces the generated image.

This concludes our introduction to DDPMs for now. However, we'll revisit them in *Chapter 9*, but in the context of Stable Diffusion.

Summary

In this chapter, we discussed some advanced computer vision tasks. We started with TL, a technique that makes it possible to bootstrap our experiments with the help of pre-trained models. We also introduced object detection and semantic segmentation models, which benefit from TL. Finally, we focused on generative models and DDPM in particular.

In the next chapter, we'll introduce language modeling and recurrent networks.

Part 3: Natural Language Processing and Transformers

We'll start this part with an introduction to natural language processing, which will serve as a backdrop for our discussion on recurrent networks and transformers. Transformers will be the main focus of this section because they represent one of the most significant deep learning advances in recent years. They are the foundation of **large language models** (**LLM**), such as ChatGPT. We'll discuss their architecture and their core element – the attention mechanism. Then, we'll discuss the properties of LLMs. Finally, we'll focus on some advanced LLM applications, such as text and image generation, and learn how to build LLM-centered applications.

This part has the following chapters:

- *Chapter 6, Natural Language Processing and Recurrent Neural Networks*
- *Chapter 7, The Attention Mechanism and Transformers*
- *Chapter 8, Exploring Large Language Models in Depth*
- *Chapter 9, Advanced Applications of Large Language Models*

6

Natural Language Processing and Recurrent Neural Networks

This chapter will introduce two different topics that nevertheless complement each other – **natural language processing (NLP)** and **recurrent neural networks (RNNs)**. NLP teaches computers to process and analyze natural language text to perform tasks such as machine translation, sentiment analysis, and text generation. Unlike images in computer vision, natural text represents a different type of data, where the order (or sequence) of the elements matters. Thankfully, RNNs are suitable for processing sequential data, such as text or time series. They help us deal with sequences of variable length by defining a recurrence relation over these sequences (hence the name). This makes NLP and RNNs natural allies. In fact, RNNs can be applied to any problem since it has been proven that they are Turing-complete – theoretically, they can simulate any program that a regular computer would not be able to compute.

However, it is not only good news, and we'll have to start with a disclaimer. Although RNNs have great theoretical properties, we now know that there are practical limitations to their use. These limitations have been mostly surpassed by a more recent **neural network (NN)** architecture called **transformer**, which we'll discuss in *Chapter 7*. In theory, the transformer has more limitations compared to RNNs. But as sometimes happens, it works better in practice. Nevertheless, I believe that this chapter will be beneficial to you. On one hand, RNNs have elegant architecture and still represent one of the major NN classes; on the other hand, the progression of knowledge presented in this and the next three chapters will closely match the real-world progression of research on these topics. So, you'll be able to apply the concepts you'll learn here in the next few chapters as well. This chapter will also allow you to fully appreciate the advantages of the newer models.

This chapter will cover the following topics:

- Natural language processing
- Introducing RNNs

Technical requirements

We'll implement the example in this chapter using Python, PyTorch, and the TorchText package (`https://github.com/pytorch/text`). If you don't have an environment set up with these tools, fret not – the example is available as a Jupyter Notebook on Google Colab. You can find the code examples in this book's GitHub repository: `https://github.com/PacktPublishing/Python-Deep-Learning-Third-Edition/tree/main/Chapter06`.

Natural language processing

NLP is a subfield of machine learning that allows computers to interpret, manipulate, and comprehend human language. This definition sounds a little dry, so, to provide a little clarity, let's start with a non-exhaustive list of the types of tasks that fall under the NLP umbrella:

- **Text classification**: This assigns a single label to the entire input text. For example, **sentiment analysis** can determine whether a product review is positive or negative.

- **Token classification**: This assigns a label for each token of the input text. A token is a building block (or a unit) of text. Words can be tokens. A popular token classification task is **named entity recognition**, which assigns each token to a list of predefined classes such as place, company, or person. **Part-of-speech** (**POS**) tagging assigns each word to a particular part of speech, such as a noun, verb, or adjective.

- **Text generation**: This uses the input text to generate new text with arbitrary length. Text generation tasks include machine translation, question answering, and text summarization (creating a shorter version of the original text while preserving its essence).

Solving NLP problems is not trivial. To understand why, let's go back to computer vision (*Chapter 4*), where the input images are represented as 2D tensors of pixel intensities with the following properties:

- The image is composed of pixels and doesn't have any other explicitly defined structure

- The pixels form implicit hierarchical structures of larger objects, based on their proximity to each other

- There is only one type of pixel, which is defined only by its scalar intensity

Thanks to its homogenous structure, we can feed the (almost) raw image to a **convolutional neural network** (**CNN**) and let it do its magic with relatively little data pre-processing.

Now, let's return to text data, which has the following properties:

- There are different types of characters with different semantic meanings, such as letters, digits, and punctuation marks. In addition, we might encounter previously unknown symbols.

- The natural text has an explicit hierarchy in the form of characters, words, sentences, and paragraphs. We also have quotes, titles, and a hierarchy of headings.

- Some parts of the text may be related to distant parts of the sequence, rather than their immediate context. For example, a fictional story can introduce a person by their name but later refer to them only as *he* or *she*. These references can be separated by long text sequences, yet, we still have to be able to find this relation.

The complexity of natural text requires several pre-processing steps before the actual NN model comes into play. The first step is **normalization**, which involves operations such as removing extra whitespace and converting all letters into lowercase. The next steps are not as straightforward, so we'll dedicate the next two sections to them.

Tokenization

One intuitive way to approach an NLP task is to split the corpus into words, which will represent the basic input units of our model. However, using words as input is not set in stone and we can use other elements, such as individual characters, phrases, or even whole sentences. The generic term for these units is **tokens**. A token refers to a text corpus in the same way as a pixel refers to an image. The process of splitting the corpus into tokens is called **tokenization** (what a surprise!). The entity (for example, an algorithm) that performs this tokenization is called a **tokenizer**.

> **Note**
> The tokenizers we'll discuss in this section are universal in the sense that they can work with different NLP ML algorithms. Therefore, the pre-processing algorithms in this section are commonly used with transformer models, which we'll introduce in *Chapter 7*.

With that, let's discuss the types of tokenizers:

- **Word-based**: Each word represents a unique token. This is the most intuitive type of tokenization, but it has serious drawbacks. For example, the words *don't* and *do not* will be represented by different tokens, but they mean the same thing. Another example is the words *car* and *cars* or *ready* and *readily*, which will be represented by different tokens, whereas a single token would be more appropriate. Because natural language is so diverse, there are many corner cases like these. The issue isn't just that semantically similar words will have unrelated tokens, but also the large number of unique tokens that come out of this. This will make the model computationally inefficient. It will also produce many tokens with a small number of occurrences, which will prove challenging for the model to learn. Finally, we might encounter unknown words in a new text corpus.

- **Character-based**: Each character (letter, digit, punctuation, and so on) in the text is a unique token. In this way, we have fewer tokens, as the total number of characters is limited and finite. Since we know all the characters in advance, we cannot encounter unknown symbols.

However, this tokenization is less intuitive than the word-based model because a context composed of characters is less meaningful than a context based on words. While the number of unique tokens is relatively small, the total number of tokens in the corpus will be very large (equal to the total number of characters).

- **Subword tokenization**: This is a two-step process that starts by splitting the corpus into words. The most obvious way to split the text is on whitespace. In addition, we can also split it on whitespace *and punctuation marks*. In NLP parlance, this step is known as **pre-tokenization** (the prefix implies that tokenization will follow). Then, it preserves the frequently used words and decomposes the rare words into meaningful subwords, which are more frequent. For example, we can decompose the word *tokenization* into the core word *token* and the suffix *ization*, each with its own token. Then, when we encounter the word *carbonization*, we can decompose it into *carbon* and *ization*. In this way, we'll have two occurrences of *ization* instead of a single occurrence of *tokenization* and *carbonization*. Subword tokenization also makes it possible to decompose unknown words into known tokens.

Special service tokens

For the concept of tokenization to work, it introduces some service tokens. These include the following:

- **UNK**: Replaces unknown tokens in the corpus (think of rare words such as alphanumeric designations)
- **EOS**: An end-of-sentence (or sequence) token
- **BOS**: A beginning-of-sentence (or sequence) token
- **SEP**: This separates two semantically different text sequences, such as question and answer
- **PAD**: This is a padding token that is appended to an existing sequence so that it can reach some predefined length and fit in a fixed-length mini-batch.

For example, we can tokenize the sentence *I bought a product called FD543C* into *BOS I bought a product called UNK EOS PAD PAD* to fit a fixed input with a length of 10.

Subword tokenization is the most popular type of tokenization because it combines the best features of character-based (smaller vocabulary size) and word-based (meaningful context) tokenization. In the next few sections, we'll discuss some of the most popular subword tokenizers.

Byte-Pair Encoding and WordPiece

Byte-Pair Encoding (**BPE**, Neural Machine Translation of Rare Words with Subword Units, `https://arxiv.org/abs/1508.07909`) is a popular subword tokenization algorithm. As with other such tokenizers, it begins with pre-tokenization, which splits the corpus into words.

Using this dataset as a starting point, BPE works in the following way:

1. Start with an initial **base** (or **seed**) **vocabulary**, which consists of the individual characters of all words in the text corpus. Therefore, each word is a sequence of single-character tokens.

2. Repeat the following until the size of the token vocabulary reaches a certain maximum threshold:

   I. Find the pair of tokens (initially, these are single characters) that occur together most frequently and merge them into a new composite token.

   II. Extend the existing token vocabulary with the new composite token.

   III. Update the tokenized text corpus with the new token structure.

To understand BPE, let's assume that our corpus consists of the following (imaginary) words: {dab: 5, adab: 4, aab: 7, bub: 9, bun: 2}. The digit following each word indicates the number of occurrences of that word in the text. And here is the same corpus, but split into tokens (that is, characters): {(d, a, b): 5, (a, d, a, b): 4, (a, a, b): 7, (b, u, b): 9, (b, u, c): 2}. Based on this, we can build our initial token vocabulary with occurrence counts for each token: {b: 36, a: 27, u: 11, d: 9, c: 2}. The following list illustrates the first four merge operations:

1. The most common pair of tokens is (a, b), which occurs freq((a, b)) = 5 + 4 + 7 = 16 times. Therefore, we merge them, and the corpus becomes {(d, **ab**): 5, (a, d, **ab**): 4, (a, **ab**): 7, (b, u, b): 9, (b, u, c): 2}. The new token vocabulary is {b: 20, **ab**: 16, a: 11, u: 11, d: 9, c: 2}.

2. The new most common token pair is (b, u) with freq((b, u)) = 9 + 2 = 11 occurrences. Again, we proceed to combine them in a new token: {(d, ab): 5, (a, d, ab): 4, (a, ab): 7, (**bu**, b): 9, (**bu**, c): 2}. The updated token vocabulary is {ab: 16, a: 11, **bu**: 11, b: 9, d: 9, c: 2}.

3. The next token pair is (d, ab) and it occurs freq((d, ab)) = 5 + 4 = 9 times. After combining them, the tokenized corpus becomes {(**dab**): 5, (a, **dab**): 4, (a, ab): 7, (bu, b): 9, (bu, c): 2}. The new token vocabulary is {a: 11, bu: 11, b: 9, **dab**: 9, ab: 7, c: 2}.

4. The new pair of tokens is (bu, b) with nine occurrences. After merging them, the corpus becomes {(dab): 5, (a, dab): 4, (a, ab): 7, (**bub**): 9, (bu, c): 2}, and the token vocabulary becomes {a: 11, **bub**: 9, **dab**: 9, ab: 7, bu: 2, c: 2}.

BPE stores all token-merge rules and their order and not just the final token vocabulary. During model inference, it applies the rules in the same order on the new unknown text to tokenize it.

End-of-word tokens

The original BPE implementation appends a special end-of-word token, <w/>, at the end of each word – for example, the word aab becomes aab<w/>. Other implementations can place the special token at the beginning of the word, instead of the end. This makes it possible for the algorithm to distinguish between, say, the token ab, as presented in the word **ab**ca<w/>, and the same token in a**ab**<w/>. In this way, the algorithm can restore the original corpus from the tokenized one (**de-tokenization**), which wouldn't be possible otherwise. In this section, we have omitted the end-of-word token for clarity.

Let's recall that our base vocabulary includes all characters of the text corpus. If these are Unicode characters (which is the usual case), we could end up with a vocabulary of up to 150,000 tokens. And this is before we even start the token-merge process. One trick to solve this issue is with the help of **byte-level BPE**. Each Unicode character can be encoded with multiple (up to 4) bytes. Byte-level BPE initially splits the corpus into a sequence of bytes, instead of full-fledged Unicode characters. If a character is encoded with n bytes, the tokenizer will treat it as a sequence of n one-byte tokens. In this way, the size of the base vocabulary will always be 256 (the maximum unique values that we can store in a byte). In addition, byte-level BPE guarantees that we won't encounter unknown tokens.

WordPiece (https://arxiv.org/abs/1609.08144) is another subword tokenization algorithm. It is similar to BPE but with one main difference. Like BPE, it starts with a base vocabulary of individual characters and then proceeds to merge them into new composite tokens. However, it defines the merge order based on a score, computed with the following formula (unlike BPE, which uses frequent co-occurrence):

$$score\big((token_1, token_2)\big) = \frac{freq\big((token_1, token_2)\big)}{freq(token_1) \times freq(token_2)}$$

In this way, the algorithm prioritizes the merging of pairs where the individual tokens are less frequent in the corpus. Let's compare this approach with BPE, which merges tokens based only on the potential gains of the new token. In contrast, WordPiece balances the gain (the nominator in the formula) with the potential loss of the existing tokens (the denominator). This makes sense because the new token will exist instead of the old pair of tokens, rather than alongside them.

In-word tokens

WordPiece adds a special ## prefix to all tokens inside a word, except for the first. For example, it will tokenize the word *aab* as [a, ##a, ##b]. The token merge removes the ## between the tokens. So, when we merge ##a and ##b, *aab* becomes [a, ##ab].

Unlike BPE, WordPiece only stores the final token vocabulary. When it tokenizes a new word, it finds the longest matching subword in the vocabulary and splits the word on it. For example, let's assume that we want to split the word *abcd* with a token vocabulary of [a, ##b, ##c, ##d, ab,

`##cd, ##bcd]`. Following the new rule, WordPiece will first select the longest subword, *bcd*, and it will tokenize *abcd* as `[a, ##bcd]`.

BPE and WordPiece are greedy algorithms – they will always merge tokens deterministically, based on frequency criteria. However, encoding the same text sequence with different tokens might be possible. This could act as regularization for a potential NLP algorithm. Next, we'll introduce a tokenization technique that takes advantage of this.

Unigram

Unlike BPE and WordPiece, the **Unigram** (*Subword Regularization: Improving Neural Network Translation Models with Multiple Subword Candidates*, https://arxiv.org/abs/1804.10959) algorithm starts with a large base vocabulary and progressively tries to reduce it. The initial base vocabulary is a union of all unique characters and the most common substrings of the corpus. One way to find the most common substrings is with BPE. The algorithm assumes that each token, x_i, occurs independently (hence the Unigram name). Because of this assumption, the probability of a token, x_i, $P(x_i)$, is just the number of its occurrences divided by the total size of the rest of the corpus. Then, the probability of a sequence of tokens with length M, $X = (x_1...x_M)$, is as follows:

$$P(x) = \prod_{i=1}^{M}P(x_i), \forall x_i \in V, \sum_{x_i \in V}P(x_i) = 1$$

Here, V is the full token vocabulary.

Say that we have the same token sequence, X, and multiple token segmentation candidates, $x \in S(X)$, for that sequence. The most probable segmentation candidate, x^*, for X is as follows:

$$x^* = argmaxP(x), \ x \in S(X)$$

Let's clarify this with an example. We'll assume that our corpus consists of some (imaginary) words, `{dab: 5, aab: 7, bun: 4}`, where the digits indicate the number of occurrences of that word in the text. Our initial token vocabulary is a union of all unique characters and all possible substrings (the numbers indicate frequency): `{a: 19, b: 16, ab: 12, aa: 7, da: 5, d: 5, bu: 4, un: 4}`. The sum of all token frequencies is $19 + 16 + 12 + 7 + 5 + 5 + 4 + 4 = 72$. Then, the independent probability for each token is $P(x_i) = count(x_i)/72$ – for example, $P(a) = 19/72 = 0.264$, $P(ab) = 12/72 = 0.167$, and so on.

Our extended vocabulary presents us with the possibility to tokenize each sequence (we'll focus on words for simplicity) in multiple ways. For example, we can represent *dab* as either `{d, a, b}`, `{da, b}`, or `{d, ab}`. Here, the probabilities for each candidate are $P(\{d, a, b\}) = P(d) * P(a) * P(b) = 0.07 * 0.264 * 0.222 = 0.0041$; $P(\{da, b\}) = 0.07 * 0.222 = 0.015$; $P(\{d, ab\}) = 0.07 * 0.167 = 0.012$. The candidate with the highest probability is $x^* = \{da, b\}$.

With that, here's how Unigram tokenization works step by step:

1. Start with the initial large base vocabulary, V.

2. Repeat the following steps until the size of $|V|$ reaches some minimum threshold value:

    I.   Find the l-best tokenization candidates, x^*, for all words in the corpus with the help of the **Viterbi** algorithm (https://en.wikipedia.org/wiki/Viterbi_algorithm – using this algorithm is necessary because this is a computationally intensive task). Taking l candidates, instead of one, makes it possible to sample different token sequences over the same text. You can think of this as a data augmentation technique over the input data, which provides additional regularization to the NLP algorithm. Once we have a tokenized corpus in this way, we can estimate the probabilities, $P(x_i), x_i \in V$, for all tokens of the current token vocabulary, V, with the help of an **expectation-minimization** algorithm (https://en.wikipedia.org/wiki/Expectation%E2%80%93maximization_algorithm).

    II.   For each token, x_i, compute a special loss function, $loss_i$, which determines how the likelihood of the corpus is reduced if we remove x_i from the token vocabulary.

    III.   Sort the tokens by their $loss_i$ and preserve only the top n % of the tokens (for example, $n = 80$). Always preserve the individual characters to avoid unknown tokens.

This concludes our introduction to tokenization. Some of these techniques were developed alongside the transformer architecture and we'll make the most use of them in the following chapters. But for now, let's focus on another fundamental technique in the NLP pipeline.

Introducing word embeddings

Now that we've learned how to tokenize the text corpus, we can proceed to the next step in the NLP data processing pipeline. For the sake of clarity, we'll assume that we've tokenized the corpus into words, rather than subwords or characters (in this section, *word* and *token* are interchangeable).

One way to feed the words of the sequence as input to the NLP algorithm is with one-hot encoding. Our input vector will have the same size as the number of tokens in the vocabulary and each token will have a unique one-hot encoded representation. However, this approach has a few drawbacks, as follows:

- **Sparse inputs**: The one-hot representation consists of mostly zeros and a single value. If our NLP algorithm is an NN (and it is), this type of input will activate only a small portion of its weights per word. Because of this, we'll need a large training set to include a sufficient number of training samples of each word of the vocabulary.

- **Computational intensity**: The large size of the vocabulary will result in large input tensors, which require large NNs and more computational resources.

- **Impracticality**: Every time we add a new word to the vocabulary, we'll increase its size. However, the size of the one-hot encoded input will also increase. Therefore, we'll have to change the structure of our NN to accommodate the new size and we'll perform additional training.

- **Lack of context**: Words such as *dog* and *wolf* are semantically similar, but the one-hot representation lacks a way to convey this similarity.

In this section, we'll try to solve these issues with the help of a lower-dimensional distributed representation of the words, known as **word embeddings** (*A Neural Probabilistic Language Model*, `http://www.jmlr.org/papers/volume3/bengio03a/bengio03a.pdf`). The distributed representation is created by learning an embedding function that transforms the one-hot encoded words into a lower-dimensional space of word embeddings, as follows:

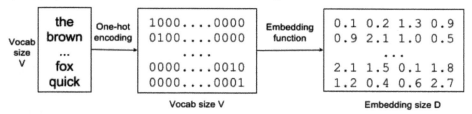

Figure 6.1 – Words -> one-hot encoding -> word embedding vectors

Words from the vocabulary with size V are transformed into one-hot encoding vectors of size V. Then, an **embedding function** transforms this V-dimensional space into a distributed representation (vector) of a **fixed** size, D (here, $D=4$). This vector serves as input to the NLP algorithm. We can see that the fixed and smaller vector size solves the issues of sparsity, computational intensity, and impracticality we just described. Next, we'll see how it solves the context issue.

The embedding function learns semantic information about the words. It maps each word in the vocabulary to a continuous-valued vector representation – that is, the word embedding. Each word corresponds to a point in this embedding space, and different dimensions correspond to the grammatical or semantic properties of these words. The concept of embedding space is similar to the latent space representation, which we first discussed in the context of diffusion models in *Chapter 5*.

The goal is to ensure that the words close to each other in the embedding space have similar meanings. By *close to each other*, we mean a high value of the dot product (similarity) of their embedding vectors. In this way, the information that some words are semantically similar can be exploited by the ML algorithm. For example, it might learn that *fox* and *cat* are semantically related and that both *the quick brown fox* and *the quick brown cat* are valid phrases. A sequence of words can then be replaced with a sequence of embedding vectors that capture the characteristics of these words. We can use this sequence as a base for various NLP tasks. For example, a classifier trying to classify the sentiment of an article might be trained on previously learned word embeddings, instead of one-hot encoding vectors. In this way, the semantic information of the words becomes readily available for the sentiment classifier.

The mapping between one-hot representation and embedding vectors

Let's assume that we have already computed the embedding vectors of each token. One way to implement the mapping between the one-hot representation and the actual embedding vector is with the help of a $V \times D$-shaped matrix, $\mathbf{W}_{V \times D}$. We can think of the matrix rows as a lookup table, where each row represents one word embedding vector. It works thanks to the one-hot encoded input word, which is a vector of all zeros, except for the index of the word itself. Because of this, the input word, w_i, will only activate its unique row (vector) of weights, $\mathbf{w}_i$, in $\mathbf{W}_{V \times D}$. So, for each input sample (word), only the word's embedding vector will participate. We can also think of $\mathbf{W}_{V \times D}$ as a weight matrix of a **fully connected** (**FC**) NN layer. In this way, we can embed the embeddings (get it?) as the first NN layer – that is, the NN takes the one-hot encoded token as input and the embedding layer transforms it into a vector. Then, the rest of the NN uses the embedding vector instead of the one-hot representation. This is a standard implementation across all deep learning libraries.

The concept of word embeddings was first introduced more than 20 years ago but remains one of the central paradigms in NLP today. **Large language models** (**LLMs**), such as ChatGPT, use improved versions of word embeddings, which we'll discuss in *Chapter 7*.

Now that we are familiar with embedding vectors, we'll continue with the algorithm to obtain and compute them.

Word2Vec

A lot of research has gone into creating better word embedding models, in particular by omitting to learn the probability function over sequences of words. One of the most popular ways to do this is with **Word2Vec** (`http://papers.nips.cc/paper/5021-distributed-representations-of-words-and-phrases-and-their-compositionality.pdf`, `https://arxiv.org/abs/1301.3781`, and `https://arxiv.org/abs/1310.4546`). It creates embedding vectors based on the context (surrounding words) of the word in focus. More specifically, the context is the n preceding and the n following words of the focus word. The following figure shows the context window as it slides across the text, surrounding different focus words:

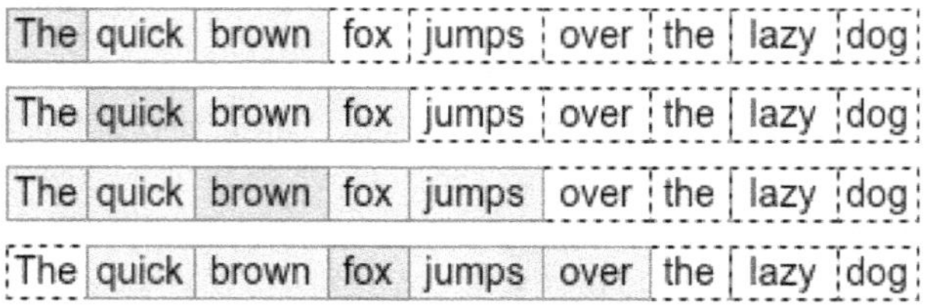

Figure 6.2 – A Word2Vec sliding context window with n=2. The same type
of context window applies to both CBOW and skip-gram

Word2vec comes in two flavors: **Continuous Bag of Words (CBOW)** and **skip-gram**. We'll start with CBOW and then we'll continue with skip-gram.

CBOW

CBOW predicts the most likely word given its context (surrounding words). For example, given the sequence *the quick _____ fox jumps*, the model will predict *brown*. It takes all words within the context window with equal weights and doesn't consider their order (hence the *bag* in the name). We can train the model with the help of the following simple NN with input, hidden, and output layers:

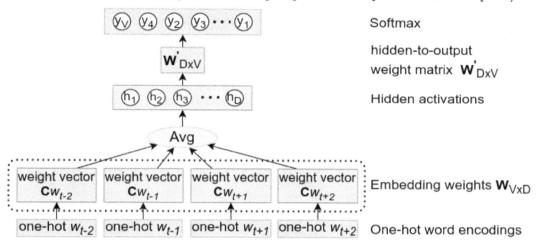

Figure 6.3 – A CBOW model NN

Here's how the model works:

- The input is the one-hot-encoded word representation (its length is equal to the vocabulary size, V).

- The embedding vectors are represented by the *input-to-hidden* matrix, $\mathbf{W}_{V \times D}$.

- The embedding vectors of all context words are averaged to produce the output of the hidden network layer (there is no activation function).

- The hidden activations serve as input to the output **Softmax** layer of size V (with the hidden-to-output weight matrix, $\mathbf{W}'_{D \times V}$), which predicts the most likely word to be found in the context (proximity) of the input words. The index with the highest activation represents the one-hot-encoded related word.

We'll train the NN with gradient descent and backpropagation. The training set consists of (context and label) one-hot encoded pairs of words that appear close to each other in the text. For example, if part of the text is [the, quick, brown, fox, jumps] and *n=2*, the training tuples will include ([quick, brown], the), ([the, brown, fox], quick), ([the, quick, `fox jumps]`, `brown)`, and so on. Since we are only interested in the embeddings, $\mathbf{W}_{V \times D}$, we'll discard the output NN weights, $\mathbf{W}'$, when the training is finished.

CBOW will tell us which word is most likely to appear in a given context. This could be a problem for rare words. For example, given the context *The weather today is really* _____, the model will predict the word *beautiful* rather than *fabulous* (hey, it's just an example). CBOW is several times faster to train than the skip-gram and achieves slightly better accuracy for frequent words.

Skip-gram

The skip-gram model can predict the context of a given input word (the opposite of CBOW). For example, the word *brown* will predict the words *The quick fox jumps*. Unlike CBOW, the input is a single one-hot encoded word vector. But how do we represent the context words in the output? Instead of trying to predict the whole context (all surrounding words) simultaneously, skip-gram transforms the context into multiple training pairs, such as `(fox, the)`, `(fox, quick)`, `(fox, brown)`, and `(fox, jumps)`. Once again, we can train the model with a simple single-layer NN:

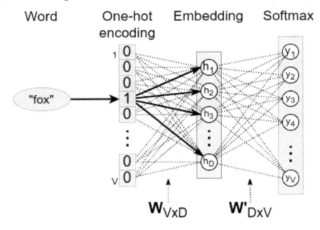

Figure 6.4 – A skip-gram model NN

As with CBOW, the output is a softmax, which represents the one-hot-encoded most probable context word. The input-to-hidden weights, $\mathbf{W}_{V \times D}$, represent the word embeddings lookup table, and the hidden-to-output weights, $\mathbf{W}'$, are only relevant during training. The hidden layer doesn't have an activation function (that is, it uses linear activation).

We'll train the model with backpropagation (no surprises here). Given a sequence of words, $w_1 \ldots w_M$, the objective of the skip-gram model is to maximize the average log probability, where n is the window size:

$$\frac{1}{M} \sum_{m=1}^{M} \sum_{i=-n, i \neq 0}^{i=n} log P\left(w_{m+i} | w_m\right)$$

The model defines the probability, $P(w_{m+i}|w_m)$, as the following softmax formula:

$$P(w_O|w_I) = \frac{exp\left(\mathbf{v}'^{\top}_{w_O}\mathbf{v}_{w_I}\right)}{\sum_{w=1}^{V}exp\left(\mathbf{v}'^{\top}_{w}\mathbf{v}_{w_I}\right)}$$

In this example, w_I and w_O are the input and output words, and $\mathbf{v}$ and $\mathbf{v}'$ are the corresponding word vectors in the input and output weight matrices, $\mathbf{W}_{V \times D}$ and $\mathbf{W}'_{D \times V}$, respectively (we keep the original notation of the paper). Since the NN doesn't have a hidden activation function, its output value for one input/output word pair is simply the multiplication of the input word vector, $\mathbf{v}_{w_I}$, and the output word vector, $\mathbf{v}'_{w_O}$ (hence the transpose operation).

The authors of the Word2Vec paper note that word representations cannot represent idiomatic phrases that are not compositions of individual words. For example, *New York Times* is a newspaper, and not just a natural combination of the meanings of *New*, *York*, and *Times*. To overcome this, the model can be extended to include whole phrases. However, this significantly increases the vocabulary size. And, as we can see from the preceding formula, the softmax denominator needs to compute the output vectors for all words of the vocabulary. Additionally, every weight of the $\mathbf{W}'_{D \times V}$ matrix is updated on every training step, which slows the training.

To solve this, we can replace the softmax operation with the so-called **negative sampling (NEG)**. For each training sample, we'll take the positive training pair (for example, (fox, brown)), as well as k additional negative pairs (for example, (fox, puzzle)), where k is usually in the range of [5,20]. Instead of predicting the word that best matches the input word (softmax), we'll simply predict whether the current pair of words is true or not. In effect, we convert the multinomial classification problem (classified as one of many classes) into a binary logistic regression (or binary classification) problem. By learning the distinction between positive and negative pairs, the classifier will eventually learn the word vectors in the same way, as with multinomial classification. In Word2Vec, the words for the negative pairs are drawn from a special distribution, which draws less frequent words more often, compared to more frequent ones.

Some of the most frequent words to occur carry less information value compared to the rare words. Examples of such words are the definite and indefinite articles *a*, *an*, and *the*. The model will benefit more from observing the pairs *London* and *city* compared to *the* and *city* because almost all words co-occur frequently with *the*. The opposite is also true – the vector representations of frequent words do not change significantly after training on many examples. To counter the imbalance between the rare and frequent words, the authors of the paper propose a subsampling approach, where each word, w_i, of the training set is discarded with some probability, computed by the heuristic formula where $f(w_i)$ is the frequency of word w_i and t is a threshold (usually around 10^{-5}):

$$P(w_i) = 1 - \sqrt{\frac{t}{f(w_i)}}$$

It aggressively subsamples words with a frequency greater than t but also preserves the ranking of the frequencies.

We can say that, in general, skip-gram performs better on rare words than CBOW, but it takes longer to train.

Now that we've learned about embedding vectors, let's learn how to visualize them.

Visualizing embedding vectors

A successful word embedding function will map semantically similar words to vectors with high dot product similarity in the embedding space. To illustrate this, we'll implement the following steps:

1. Train a Word2Vec skip-gram model on the `text8` dataset, which consists of the first 100,000,000 bytes of plain text from Wikipedia (`http://mattmahoney.net/dc/textdata.html`). Each embedding vector is 100-dimensional, which is the default value for this type of model.

2. Select a list of *seed* words. In this case, the words are *mother, car, tree, science, building, elephant,* and *green*.

3. Compute the dot-product similarity between the Word2Vec embedding vector of each seed word and the embedding vectors of all other words in the vocabulary. Then, select a cluster of the top-*k* (in our case, *k=5*) similar words (based on their dot-product similarity) for each seed word.

4. Visualize the similarity between the seed embeddings and the embeddings of their respective clusters of similar words in a 2D plot. Since the embeddings are 100-dimensional, we'll use the t-SNE (`https://en.wikipedia.org/wiki/T-distributed_stochastic_neighbor_embedding`) dimensionality-reduction algorithm. It maps each high-dimensional embedding vector on a two- or three-dimensional point in a way where similar objects are modeled on nearby points and dissimilar objects are modeled on distant points with a high probability. We can see the result in the following scatterplot:

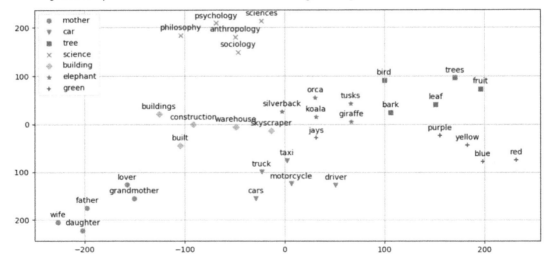

Figure 6.5 – t-SNE visualization of the seed words and their clusters of the most similar words

This graph proves that the obtained word vectors contain relevant information for the words.

Word2Vec (and similar models) create **static** (or **context independent**) **embeddings**. Each word has a single embedding vector, based on all occurrences (that is, all contexts) of that word in the text corpus. This imposes some limitations. For example, *bank* has a different meaning in different contexts, such as *river bank*, *savings bank*, and *bank holiday*. Despite this, it is represented with a single embedding. In addition, the static embedding doesn't take into account the word order in the context. For example, the expressions *I like apples, but I don't like oranges* and *I like oranges, but I don't like apples* have opposite meanings, but Word2Vec interprets them as the same. We can solve these problems with the so-called **dynamic** (**context dependent**) **embeddings**, which we'll discuss in *Chapter 7*.

So far, we've focused on single words (or tokens). Next, we'll expand our scope to text sequences.

Language modeling

A word-based **language model** (**LM**) defines a probability distribution over sequences of **tokens**. For this section, we'll assume that the tokens are words. Given a sequence of words of length m (for example, a sentence), an LM assigns a probability, $P(w_1...w_m)$, that the full sequence of words could exist. One application of these probabilities is a generative model to create new text – a word-based LM can compute the likelihood of the next word, given an existing sequence of words that precede it. Once we have this new word, we can append it to the existing sequence and predict yet another new word, and so on. In this way, we can generate new text sequences with arbitrary length. For example, given the sequence *the quick brown*, the LM might predict *fox* as the next most likely word. Then, the sequence becomes *the quick brown fox*, and we task the LM to predict the new most likely word based on the updated sequence. A model whose output depends on its previous values, as well as its stochastic (that is, with some randomness) output (new value), is called an **autoregressive model**.

Next, we'll focus on the properties of the word sequence, rather than the model.

> **Note**
>
> Even the most advanced LLMs, such as ChatGPT, are autoregressive models – they just predict the next word, one word at a time.

Understanding N-grams

The inference of the probability of a long sequence, say $w_1...w_m$, is typically infeasible. To understand why, let's note that we can calculate the joint probability of $P(w_1...w_m)$ with the chain rule of joint probability (*Chapter 2*):

$$P(w_1...w_m) = P(w_m|w_1...w_{m-1})...P(w_3|w_1,w_2)P(w_2|w_1)P(w_1)$$

The probability of the later words given the earlier words would be especially difficult to estimate from the data. That's why this joint probability is typically approximated by an independence assumption that the i-th word is only dependent on the n-1 previous words. We'll only model the joint probabilities of combinations of n sequential words, called n-grams. For example, in the phrase *the quick brown fox*, we have the following n-grams:

- **1-gram (unigram)**: *the*, *quick*, *brown*, and *fox* (this is where Unigram tokenization takes its name)

- **2-gram (bigram)**: *the quick*, *quick brown*, and *brown fox*

- **3-gram (trigram)**: *the quick brown* and *quick brown fox*

- **4-gram**: *the quick brown fox*

Note

The term n-grams can refer to other types of sequences of length n, such as n characters.

The inference of the joint distribution is approximated with the help of n-gram models that split the joint distribution into multiple independent parts. If we have a large corpus of text, we can find all the n-grams up until a certain n (typically 2 to 4) and count the occurrence of each n-gram in that corpus. From these counts, we can estimate the probabilities of the last word of each n-gram, given the previous n-1 words:

- Unigram: $P(w) = \frac{count(w)}{\text{total number of words in the corpus}}$

- Bigram: $P\left(w_i \middle| w_{i-1}\right) = \frac{count(w_{i-1}, w_i)}{count(w_{i-1})}$

- n-gram: $P\left(w_{n+i} \middle| w_n, \ldots, w_{n+i-1}\right) = \frac{count(w_n, \ldots, w_{n+i-1}, w_{n+i})}{count(w_n, \ldots, w_{n+i-1})}$

The independent assumption that the i-th word is only dependent on the previous n-1 words can now be used to approximate the joint distribution.

For example, we can approximate the joint distribution for a unigram with the following formula:

$$P(w_1, \ldots, w_m) = P(w_1)P(w_2)\ldots P(w_m)$$

For a trigram, we can approximate the joint distribution with the following formula:

$$P(w_1, \ldots, w_m) = P(w_1)P\left(w_2 \middle| w_1\right)P\left(w_3 \middle| w_1, w_2\right)\ldots P\left(w_m \middle| w_{m-2}, w_{m-1}\right)$$

We can see that, based on the vocabulary size, the number of n-grams grows exponentially with n. For example, if a small vocabulary contains 100 words, then the number of possible 5-grams would be $100^5 = 10,000,000,000$ different 5-grams. In comparison, the entire works of Shakespeare contain around 30,000 different words, illustrating the infeasibility of using n-grams with a large n. Not only is there the issue of storing all the probabilities, but we would also need a very large text corpus to create decent n-gram probability estimations for larger values of n.

The curse of dimensionality

When the number of possible input variables (words) increases, the number of different combinations of these input values increases exponentially. This problem is known as the curse of dimensionality. It arises when the learning algorithm needs at least one example per relevant combination of values, which is the case in *n*-gram modeling. The larger our *n*, the better we can approximate the original distribution and the more data we would need to make good estimations of the *n*-gram probabilities.

But fret not, as the *n*-gram LM gives us some important clues on how to proceed. Its theoretical formulation is sound, but the curse of dimensionality makes it unfeasible. In addition, the *n*-gram model reinforces the importance of the word context, just as with Word2Vec. In the next few sections, we'll learn how to simulate an *n*-gram model probability distribution with the help of NNs.

Introducing RNNs

An RNN is a type of NN that can process sequential data with variable length. Examples of such data include text sequences or the price of a stock at various moments in time. By using the word *sequential*, we imply that the sequence elements are related to each other and their order matters. For example, if we take a book and randomly shuffle all the words in it, the text will lose its meaning, even though we'll still know the individual words.

RNNs get their name because they apply the same function over a sequence recurrently. We can define an RNN as a recurrence relation:

$$\mathbf{s}_t = f(\mathbf{s}_{t-1}, \mathbf{x}_t)$$

Here, f is a differentiable function, $\mathbf{s}_t$ is a vector of values called internal RNN state (at step t), and $\mathbf{x}_t$ is the network input at step t. Unlike regular NNs, where the state only depends on the current input (and RNN weights), here, $\mathbf{s}_t$ is a function of both the current input, as well as the previous state, $\mathbf{s}_{t-1}$. You can think of $\mathbf{s}_{t-1}$ as the RNN's summary of all previous inputs. The recurrence relation defines how the state evolves step by step over the sequence via a feedback loop over previous states, as illustrated in the following diagram:

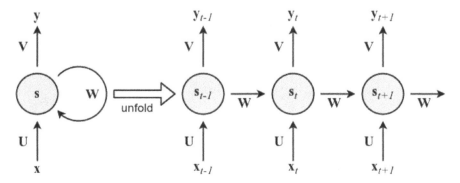

Figure 6.6 – An unfolded RNN

On the left, we have a visual illustration of the RNN recurrence relation. On the right, we have the RNN states recurrently unfolded over the sequence *t-1*, *t*, *t+1*.

The RNN has three sets of parameters (or weights), shared between all steps:

- **U**: Transforms the input, $\mathbf{x}_t$, into the state, $\mathbf{s}_t$
- **W**: Transforms the previous state, $\mathbf{s}_{t-1}$, into the current state, $\mathbf{s}_t$
- **V**: Maps the newly computed internal state, $\mathbf{s}_t$, to the output, $\mathbf{y}_t$

U, **V**, and **W** apply linear transformation over their respective inputs. The most basic case of such a transformation is the familiar FC operation we know and love (therefore, **U**, **V**, and **W** are weight matrices). We can now define the internal state and the RNN output as follows:

$$\mathbf{s}_t = f(\mathbf{s}_{t-1}\mathbf{W} + \mathbf{x}_t\mathbf{U})$$

$$\mathbf{y}_t = \mathbf{s}_t\mathbf{V}$$

Here, *f* is the non-linear activation function (such as tanh, sigmoid, or ReLU).

For example, in a word-level LM, the input, *x*, will be a sequence of word embedding vectors $(\mathbf{x}_1...\mathbf{x}_t)$. The state, *s*, will be a sequence of state vectors $(\mathbf{s}_1...\mathbf{s}_t)$. Finally, the output, *y*, will be a sequence of probability vectors $(\mathbf{y}_1...\mathbf{y}_t)$ of the next words in the sequence.

Note that in an RNN, each state is dependent on all previous computations via this recurrence relation. An important implication of this is that RNNs have memory over time because the states, $\mathbf{s}_t$, contain information based on the previous steps. In theory, RNNs can remember information for an arbitrarily long period, but in practice, they are limited to looking back only a few steps. We will address this issue in more detail in the *Vanishing and exploding gradients* section.

The RNN we described is somewhat equivalent to a single-layer regular NN (with an additional recurrence relation). But as with regular NNs, we can stack multiple RNNs to form a **stacked RNN**. The cell state, $\mathbf{s}_t^l$, of an RNN cell at level *l* at time *t* will take the output, $\mathbf{y}_t^{l-1}$, of the RNN cell from level *l-1* and the previous cell state, $\mathbf{s}_{t-1}^l$, of the cell at the same level *l* as the input:

$$\mathbf{s}_t^l = f(\mathbf{s}_{t-1}^l, \mathbf{y}_t^{l-1})$$

In the following diagram, we can see an unfolded, stacked RNN:

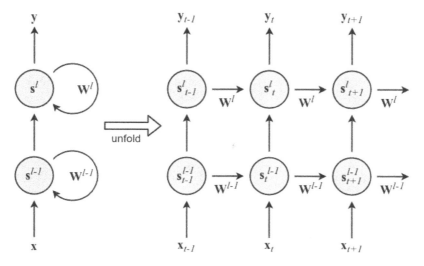

Figure 6.7 – Stacked RNN

Because RNNs are not limited to processing fixed-size inputs, they expand the possibilities of what we can compute with NNs. We can identify several types of tasks, based on the relationship between the input and output sizes:

- **One-to-one**: Non-sequential processing, such as feedforward NNs and CNNs. There isn't much difference between a feedforward NN and applying an RNN to a single time step. An example of one-to-one processing is image classification.

- **One-to-many**: This generates a sequence based on a single input – for example, caption generation from an image (*Show and Tell: A Neural Image Caption Generator*, https://arxiv.org/abs/1411.4555).

- **Many-to-one**: This outputs a single result based on a sequence – for example, sentiment classification of text.

- **Many-to-many indirect**: A sequence is encoded into a state vector, after which this state vector is decoded into a new sequence – for example, language translation (*Learning Phrase Representations using RNN Encoder-Decoder for Statistical Machine Translation*, https://arxiv.org/abs/1406.1078 and *Sequence to Sequence Learning with Neural Networks*, http://papers.nips.cc/paper/5346-sequence-to-sequence-learning-with-neural-networks.pdf).

- **Many-to-many direct**: Outputs a result for each input step – for example, frame phoneme labeling in speech recognition.

> **Note**
> The many-to-many models are often referred to as **sequence-to-sequence (seq2seq)** models.

The following is a graphical representation of the preceding input-output combinations:

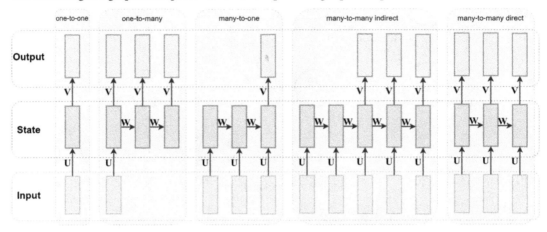

Figure 6.8 – RNN input-output combinations, inspired by http://karpathy.
github.io/2015/05/21/rnn-effectiveness/

Now that we've introduced RNNs, let's improve our knowledge by implementing a simple RNN example.

RNN implementation and training

In the preceding section, we briefly discussed what RNNs are and what problems they can solve. Let's dive into the details of an RNN and how to train it with a very simple toy example: counting ones in a sequence.

We'll teach a basic RNN how to count the number of ones in the input and then output the result at the end of the sequence. This is an example of a many-to-one relationship, which we defined in the previous section.

We'll implement this example with Python (no DL libraries) and numpy. An example of the input and output is as follows:

```
In: (0, 0, 0, 0, 1, 0, 1, 0, 1, 0)
Out: 3
```

The RNN we'll use is illustrated in the following diagram:

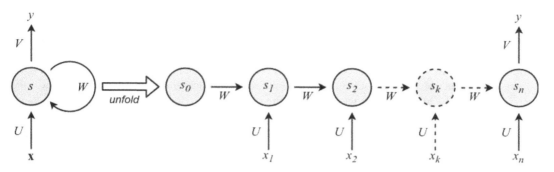

Figure 6.9 – Basic RNN for counting ones in the input

> **Note**
>
> Since s, x, U, W, and y are scalar values (**x** remains a vector), we won't use the matrix notation (bold capital letters) in the RNN implementation and training section and its subsections. We'll use italic notation instead. In the code sections, we'll denote them as variables. However, note that the generic versions of these formulas use matrix and vector parameters.

The RNN will have only two parameters: an input weight, U, and a recurrence weight, W. The output weight, V, is set to 1 so that we just read out the last state as the output, y.

First, let's add some code so that our example can be executed. We'll import numpy and define our training set – inputs, **x**, and labels, y. **x** is two-dimensional since the first dimension represents the sample in the mini-batch. y is a single numerical value (it still has a batch dimension). For the sake of simplicity, we'll use a mini-batch with a single sample:

```
import numpy as np

# The first dimension represents the mini-batch
x = np.array([[0, 0, 0, 0, 1, 0, 1, 0, 1, 0]])
y = np.array([3])
```

The recurrence relation defined by this RNN is $s_t = f(s_{t-1} W + x_t U)$. Note that this is a linear model since we don't apply a non-linear function in this formula. We can implement a recurrence relationship in the following way:

```
def step(s_t, x_t, U, W):
    return x_t * U + s_t * W
```

The states, `s_t`, and the weights, `W` and `U`, are single scalar values. `x_t` represents a single element of the input sequence (in our case, one or zero).

> **Note**
>
> One solution to this task is to just get the sum of the elements of the input sequence. If we set U=1, then whenever input is received, we will get its full value. If we set W=1, then the value we would accumulate would never decay. So, for this example, we would get the desired output: 3. Nevertheless, let's use this simple example to explain the training and implementation of the RNN. This will be interesting, as we will see in the rest of this section.

We can think of an RNN as a special type of regular NN by unfolding it through time for a certain number of time steps (as illustrated in the preceding diagram). This regular NN has as many hidden layers as the size of the elements of the input sequence. In other words, one hidden layer represents one step through time. The only difference is that each layer has multiple inputs: the previous state, s_{t-1}, and the current input, x_t. The parameters, U and W, are shared between all of the hidden layers.

The forward pass unfolds the RNN along the sequence and builds a stack of states for each step. In the following code block, we can see an implementation of the forward pass, which returns the activation, s, for each recurrent step and each sample in the batch:

```
def forward(x, U, W):
    # Number of samples in the mini-batch
    number_of_samples = len(x)
    # Length of each sample
    sequence_length = len(x[0])
    # Initialize the state activation for each sample along the
sequence
    s = np.zeros((number_of_samples, sequence_length + 1))
    # Update the states over the sequence
    for t in range(0, sequence_length):
        s[:, t + 1] = step(s[:, t], x[:, t], U, W)  # step function
    return s
```

Now that we have the RNN forward pass, let's look at how to train our unfolded RNN.

Backpropagation through time

Backpropagation through time (BPTT) is the typical algorithm we use to train RNNs (*Backpropagation Through Time: What It Does and How to Do It*, http://axon.cs.byu.edu/~martinez/classes/678/Papers/Werbos_BPTT.pdf). As its name suggests, it's an adaptation of the backpropagation algorithm we discussed in *Chapter 2*.

Let's assume that we'll use the **mean squared error** (MSE) cost function. Now that we also have our forward step implementation, we can define how the gradient is propagated backward. Since the unfolded RNN is equivalent to a regular feedforward NN, we can use the backpropagation chain rule we introduced in *Chapter 2*.

Because the weights, W and U, are shared across the layers, we'll accumulate the error derivatives for each recurrent step, and in the end, we'll update the weights with the accumulated value.

First, we need to get the gradient of the output, s_t, concerning the loss function, J, $\partial J/\partial s$. Once we have it, we'll propagate it backward through the stack of activities we built during the forward step. This backward pass pops activities off of the stack to accumulate their error derivatives at each time step. The recurrence relation that propagates this gradient through the RNN can be written as follows (chain rule):

$$\frac{\partial J}{\partial s_{t-1}} = \frac{\partial J}{\partial s_t}\frac{\partial s_t}{\partial s_{t-1}} = \frac{\partial J}{\partial s_t}W$$

The gradients of the weights, U and W, are accumulated as follows:

$$\frac{\partial J}{\partial U} = \sum_{t=0}^{n}\frac{\partial J}{\partial s_t}x_t$$

$$\frac{\partial J}{\partial W} = \sum_{t=0}^{n}\frac{\partial J}{\partial s_t}s_{t-1}$$

Armed with this knowledge, let's implement the backward pass:

1. Accumulate the gradients for U and W in gU and gW, respectively:

```python
def backward(x, s, y, W):
    sequence_length = len(x[0])
    # The network output is just the last activation of sequence
    s_t = s[:, -1]
    # Compute the gradient of the output w.r.t. MSE cost
function at final state
    gS = 2 * (s_t - y)
    # Set the gradient accumulations to 0
    gU, gW = 0, 0
    # Accumulate gradients backwards
    for k in range(sequence_length, 0, -1):
        # Compute the parameter gradients and accumulate the
results.
        gU += np.sum(gS * x[:, k - 1])
        gW += np.sum(gS * s[:, k - 1])
        # Compute the gradient at the output of the previous
layer
        gS = gS * W
    return gU, gW
```

2. Use gradient descent to optimize our RNN. Compute the gradients (using MSE) with the help of the backward function and use them to update the weights value:

```
def train(x, y, epochs, learning_rate=0.0005):
    # Set initial parameters
    weights = (-2, 0)  # (U, W)
    # Accumulate the losses and their respective weights
    losses, gradients_u, gradients_w = list(), list(), list()
    # Perform iterative gradient descent
    for i in range(epochs):
        # Perform forward and backward pass to get the gradients
        s = forward(x, weights[0], weights[1])
        # Compute the loss
        loss = (y[0] - s[-1, -1]) ** 2
        # Store the loss and weights values for later display
        losses.append(loss)
        gradients = backward(x, s, y, weights[1])
        gradients_u.append(gradients[0])
        gradients_w.append(gradients[1])
        # Update each parameter `p` by p = p - (gradient *
learning_rate).
        # `gp` is the gradient of parameter `p`
        weights = tuple((p - gp * learning_rate) for p, gp in
zip(weights, gradients))
    return np.array(losses), np.array(gradients_u),
np.array(gradients_w)
```

3. Run the training for 150 epochs:

```
losses, gradients_u, gradients_w = train(x, y,
    epochs=150)
```

4. Finally, display the loss function and the gradients for each weight over the epochs. We'll do this with the help of the plot_training function, which is not implemented here but is available in the full example on GitHub. plot_training produces the following graph:

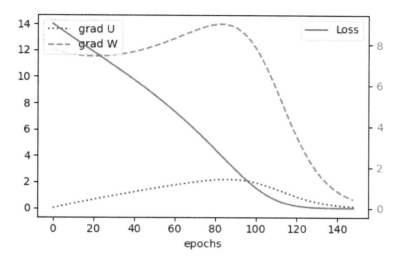

Figure 6.10 – The RNN loss – uninterrupted line – loss value;
dashed lines – the weight gradients during training

Now that we've learned about backpropagation through time, let's discuss how the familiar vanishing and exploding gradient problems affect it.

Vanishing and exploding gradients

The preceding example has an issue. To illustrate it, let's run the training process with a longer sequence:

```
x = np.array([[0, 0, 0, 0, 1, 0, 1, 0, 1, 0, 0, 0, 0, 0, 1, 0, 1, 0,
1, 0, 0, 0, 0, 0, 1, 0, 1, 0, 1, 0, 0, 0, 0, 0, 1, 0, 1, 0, 1, 0]])
y = np.array([12])
losses, gradients_u, gradients_w = train(x, y, epochs=150)
plot_training(losses, gradients_u, gradients_w)
```

The output is as follows:

```
RuntimeWarning: overflow encountered in multiply
  return x * U + s * W
RuntimeWarning: invalid value encountered in multiply
  gU += np.sum(gS * x[:, k - 1])
RuntimeWarning: invalid value encountered in multiply
  gW += np.sum(gS * s[:, k - 1])
```

The reason for these warnings is that the final parameters, U and W, end up as **Not a Number (NaN)**. To display the gradients properly, we'll need to change the scale of the gradient axis in the plot_ training function to produce the following result:

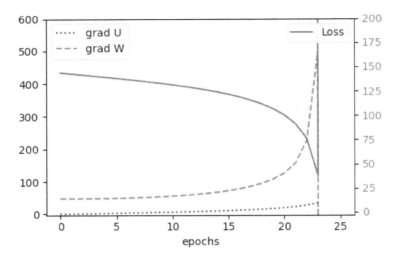

Figure 6.11 – Parameters and loss function during an exploding gradients scenario

In the initial epochs, the gradients slowly increase, similar to the way they increased for the shorter sequence. However, when they get to epoch 23 (the exact epoch is unimportant, though), the gradient becomes so large that it goes out of the range of the float variable and becomes NaN (as illustrated by the jump in the plot). This problem is known as **exploding gradients**. We can stumble upon exploding gradients in a regular feedforward NN, but it is especially pronounced in RNNs. To understand why, let's recall the recurrent gradient propagation chain rule for the two consecutive sequence steps we defined in the *Backpropagation through time* section:

$$\frac{\partial J}{\partial s_{t-1}} = \frac{\partial J}{\partial s_t}\frac{\partial s_t}{\partial s_{t-1}} = \frac{\partial J}{\partial s_t}W$$

Depending on the sequence's length, an unfolded RNN can be much deeper compared to a regular NN. At the same time, the weights, W, of an RNN are shared across all of the steps. Therefore, we can generalize this formula to compute the gradient between two non-consecutive steps of the sequence. Because W is shared, the equation forms a geometric progression:

$$\frac{\partial s_t}{\partial s_{t-k}} = \frac{\partial s_t}{\partial s_{t-1}}\frac{\partial s_{t-1}}{\partial s_{t-2}}...\frac{\partial s_{t-k+1}}{\partial t-k} = \prod_{j=1}^{k}\frac{\partial s_{t-j+1}}{\partial s_{t-j}}W$$

In our simple linear RNN, the gradient grows exponentially if $|W|>1$ (exploding gradient), where W is a single scalar weight – for example, 50 time steps over $W=1.5$ is $W^{(50)} \approx 637{,}621{,}500$. The gradient shrinks exponentially if $|W|<1$ (vanishing gradient), for example, 10 time steps over $W=0.6$ is $W^{(10)} = 0.00097$. If the weight parameter, W, is a matrix instead of a scalar, this exploding or vanishing gradient is related to the largest eigenvalue, ρ, of W (also known as a spectral radius). It is sufficient for $\rho<1$ for the gradients to vanish, and it is necessary for $\rho>1$ for them to explode.

The vanishing gradients problem, which we first mentioned in *Chapter 3*, has another more subtle effect in RNNs: the gradient decays exponentially over the number of steps to a point where it becomes extremely small in the earlier states. In effect, they are overshadowed by the larger gradients from more recent time steps, and the RNN's ability to retain the history of these earlier states vanishes. This problem is harder to detect because the training will still work, and the NN will produce valid outputs (unlike with exploding gradients). It just won't be able to learn long-term dependencies.

With that, we are familiar with some of the problems surrounding RNNs. This knowledge will serve us well because, in the next section, we'll discuss how to solve these problems with the help of a special type of RNN cell.

Long-short term memory

Hochreiter and Schmidhuber studied the problems of vanishing and exploding gradients extensively and came up with a solution called **long short-term memory** (**LSTM** – https:// www.bioinf.jku.at/publications/older/2604.pdf and *Learning to Forget: Continual Prediction with LSTM*, https://citeseerx.ist.psu.edu/viewdoc/ download?doi=10.1.1.55.5709&rep=rep1&type=pdf). LSTMs can handle long-term dependencies due to a specially crafted memory cell. They work so well that most of the current accomplishments in training RNNs on a variety of problems are due to the use of LSTMs. In this section, we'll explore how this memory cell works and how it solves the vanishing gradients issue.

The following is a diagram of an LSTM cell:

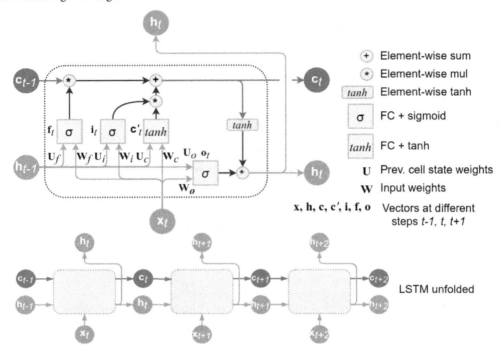

Figure 6.12 – LSTM cell (top); unfolded LSTM cell (bottom). Inspired by `http://colah.github.io/posts/2015-08-Understanding-LSTMs/`

The key idea of LSTM is the cell state, c_t (in addition to the hidden RNN state, h_t), where the information can only be explicitly written in or removed so that the state stays constant if there is no outside interference. The cell state can only be modified by specific gates, which are a way to let information pass through. A typical LSTM is composed of three gates: a **forget gate**, an **input gate**, and an **output gate**. The cell state, input, and output are all vectors so that the LSTM can hold a combination of different information blocks at each time step.

LSTM notations

x_t, c_t, and h_t are the LSTM's input, cell memory state, and output (or hidden state) vectors in moment t. c'_t is the candidate cell state vector (more on that later). The input, x_t, and the previous cell output, h_{t-1}, are connected to each gate and the candidate cell vector with sets of FC weights, $\mathbf{W}$ and $\mathbf{U}$, respectively. f_t, i_t, and o_t are the forget, input, and output gates of the LSTM cell (the gates use vector notation as well).

The gates are composed of FC layers, sigmoid activations, and element-wise multiplication (denoted with $\odot$). Because the sigmoid only outputs values between 0 and 1, the multiplication can only reduce the value running through the gate. Let's discuss them in order:

- **Forget gate, $\mathbf{f}_t$:** It decides whether we want to erase parts of the existing cell state or not. It bases its decision on the weighted vector sum of the output of the previous cell, $\mathbf{h}_{t-1}$, and the current input, $\mathbf{x}_t$:

$$\mathbf{f}_t = \sigma\left(\mathbf{W}_f \mathbf{x}_t + \mathbf{U}_f \mathbf{h}_{t-1}\right)$$

  From the preceding formula, we can see that the forget gate applies element-wise sigmoid activations to each element of the previous state vector, $\mathbf{c}_{t-1}$: $\mathbf{f}_t \odot \mathbf{c}_{t-1}$ (note the circle-dot notation). Since the operation is elementwise, the values of this vector are squashed in the [0, 1] range. An output of 0 erases a specific $c_{t-1,j}$ cell block completely and an output of 1 allows the information in that cell block to pass through. In this way, the LSTM can get rid of irrelevant information in its cell state vector.

- **Input gate, $\mathbf{i}_t$:** It decides what new information is going to be added to the memory cell in a multi-step process. The first step determines whether any information is going to be added. As in the forget gate, its decision is based on $\mathbf{h}_{t-1}$ and $\mathbf{x}_t$: it outputs 0 or 1 through the sigmoid function for each cell of the candidate state vector. An output of 0 means that no information is added to that cell block's memory. As a result, the LSTM can store specific pieces of information in its cell state vector:

$$\mathbf{i}_t = \sigma\left(\mathbf{W}_i \mathbf{x}_t + \mathbf{U}_i \mathbf{h}_{t-1}\right)$$

  In the next step of the input gate sequence, we compute the new candidate cell state, $\mathbf{c}'_t$. It is based on the previous output, $\mathbf{h}_{t-1}$, and the current input, $\mathbf{x}_t$, and is transformed via a tanh function:

$$\mathbf{c}'_t = tanh\left(\mathbf{W}_c \mathbf{x}_t + \mathbf{U}_c \mathbf{h}_{t-1}\right)$$

  Then, we combine $\mathbf{c}'_t$ with the sigmoid outputs of the input gate via element-wise multiplication:

$$\mathbf{i}_t \odot \mathbf{c}'_t.$$

  To recap, the forget and input gates decide what information to forget and include from the previous and candidate cell states, respectively. The final version of the new cell state, $\mathbf{c}_t$, is just an element-wise sum between these two components:

$$\mathbf{c}_t = \mathbf{f}_t \odot \mathbf{c}_{t-1} \oplus \mathbf{b}_t \odot \mathbf{c}'_t$$

- **Output gate, $\mathbf{o}_t$:** It decides what the total cell output is going to be. It takes $\mathbf{h}_{t-1}$ and $\mathbf{x}_t$ as inputs. It outputs a value in the (0, 1) range (via the sigmoid function) for each block of the cell's memory. Like before, 0 means that the block doesn't output any information and 1 means that the block can pass through as a cell's output. Therefore, the LSTM can output specific blocks of information from its cell state vector:

$$\mathbf{o}_t = \sigma\left(\mathbf{W}_o \mathbf{x}_t + \mathbf{U}_o \mathbf{h}_{t-1}\right)$$

Finally, the LSTM cell's output is transferred by a tanh function:

$$\mathbf{h}_t = \mathbf{o}_t \odot \tanh(\mathbf{c}_t)$$

Because all these formulas are derivable, we can chain LSTM cells together, just like when we chain simple RNN states together and train the network via backpropagation through time.

But how does the LSTM protect us from vanishing gradients? Let's start with the forward phase. Notice that the cell state is copied identically from step to step if the forget gate is 1 and the input gate is 0: $\mathbf{c}_t = \mathbf{f}_t \odot \mathbf{c}_{t-1} \oplus \mathbf{i}_t \odot \mathbf{c}'_t = 1 \odot \mathbf{c}_{t-1} \oplus 0 \odot \mathbf{c}'_t = \mathbf{c}_{t-1}$. Only the forget gate can completely erase the cell's memory. As a result, the memory can remain unchanged over a long period. Also, note that the input is a tanh activation that's been added to the current cell's memory. This means that the cell's memory doesn't blow up and is quite stable.

Let's use an example to demonstrate how an LSTM cell is unfolded. For the sake of simplicity, we'll assume that it has one-dimensional (single scalar value) input, state, and output vectors. Because the values are scalar, we won't use vector notation for the rest of this example:

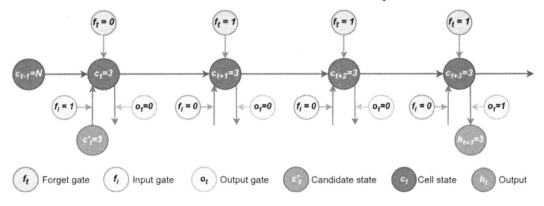

Figure 6.13 – Unrolling an LSTM through time

The process is as follows:

1. First, we have a value of 3 as a candidate state. The input gate is set to $f_i = 1$ and the forget gate is set to $f_t = 0$. This means that the previous state, $c_{t-1} = N$, is erased and replaced with the new state, $c_t = 0 \odot N \oplus 1 \odot 3 = 3$.

2. For the next two time steps, the forget gate is set to 1, while the input gate is set to 0. By doing this, all the information is kept throughout these steps and no new information is added because the input gate is set to 0: $c_{t+1} = 1 \odot 3 \oplus 0 \odot c'_{t+1} = 3$.

3. Finally, the output gate is set to $o_t = 1$ and 3 is output and remains unchanged. We have successfully demonstrated how the internal state is stored across multiple steps.

Next, let's focus on the backward phase. The cell state, c_t, can mitigate the vanishing/exploding gradients as well with the help of the forget gate, f_t. Like the regular RNN, we can use the chain rule to compute the partial derivative, $\partial c_t / \partial c_{t-1}$, for two consecutive steps. Following the formula $c_t = f_t \odot c_{t-1} \oplus i_t \odot c'_t$ and without going into details, its partial derivative is as follows:

$$\frac{\partial c_t}{\partial c_{t-1}} \approx f_t$$

We can also generalize this to non-consecutive steps:

$$\frac{\partial c_t}{\partial c_{t-k}} = \frac{\partial c_t}{\partial c_{t-1}} \frac{\partial c_{t-1}}{\partial c_{t-2}} \cdots \frac{\partial c_{t-k+1}}{\partial c_{t-k}} \approx \prod_{j=1}^{k} f_{t-j+1}$$

If the forget gate values are close to 1, gradient information can pass back through the network states almost unchanged. This is because f_t uses sigmoid activation and information flow is still subject to the vanishing gradient that's specific to sigmoid activations. But unlike the gradients in the regular RNN, f_t has a different value at each time step. Therefore, this is not a geometric progression, and the vanishing gradient effect is less pronounced.

Next, we'll introduce a new type of lightweight RNN cell that still preserves the properties of LSTM.

Gated recurrent units

A **gated recurrent unit** (**GRU**) is a type of recurrent block that was introduced in 2014 (*Learning Phrase Representations using RNN Encoder-Decoder for Statistical Machine Translation*, https:// arxiv.org/abs/1406.1078 and *Empirical Evaluation of Gated Recurrent Neural Networks on Sequence Modeling*, https://arxiv.org/abs/1412.3555) as an improvement over LSTM. A GRU unit usually has similar or better performance than an LSTM, but it does so with fewer parameters and operations:

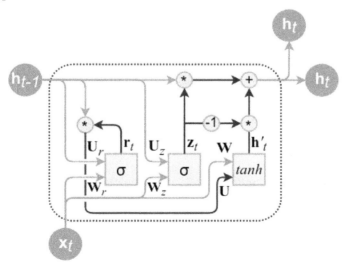

Figure 6.14 – A GRU cell diagram

Similar to the classic RNN, a GRU cell has a single hidden state, $\mathbf{h}_t$. You can think of it as a combination of the hidden and cell states of an LSTM. The GRU cell has two gates:

- **Update gate**, $\mathbf{z}_t$: Combines the input and forget LSTM gates. It decides what information to discard and what new information to include in its place based on the network input, $\mathbf{x}_t$, and the previous hidden state, $\mathbf{h}_{t-1}$. By combining the two gates, we can ensure that the cell will forget information, but only when we are going to include new information in its place:

$$\mathbf{z}_t = \sigma\left(\mathbf{W}_z\mathbf{x}_t + \mathbf{U}_z\mathbf{h}_{t-1}\right)$$

- **Reset gate**, $\mathbf{r}_t$: Uses the previous hidden state, $\mathbf{h}_{t-1}$, and the network input, $\mathbf{x}_t$, to decide how much of the previous state to pass through:

$$\mathbf{r}_t = \sigma\left(\mathbf{W}_r\mathbf{x}_t + \mathbf{U}_r\mathbf{h}_{t-1}\right)$$

Next, we have the candidate state, $\mathbf{h}_t'$:

$$\mathbf{h}_t' = tanh\left(\mathbf{W}\mathbf{x}_t + \mathbf{U}\left(\mathbf{r}_t\odot\mathbf{h}_{t-1}\right)\right)$$

Finally, the GRU output, $\mathbf{h}_t$, at time t is an element-wise sum between the previous output, $\mathbf{h}_{t-1}$, and the candidate output, $\mathbf{h}_t'$:

$$\mathbf{h}_t = \mathbf{z}_t\odot\mathbf{h}_{t-1} \oplus \left(1 - \mathbf{z}_t\right)\odot\mathbf{h}_t'$$

Since the update gate allows us to both forget and store data, it is directly applied to the previous output, $\mathbf{h}_{t-1}$, and applied over the candidate output, $\mathbf{h}_t'$.

We'll conclude our introduction to RNNs by returning to the disclaimer at the start of this chapter – the practical limitations of RNNs. We can solve one of them – the vanishing and exploding gradients – with the help of LSTM or GRU cells. However, there are two others:

- The RNN's internal state is updated after each element of the sequence – a new element requires all preceding elements to be processed in advance. Therefore, the RNN sequence processing cannot be parallelized and RNNs cannot take advantage of the GPU parallelization capabilities.

- The information of all preceding sequence elements is summarized in a single hidden cell state. The RNN doesn't have direct access to the historical sequence elements and has to rely on the cell state instead. In practice, this means that an RNN (even LSTM or GRU) can meaningfully process sequences with a maximum length of around 100 elements.

As we'll see in the next chapter, the transformer architecture successfully solves both of these limitations. But for now, let's see how to use LSTMs in practice.

Implementing text classification

In this section, we'll use LSTM to implement a sentiment analysis example over the Large Movie Review Dataset (**IMDb**, http://ai.stanford.edu/~amaas/data/sentiment/), which consists of 25,000 training and 25,000 testing reviews of popular movies. Each review has a binary label that indicates whether it is positive or negative. This type of problem is an example of a **many-to-one** relationship, which we defined in the *Recurrent neural networks (RNNs)* section.

The sentiment analysis model is displayed in the following diagram:

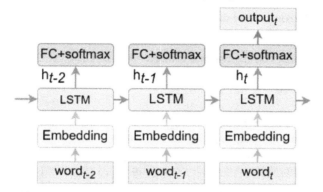

Figure 6.15 – Sentiment analysis with word embeddings and LSTM

Let's describe the model components (these are valid for any text classification algorithm):

1. Each word of the sequence is replaced with its embedding vector. These embeddings can be produced with word2vec.
2. The word embedding is fed as input to the LSTM cell.
3. The cell output, $\mathbf{h}_t$, serves as input to an FC layer with two output units and softmax. The softmax output represents the probability of the review being positive (1) or negative (0).
4. The network can be produced with Word2Vec.
5. The output for the final element of the sequence is taken as a result of the whole sequence.

To implement this example, we'll use PyTorch and the TorchText package. It consists of data processing utilities and popular datasets for natural language. We'll only include the interesting portions of the code, but the full example is available in this book's GitHub repo. With that, let's start:

1. Define the device (by default, this is GPU with a fallback on CPU):

    ```
    import torch
    device = torch.device("cuda:0" if torch.cuda.is_available() else
    "cpu")
    ```

2. Start the training and testing dataset pipeline. First, define the `basic_english` tokenizer, which splits the text on spaces (that is, word tokenization):

```
from torchtext.data.utils import get_tokenizer
tokenizer = get_tokenizer('basic_english')
```

3. Next, use `tokenizer` to build the token `vocabulary`:

```
from torchtext.datasets import IMDB
from torchtext.vocab import build_vocab_from_iterator
def yield_tokens(data_iter):
    for _, text in data_iter:
        yield tokenizer(text)
vocabulary = build_vocab_from_iterator(
    yield_tokens(IMDB(split='train')),
    specials=["<unk>"])
vocabulary.set_default_index(vocabulary["<unk>"])
```

Here, `IMDB(split='train')` provides an iterator of all movie reviews in the training set (each review is represented as a string). The `yield_tokens(IMDB(split='train'))` generator iterates over all samples and splits them into words. The result serves as input to `build_vocab_from_iterator`, which iterates over the tokenized samples and builds the token `vocabulary`. Note that the vocabulary only includes training samples. Therefore, any token that exists in the test set (but not the training one) will be replaced with the default unknown <unk> token.

4. Next, define the `collate_batch` function, which takes a `batch` of tokenized samples with varying lengths, and concatenates them in a single long sequence of tokens:

```
def collate_batch(batch):
    labels, samples, offsets = [], [], [0]
    for (_label, _sample) in batch:
        labels.append(int(_label) - 1)
        processed_text = torch.tensor(
            vocabulary(tokenizer(_sample)),
            dtype=torch.int64)
        samples.append(processed_text)
        offsets.append(processed_text.size(0))
    labels = torch.tensor(
        labels,
        dtype=torch.int64)
    offsets = torch.tensor(
        offsets[:-1]).cumsum(dim=0)
    samples = torch.cat(samples)
    return labels, samples, offsets
```

Here, the `samples` list aggregates all tokenized `_sample` instances of `batch`. In the end, they are concatenated into a single list. The `offsets` list contains the offset from the start of each concatenated sample. This information makes it possible to reverse-split the long `samples` sequence into separate items again. The purpose of the function is to create a compressed `batch` representation. This is necessary because of the varying length of each sample. The alternative would be to pad all samples to match the length of the longest one so that they can fit in the batch tensor. Fortunately, PyTorch provides us with the `offsets` optimization to avoid this. Once we feed the compressed batch to the RNN, it will automatically reverse it back into separate samples.

5. Then, we define the LSTM model:

```python
class LSTMModel(torch.nn.Module):
    def __init__(self, vocab_size, embedding_size, hidden_size,
        num_classes):

        super().__init__()
        # Embedding field
        self.embedding = torch.nn.EmbeddingBag(
            num_embeddings=vocab_size,
            embedding_dim=embedding_size)
        # LSTM cell
        self.rnn = torch.nn.LSTM(
            input_size=embedding_size,
            hidden_size=hidden_size)
        # Fully connected output
        self.fc = torch.nn.Linear(
            hidden_size, num_classes)

    def forward(self, text_sequence, offsets):
        # Extract embedding vectors
        embeddings = self.embedding(
            text_sequence, offsets)
        h_t, c_t = self.rnn(embeddings)
        return self.fc(h_t)
```

The model implements the scheme we introduced at the start of this section. As its name suggests, the `embedding` property (an instance of `EmbeddingBag`) maps the token (in our case, word) index to its embedding vector. We can see that the constructor takes the vocabulary size (`num_embeddings`) and the embedding vector size (`embedding_dim`). In theory, we could initialize `EmbeddingBag` with pre-computed Word2Vec embedding vectors. But in our case, we'll simply use random initialization and let the model learn them as part of the training. `embedding` also takes care of the compressed batch representation (hence the `offsets` parameter in the `forward` method). The embedding's output serves as input to the `rnn` LSTM cell, which, in turn, feeds the output `fc` layer.

6. Define the `train_model(model, cost_function, optimizer, data_loader)` and `test_model(model, cost_function, data_loader)` functions. These are almost the same functions that we first defined in *Chapter 3*, so we won't include them here. However, they have been adapted to the compressed batch representation and the additional `offsets` parameter.

7. Proceed with the experiment. Instantiate the LSTM model, the cross-entropy cost function, and the Adam optimizer:

```
model = LSTMModel(
    vocab_size=len(vocabulary),
    embedding_size=64,
    hidden_size=64,
    num_classes=2)
cost_fn = torch.nn.CrossEntropyLoss()
optim = torch.optim.Adam(model.parameters())
```

8. Define `train_dataloader`, `test_dataloader`, and their respective datasets (use a mini-batch size of 64):

```
from torchtext.data.functional import to_map_style_dataset
train_iter, test_iter = IMDB()
train_dataset = to_map_style_dataset(train_iter)
test_dataset = to_map_style_dataset(test_iter)
from torch.utils.data import DataLoader
train_dataloader = DataLoader(
    train_dataset, batch_size=64,
    shuffle=True, collate_fn=collate_batch)
test_dataloader = DataLoader(
    test_dataset, batch_size=64,
    shuffle=True, collate_fn=collate_batch)
```

9. Run the training for 5 epochs:

```
for epoch in range(5):
    print(f'Epoch: {epoch + 1}')
    train_model(model, cost_fn, optim, train_dataloader)
    test_model(model, cost_fn, test_dataloader)
```

The model achieves a test accuracy in the realm of 87%.

This concludes our small practical example of LSTM text classification. Coincidentally, it also concludes this chapter.

Summary

In this chapter, we introduced two complementary topics – NLP and RNNs. We discussed the tokenization technique and the most popular tokenization algorithms – BPE, WordPiece, and Unigram. Then, we introduced the concept of word embedding vectors and the Word2Vec algorithm to produce them. We also discussed the n-gram LM, which provided us with a smooth transition to the topic of RNNs. There, we implemented a basic RNN example and introduced two of the most advanced RNN architectures – LSTM and GRU. Finally, we implemented a sentiment analysis model.

In the next chapter, we'll supercharge our NLP potential by introducing the attention mechanism and transformers.

7

The Attention Mechanism and Transformers

In *Chapter 6*, we outlined a typical **natural language processing** (**NLP**) pipeline, and we introduced **recurrent neural networks** (**RNNs**) as a candidate architecture for NLP tasks. But we also outlined their drawbacks—they are inherently sequential (that is, not parallelizable) and cannot process longer sequences, because of the limitations of their internal sequence representation. In this chapter, we'll introduce the **attention mechanism**, which allows a **neural network** (**NN**) to have direct access to the whole input sequence. We'll briefly discuss the attention mechanism in the context of RNNs since it was first introduced as an RNN extension. However, the star of this chapter will be the **transformer**—a recent NN architecture that relies entirely on attention. Transformers have been one of the most important NN innovations in the past 10 years. They are at the core of all recent **large language models** (**LLMs**), such as ChatGPT (https://chat.openai.com/), and even image generation models such as Stable Diffusion (https://stability.ai/stable-diffusion). This is the second chapter in our arc dedicated to NLP and the first of three chapters dedicated to transformers.

This chapter will cover the following topics:

- Introducing **sequence-to-sequence** (**seq2seq**) models
- Understanding the attention mechanism
- Building transformers with attention

Technical requirements

We'll implement the example in this chapter using Python, PyTorch, and the Hugging Face Transformers library (https://github.com/huggingface/transformers). If you don't have an environment set up with these tools, fret not—the example is available as a Jupyter notebook on Google Colab. You can find the code examples in the book's GitHub repository: https://github.com/PacktPublishing/Python-Deep-Learning-Third-Edition/tree/main/Chapter07.

Introducing seq2seq models

In *Chapter 6*, we outlined several types of recurrent models, depending on the input/output combinations. One of them is indirect many-to-many, or **seq2seq**, where an input sequence is transformed into another, different output sequence, not necessarily with the same length as the input. One type of seq2seq task is machine translation. The input sequences are the words of a sentence in one language, and the output sequences are the words of the same sentence translated into another language. For example, we can translate the English sequence *tourist attraction* to the German *Touristenattraktion*. Not only is the output of a different length but there is no direct correspondence between the elements of the input and output sequences. One output element corresponds to a combination of two input elements.

Another type of indirect many-to-many task is conversational chatbots such as ChatGPT, where the initial input sequence is the first user query. After that, the whole conversation so far (including both user queries and bot responses) serves as an input sequence for the newly generated bot responses.

In this section, we'll focus on encoder-decoder seq2seq models (*Sequence to Sequence Learning with Neural Networks*, https://arxiv.org/abs/1409.3215; *Learning Phrase Representations using RNN Encoder-Decoder for Statistical Machine Translation*, https://arxiv.org/abs/1406.1078), first introduced in 2014. They use RNNs in a way that's especially suited for solving indirect many-to-many tasks such as these. The following is a diagram of the seq2seq model, where an input sequence [A, B, C, <EOS>] is decoded into an output sequence [W, X, Y, Z, <EOS>]:

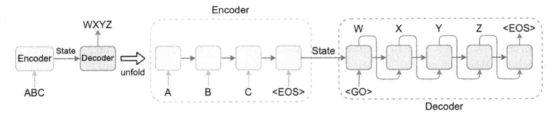

Figure 7.1 – A seq2seq model (inspired by https://arxiv.org/abs/1409.3215)

The model consists of two parts:

- **Encoder**: An RNN such as **Long Short-Term Memory (LSTM)** or **Gated Recurrent Unit (GRU)**. Taken by itself, the encoder works like a regular RNN—it reads the input sequence, one step at a time, and updates its internal state after each step. The encoder will stop reading the input sequence once a special <EOS>—end-of-sequence—token is reached. Let's assume that the input is a textual sequence using word-level tokenization. Then, we'll use word-embedding vectors as the encoder input at each step, and the <EOS> token signals the end of a sentence. The encoder output is discarded and has no role in the seq2seq model, as we're only interested in the hidden encoder state.

- **Decoder**: Once the encoder is finished, we'll signal the decoder so that it can start generating the output sequence with a special `<GO>` input signal. The encoder is also an RNN (LSTM or GRU). The link between the encoder and the decoder is the most recent encoder's internal state vector, $\mathbf{h}_t$ (also known as the **thought vector**), which is fed as the recurrence relation at the first decoder step. The decoder output y_{t+1} at step $t+1$ is one element of the output sequence. Next, we'll use y_{t+1} as a model input at step $t+2$ to generate new output, and so on (this type of model is called **autoregressive**). In the case of textual sequences, the decoder output is a softmax operation over all the words in the vocabulary. At each step, we take the word with the highest probability, and we feed it as input to the next step. Once `<EOS>` becomes the most probable symbol, the decoding is finished.

An example of an autoregressive model

Let's assume that we want to translate the English sentence *How are you today?* into Spanish. We'll tokenize it as [`how, are, you, today, ?, <EOS>`]. An autoregressive model will start with an initial sequence [`<GO>`]. Then, it will generate the first word of the translation and will append it to the existing input sequence: [`<GO>, ¿`]. The new sequence will serve as new input to the decoder, so it can produce the next element and extend the sequence again: [`<GO>, ¿, cómo`]. We'll repeat the same steps until the decoder predicts the `<EOS>` token: [`<GO>, ¿, cómo, estás, hoy, ?, <EOS>`].

The training of the model is supervised, as it needs to know both the input sequence and its corresponding target output sequence (for example, the same text in multiple languages). We feed the input sequence to the encoder, generate the thought vector, $\mathbf{h}_t$, and use it to initiate the output sequence generation from the decoder. Training the decoder uses a process called **teacher forcing**—its input at step t is always the correct word from the target sequence at step $t-1$, even if the decoder prediction of step $t-1$ is wrong. For example, let's say that the correct target sequence until step t is [`W, X, Y`], but the current decoder-generated output sequence is [`W, X, Z`]. With teacher forcing, the decoder input at step $t+1$ will be Y instead of Z. In other words, the decoder learns to generate target values [`t+1, ...`] given target values [`..., t`]. We can think of this in the following way: the decoder input is the target sequence, while its output (target values) is the same sequence but shifted one position to the right.

To summarize, the seq2seq model solves the problem of varying input/output sequence lengths by encoding the input sequence into a fixed-length state vector, $\mathbf{v}$, and then using this vector as a base to generate the output sequence. We can formalize this by saying that it tries to maximize the following probability:

$$P(y_1, \ldots, y_T | x_1, \ldots, x_T) = \prod_{t=1}^{T} P(y_t | \mathbf{v}, y_1, \ldots, y_{t-1})$$

This is equivalent to the following:

$$P(y_1, \ldots, y_T | x_1, \ldots, x_T) = P(y_1 | \mathbf{v}) P(y_2 | \mathbf{v}, y_1) \ldots P(y_T | \mathbf{v}, y_1, \ldots, y_{T-1})$$

Let's look at the elements of this formula in more detail:

- $P(y_1...y_{T'}|x_1...x_T)$: The conditional probability where $(x_1...x_T)$ is the input sequence with length T and $(y_1...y_{T'})$ is the output sequence with length T'
- $\mathbf{v}$: The fixed-length encoding of the input sequence (the thought vector)
- $P(y_{T'}|\mathbf{v}, y1...y_{T-1})$: The probability of an output word $y_{T'}$ given prior words y, as well as the thought vector, $\mathbf{v}$

The original seq2seq paper introduces a few tricks to enhance the training and performance of the model. For example, the encoder and decoder are two separate LSTMs. In the case of machine translations, this makes it possible to train different decoders for different languages with the same encoder.

Another improvement is that the input sequence is fed to the decoder in reverse. For example, [A, B, C] -> [W, X, Y, Z] would become [C, B, A] -> [W, X, Y, Z]. There is no clear explanation of why this works, but the authors have shared their intuition: since this is a step-by-step model, if the sequences were in normal order, each source word in the source sentence would be far from its corresponding word in the output sentence. If we reverse the input sequence, the average distance between input/output words won't change, but the first input words will be very close to the first output words. This will help the model to establish better communication between the input and output sequences. However, this improvement also illustrates the deficiencies of the hidden state of RNNs (even LSTM or GRU)—the more recent sequence elements suppress the available information for the older elements. In the next section, we'll introduce an elegant way to solve this issue once and for all.

Understanding the attention mechanism

In this section, we'll discuss several iterations of the attention mechanism in the order that they were introduced.

Bahdanau attention

The first attention iteration (*Neural Machine Translation by Jointly Learning to Align and Translate*, https://arxiv.org/abs/1409.0473), known as **Bahdanau** attention, extends the seq2seq model with the ability for the decoder to work with all encoder hidden states, not just the last one. It is an addition to the existing seq2seq model, rather than an independent entity. The following diagram shows how Bahdanau attention works:

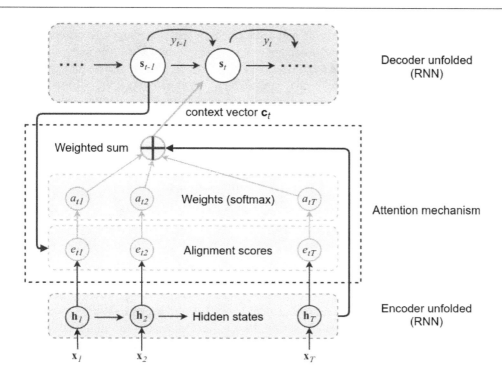

Figure 7.2 – The attention mechanism

Don't worry—it looks scarier than it is. We'll go through this diagram from top to bottom: the attention mechanism works by plugging an additional **context vector**, c_t, between the encoder and the decoder. The hidden decoder state s_t at time t is now a function not only of the hidden state and decoder output at step t-1 but also of the context vector c_t:

$$s_t = f(s_{t-1}, y_{t-1}, c_t)$$

Each decoder step has a unique context vector, and the context vector for one decoder step is just *a weighted sum of all encoder hidden states*. In this way, the encoder can access all input sequence states at each output step t, which removes the necessity to encode all information of the source sequence into a fixed-length thought vector, as the regular seq2seq model does:

$$c_t = \sum_{i=1}^{T} \alpha_{t,i} h_i$$

Let's discuss this formula in more detail:

- c_t: The context vector for a decoder output step t out of T' total output steps
- h_i: The hidden state vector of encoder step i out of T total input steps
- $\alpha_{t,i}$: The scalar weight associated with h_i in the context of the current decoder step t

Note that $\alpha_{t,i}$ is unique for both the encoder and decoder steps—that is, the input sequence states will have different weights depending on the current output step. For example, if the input and output sequences have lengths of 10, then the weights will be represented by a 10×10 matrix for a total of 100 weights. This means that the attention mechanism will focus the attention (get it?) of the decoder on different parts of the input sequence, depending on the current state of the output sequence. If $\alpha_{t,i}$ is large, then the decoder will pay a lot of attention to $\mathbf{h}_i$ at step t.

But how do we compute the weights $\alpha_{t,i}$? First, we should mention that the sum of all $\alpha_{t,i}$ weights for a decoder at step t is 1. We can implement this with a softmax operation on top of the attention mechanism:

$$\alpha_{t,i} = \frac{exp(e_{t,i})}{\sum_{j=1}^{T} exp(e_{t,j})} = \text{softmax}(e_{t,i}/e_t)$$

Here, $e_{t,i}$ is an alignment score, which indicates how well the input sequence elements around position i match (or align with) the output at position t. This score (represented by the weight $\alpha_{t,i}$) is based on the previous decoder state $\mathbf{s}_{t-1}$ (we use $\mathbf{s}_{t-1}$ because we have not computed $\mathbf{s}_t$ yet), as well as the encoder state $\mathbf{h}_i$:

$$e_{t,i} = a(\mathbf{s}_{t-1}, \mathbf{h}_i)$$

Here, a (not alpha) is a differentiable function, which is trained with backpropagation together with the rest of the system. Different functions satisfy these requirements, but the authors of the paper chose the so-called **additive attention**, which combines $\mathbf{s}_{t-1}$ and $\mathbf{h}_i$ with the help of vector addition. It exists in two flavors:

$$e_{t,i} = a(\mathbf{s}_{t-1}, \mathbf{h}_i) = \mathbf{v}^\top tanh(\mathbf{W}[\mathbf{h}_i; \mathbf{s}_{t-1}])$$

$$e_{t,i} = a(\mathbf{s}_{t-1}, \mathbf{h}_i) = \mathbf{v}^\top tanh(\mathbf{W}_1\mathbf{h}_i + \mathbf{W}_2\mathbf{s}_{t-1})$$

In the first formula, $\mathbf{W}$ is a weight matrix, applied over the concatenated vectors $\mathbf{s}_{t-1}$ and $\mathbf{h}_i$, and $\mathbf{v}$ is a weight vector. The second formula is similar, but this time we have separate **fully connected (FC)** layers (the weight matrices $\mathbf{W}_1$ and $\mathbf{W}_2$) and we sum $\mathbf{s}_{t-1}$ and $\mathbf{h}_i$. In both cases, the alignment model can be represented as a simple **feedforward network (FFN)** with one hidden layer.

Now that we know the formulas for $\mathbf{c}_t$ and $\alpha_{t,i}$, let's replace the latter with the former:

$$\mathbf{c}_t = \sum_{i=1}^{T} \alpha_{t,i} \mathbf{h}_i = \sum_{i=1}^{T} \frac{exp(e_{t,i})}{\sum_{j=1}^{T} exp(e_{t,j})} \mathbf{h}_i$$

As a conclusion, let's summarize the attention algorithm in a step-by-step manner as follows:

1. Feed the encoder with the input sequence and compute the set of hidden states, $H = \{\mathbf{h}_1, \mathbf{h}_2 ... \mathbf{h}_T\}$.

2. Compute the alignment scores, $e_{t,i} = a(\mathbf{s}_{t-1}, \mathbf{h}_i)$, that use the decoder state from the preceding step $\mathbf{s}_{t-1}$. If $t=1$, we'll use the last encoder state, $\mathbf{h}_T$, as the initial hidden state.

3. Compute the weights $\alpha_{t,i} = \text{softmax}(e_{t,i}/e_t)$.

4. Compute the context vector $\mathbf{c}_t = \sum_{i=1}^{T} \alpha_{t,i} \mathbf{h}_i$.

5. Compute the hidden state, $s_t = RNN_{decoder}([s_{t-1}; c_t], y_{t-1})$, based on the concatenated vectors s_{t-1} and c_t and the previous decoder output y_{t-1}. At this point, we can compute the final output y_t. In the case where we need to classify the next word, we'll use the softmax output, $y_t = $ softmax $(W_y s_t)$, where W_y is a weight matrix.

6. Repeat *steps 2* to *5* until the end of the sequence.

Next, we'll discuss a slightly improved version of Bahdanau attention.

Luong attention

Luong attention (*Effective Approaches to Attention-based Neural Machine Translation*, https:// arxiv.org/abs/1508.04025) introduces several improvements over Bahdanau attention. Most notably, the alignment scores depend on the decoder's hidden state s_t, as opposed to s_{t-1} in Bahdanau attention. To better understand this, let's compare the two algorithms:

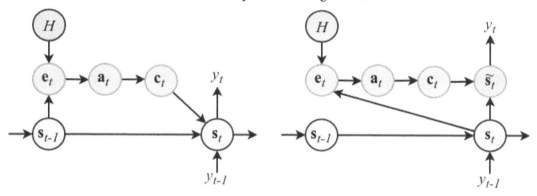

Figure 7.3 – Left: Bahdanau attention; right: Luong attention

We'll go through a step-by-step execution of Luong attention:

1. Feed the encoder with the input sequence and compute the set of encoder hidden states $H = \{h_1, h_2 ... h_T\}$.

2. $s_t = RNN_{decoder}(s_{t-1}, y_{t-1})$: Compute the decoder's hidden state based on the previous decoder's hidden state s_{t-1} and the previous decoder's output y_{t-1} (not the context vector, though).

3. $e_{t,i} = a(s_t, h_i)$: Compute the alignment scores, which use the decoder state from the current step, s_t. Besides additive attention, the Luong attention paper also proposes two types of **multiplicative attention**:

    ▪ $e_{t,i} = s_t^T h_i$: **Dot product** without any parameters. In this case, the vectors s and h (represented as column and row matrices) need to have the same sizes.

    ▪ $e_{t,i} = s_t^T W_m h_i$: Here, W_m is a trainable weight matrix of the attention layer.

The multiplication of the vectors as an alignment score measurement has an intuitive explanation—as we mentioned in *Chapter 2*, the dot product acts as a similarity measure between vectors. Therefore, if the vectors are similar (that is, aligned), the result of the multiplication will be a large value and the attention will be focused on the current *t,i* relationship.

4. $\alpha_{t,i} = \text{softmax}(e_{t,i}/e_t)$: Compute the weights.

5. $\mathbf{c}_t = \sum_{i=1}^{T}\alpha_{t,i}\mathbf{h}_i$: Compute the context vector.

6. $\widetilde{\mathbf{s}}_t = \text{tanh}(\mathbf{W}_c[\mathbf{c}_t;\mathbf{s}_t])$: Compute the intermediate vector based on the concatenated vectors $\mathbf{c}_t$ and $\mathbf{s}_t$. At this point, we can compute the final output y_t. In the case of classification, we'll use softmax, $\mathbf{y}_t = \text{softmax}(\mathbf{W}_y\widetilde{\mathbf{s}}_t)$, where $\mathbf{W}_y$ is a weight matrix.

7. Repeat *steps 2* to *6* until the end of the sequence.

Next, we'll use Bahdanau and Luong attention as a stepping stone to a generic attention mechanism.

General attention

Although we've discussed the attention mechanism in the context of seq2seq with RNNs, it is a general **deep learning** (**DL**) technique in its own right. To understand it, let's start with the following diagram:

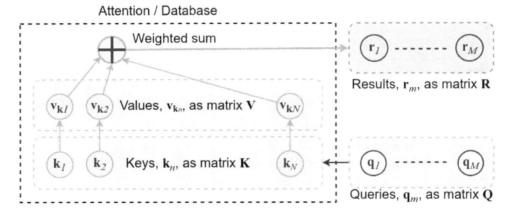

Figure 7.4 – General attention

It starts with a query, $\mathbf{q}$, executed against a database of key-value pairs, $\mathbf{k}$ and $\mathbf{v}_k$, respectively. Each key, $\mathbf{k}_n$, has a single corresponding value, $\mathbf{v}_k$. The query, keys, and values are vectors. Because of this, we can represent the key-value store as two matrices, $\mathbf{K}$ and $\mathbf{V}$. If we have multiple queries, $\mathbf{q}_m$, we can also represent them as a matrix, $\mathbf{Q}$. Hence, these are often abbreviated as $\mathbf{Q}$, $\mathbf{K}$, and $\mathbf{V}$.

> **Differences between general attention and Bahdanau/Luong attention**
>
> Unlike general attention, the keys **K** and the values **V** of Bahdanau and Luong attention are the same thing—that is, these attention models are more like **Q/V**, rather than **Q/K/V**. Having separate keys and values provides more flexibility to the general attention—the keys specialize in matching the input queries, and the values carry the actual information. We can think of the Bahdanau vector s_{t-1} (or s_t in Luong attention) as the query, q_m, executed against the database of key-value pairs, where the keys/values are the hidden states h_i.

General attention uses a multiplicative, rather than additive, mechanism (like Luong attention). Here's how it works:

1. The starting point is one of the input query vectors, q_m.

2. $e_{q_m,k_n} = q_m^\top k_n$: Compute the alignment scores, using the dot product between the query, q_m, and each key vector, k_n. As we mentioned in the *Bahdanau attention* section, the dot product acts as a similarity measure, and it makes sense to use it on this occasion.

3. $\alpha_{q_m,k_n} = \frac{exp(e_{q_m,k_n})}{\sum_{j=1}^{N} exp(e_{q_m,k_j})}$: Compute the final weights of each value vector against the query with the help of softmax.

4. The final attention vector is the weighted addition (that is, an element-wise sum) of all value vectors, v_{k_i}:

$$\text{Attention}(q_m, Q, V) = \sum_{n=1}^{N} \alpha_{q_m,k_n} v_{k_n} = \sum_{n=1}^{N} \frac{exp(e_{q_m,k_n})}{\sum_{j=1}^{N} exp(e_{q_m,k_j})} v_{k_i}$$

To better understand the attention mechanism, we'll use the numerical example displayed in the following diagram:

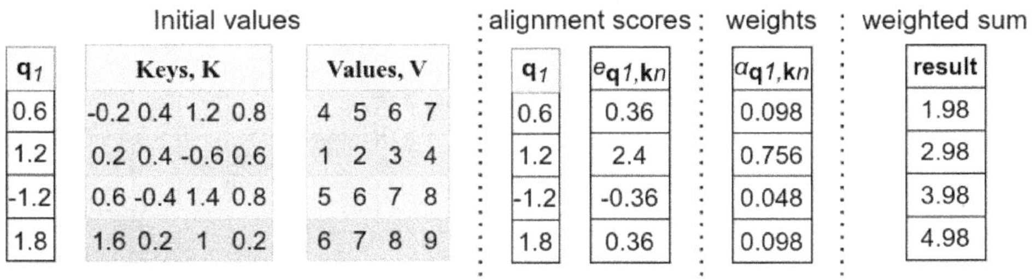

q_1	Keys, K	Values, V	q_1	$e_{q1,kn}$	$\alpha_{q1,kn}$	result
0.6	-0.2 0.4 1.2 0.8	4 5 6 7	0.6	0.36	0.098	1.98
1.2	0.2 0.4 -0.6 0.6	1 2 3 4	1.2	2.4	0.756	2.98
-1.2	0.6 -0.4 1.4 0.8	5 6 7 8	-1.2	-0.36	0.048	3.98
1.8	1.6 0.2 1 0.2	6 7 8 9	1.8	0.36	0.098	4.98

Initial values : alignment scores : weights : weighted sum

Figure 7.5 – An attention example with a four-dimensional query
executed against a key-value store with four vectors

Let's track it step by step:

1. Execute a four-dimensional query vector, $q_1 = [0.6, 1.2, -1.2, 1.8]$, against a key-value store of four four-dimensional vectors.

2. Compute the alignment scores. For example, the first score is e_{q_1,k_1} = 0.6 × (−0.2) + 1.2 × 0.4 + (−1.2) × 1.2 + 1.8 × 0.8 = 0.36. The rest of the scores are displayed in *Figure 7.5*. We have intentionally selected the query, q_1 = [0.6,1.2,−1.2,1.8], to be relatively similar to the second key vector, k_2 = [0.2,0.4,−0.6,0.6]. In this way, k_2 has the largest alignment score, e_{q_1,k_2} = 2.4, and it should have the largest influence over the final result.

3. Compute the weights, α_{q_1,k_2}, with the help of softmax—for example, α_{q_1,k_2} = $exp(2.4)/(exp$ $(0.36) + exp(2.4) + exp(0.36) + exp(0.36))$ = 0.756. The key vector, k_2, has the largest weight because of its large alignment score. The softmax function exaggerates the differences between the inputs, hence the final weight of k_2 is even higher compared to the ratio of the input alignment scores.

4. Compute the final result, r = [1.98,2.98,3.98,4.98], which is the weighted element-wise sum of the value vectors, v_k. For example, we can compute the first element of the result as r_1 = 0.098 × 4 + 0.756 × 1 + 0.048 × 5 + 0.098 × 6 = 1.98. We can see that the values of the result are closest to the value vector, v_{k_2}, which, again, reflects the large alignment between the key vector, k_2, and the input query, q_1.

I hope that this example has helped you understand the attention mechanism, as this is one of the major DL innovations in the past 10 years. Next, we'll discuss an even more advanced attention version.

Transformer attention

In this section, we'll discuss the attention mechanism, as it appears in the transformer NN architecture (*Attention Is All You Need*, https://arxiv.org/abs/1706.03762). Don't worry—you don't need to know about transformers yet, as **transformer attention** (**TA**) is an independent self-sufficient building block of the entire model. It is displayed in the following diagram:

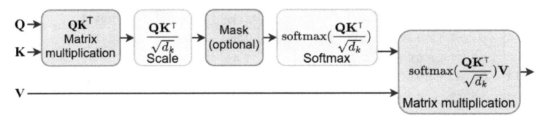

Figure 7.6 – Scaled dot product (multiplicative) TA (inspired by https://arxiv.org/abs/1706.03762)

The TA uses dot product (multiplicative) similarity and follows the general attention procedure we introduced in the *General attention* section (as we have already mentioned, it is not restricted to RNN models). We can define it with the following formula:

$$\text{Attention}(\mathbf{Q}, \mathbf{K}, \mathbf{V}) = \text{softmax}\left(\frac{\mathbf{Q}\mathbf{K}^\top}{\sqrt{d_k}}\right)\mathbf{V}$$

In practice, we'll compute the TA function over a set of queries simultaneously, packed in a matrix **Q** (the keys **K**, the values **V**, and the result are also matrices). Let's discuss the steps of the formula in more detail:

1. Match the queries, **Q**, against the database (keys **K**) with matrix multiplication to produce the alignment scores, $\mathbf{Q}\mathbf{K}^\mathsf{T}$. Matrix multiplication is equivalent to applying dot product similarity between each unique pair of query and key vectors. Let's assume that we want to match m different queries to a database of n values and the query-key vector length is d_k. Then, we have the query matrix, $\mathbf{Q} \in \mathbb{R}^{m \times d_k}$, with one d_k-dimensional query per row for m total rows. Similarly, we have the key matrix, $\mathbf{K} \in \mathbb{R}^{n \times d_k}$, with one d_k-dimensional key vector per row for n total rows (its transpose is $\mathbf{K}^\mathsf{T} \in \mathbb{R}^{d_k \times n}$). Then, the output matrix will be $\mathbf{Q}\mathbf{K}^\mathsf{T} \in \mathbb{R}^{m \times n}$, where one row contains the alignment scores of a single query against all keys of the database:

$$\mathbf{Q}\mathbf{K}^\mathsf{T} = \underbrace{\begin{bmatrix} q_{11} & \cdots & q_{1d_k} \\ \vdots & \ddots & \vdots \\ q_{m1} & \cdots & q_{md_k} \end{bmatrix}}_{Q} \cdot \underbrace{\begin{bmatrix} k_{11} & \cdots & k_{1n} \\ \vdots & \ddots & \vdots \\ k_{d_k1} & \cdots & k_{d_kn} \end{bmatrix}}_{K^\mathsf{T}} = \underbrace{\begin{bmatrix} e_{11} & \cdots & e_{1n} \\ \vdots & \ddots & \vdots \\ e_{m1} & \cdots & e_{mn} \end{bmatrix}}_{QK^\mathsf{T}}$$

    In other words, we can match multiple queries against multiple database keys in a single matrix-matrix multiplication. For example, in the context of translation, we can compute the alignment scores of all words of the target sentence over all words of the source sentence in the same way.

2. Scale the alignment scores with $1/\sqrt{d_k}$, where d_k is the same vector size as the key vectors in the matrix **K**, which is also equal to the size of the query vectors in **Q** (analogously, d_v is the vector size of the value vectors **V**). The authors of the paper suspect that for large values of d_k, the dot product grows large in magnitude and pushes the softmax in regions with extremely small gradients. This, in turn, leads to the vanishing gradients problem, hence the need to scale the results.

3. Compute the attention scores with the softmax operation along the rows of the matrix (we'll talk about the **mask** operation later):

$$\text{softmax}\left(\frac{\mathbf{Q}\mathbf{K}^\mathsf{T}}{\sqrt{d_k}}\right) = \begin{bmatrix} \text{softmax}\left(\left[e_{11}/\sqrt{d_k}, \quad e_{12}/\sqrt{d_k} \quad \cdots \quad e_{1n}/\sqrt{d_k}\right]\right) \\ \text{softmax}\left(\left[e_{21}/\sqrt{d_k}, \quad e_{22}/\sqrt{d_k} \quad \quad e_{2n}/\sqrt{d_k}\right]\right) \\ \vdots \quad \quad \ddots \quad \vdots \\ \text{softmax}\left(\left[e_{m1}/\sqrt{d_k}, \quad e_{m2}/\sqrt{d_k} \quad \cdots \quad e_{mn}/\sqrt{d_k}\right]\right) \end{bmatrix}$$

4. Compute the final attention vector by multiplying the attention scores with the values **V**:

$$\text{softmax}\left(\frac{\mathbf{Q}\mathbf{K}^\mathsf{T}}{\sqrt{d_k}}\right)\mathbf{V} = \begin{bmatrix} \text{softmax}\left(\left[e_{11}/\sqrt{d_k}, \quad e_{12}/\sqrt{d_k} \quad \cdots \quad e_{1n}/\sqrt{d_k}\right]\right) \\ \text{softmax}\left(\left[e_{21}/\sqrt{d_k}, \quad e_{22}/\sqrt{d_k} \quad \quad e_{2n}/\sqrt{d_k}\right]\right) \\ \vdots \quad \quad \ddots \quad \vdots \\ \text{softmax}\left(\left[e_{m1}/\sqrt{d_k}, \quad e_{m2}/\sqrt{d_k} \quad \cdots \quad e_{mn}/\sqrt{d_k}\right]\right) \end{bmatrix} \cdot \begin{bmatrix} v_{11} & \cdots & v_{1d_v} \\ \vdots & \ddots & \vdots \\ v_{n1} & \cdots & v_{nd_v} \end{bmatrix} = \mathbf{A} \in \mathbb{R}^{m \times d_v}$$

The full TA uses a collection of such attention blocks and is known as **multi-head attention (MHA)**, as displayed in the following diagram:

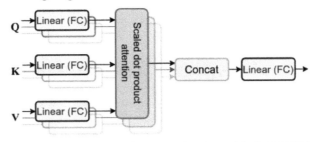

Figure 7.7 – MHA (inspired by https://arxiv.org/abs/1706.03762)

Instead of a single attention function with d_{model}-dimensional keys, we linearly project the keys, queries, and values h times to produce h different d_k-, d_k-, and d_v-dimensional projections of these values. Then, we apply separate parallel attention blocks (or **heads**) over the newly created vectors, which yield a single d_v-dimensional output for each head. Next, we concatenate the head outputs and linearly project them to produce the final attention result.

> **Note**
>
> By linear projection, we mean applying an FC layer. That is, initially we branch the **Q/K/V** matrices with the help of separate FC operations. In the end, we use an FC layer to combine and compress the concatenated head outputs. In this case, we follow the terminology used in the original paper.

MHA allows each head to attend to different elements of the sequence. At the same time, the model combines the outputs of the heads in a single cohesive representation. More precisely, we can define this with the following formula:

$$\text{MultiHead}(\mathbf{Q}, \mathbf{K}, \mathbf{V}) \;=\; \text{Concat}(\mathbf{head}_1, \mathbf{head}_2 \ldots \mathbf{head}_h)\mathbf{W}^o$$

Here, $\mathbf{head}_i \;=\; \text{Attention}(\mathbf{Q}\mathbf{W}_i^Q, \mathbf{K}\mathbf{W}_i^K, \mathbf{V}\mathbf{W}_i^V)$.

Let's look at this in more detail, starting with the heads:

- Each head receives the linearly projected versions of the initial **Q**, **K**, and **V** matrices. The projections are computed with the learnable weight matrices $\mathbf{W}_i^Q$, $\mathbf{W}_i^K$, and $\mathbf{W}_i^V$ respectively (again, the projections are FC layers). Note that we have a separate set of weights for each component (**Q**, **K**, **V**) and for each head, i. To satisfy the transformation from d_{model} to d_k and d_v, the dimensions of these matrices are $\mathbf{W}_i^Q \in \mathbb{R}^{d_{model} \times d_k}$, $\mathbf{W}_i^K \in \mathbb{R}^{d_{model} \times d_k}$, and $\mathbf{W}_i^V \in \mathbb{R}^{d_{model} \times d_v}$.

- Once **Q**, **K**, and **V** are transformed, we can compute the attention of each head using the regular attention block we described at the beginning of this section.

- The final attention result is the linear projection (FC layer with a weight matrix $\mathbf{W}^O$ of learnable weights) over the concatenated head outputs $\mathbf{head}_i$.

So far, we've assumed that the attention works for different input and output sequences. For example, in translation, each word of the translated sentence *attends* to the words of the source sentence. However, there is another valid attention use case. The transformer also relies on **self-attention** (or **intra-attention**), where the queries, **Q**, belong to the same dataset as the keys, **K**, and values, **V**, of the query database. In other words, in self-attention, the source and the target are the same sequence (in our case, the same sentence). The benefit of self-attention is not immediately obvious, as there is no direct task to apply it to. On an intuitive level, it allows us to see the relationship between words of the same sequence. To understand why this is important, let's recall the word2vec model (*Chapter 6*), where we use the context of a word (that is, its surrounding words) to learn an embedding vector of said word. One of the limitations of word2vec is that the embedding is static (or context independent)—we have a single embedding vector for all contexts of the word in the whole training corpus. For example, the word *new* will have the same embedding vector, regardless of whether we use it in the phrase *new shoes* or *New York*. Self-attention allows us to solve this problem by creating a **dynamic embedding** (or **context dependent**) of that word. We won't go into too much detail just yet (we'll do this in the *Building transformers with attention* section), but the dynamic embedding works in the following way: we feed the current word into the attention block, but also its current immediate surrounding (context). The word is the query, **q**, and the context is the **K/V** key-value store. In this way, the self-attention mechanism allows the model to produce a dynamic embedding vector, unique to the current context of the word. This vector serves as an input for a variety of downstream tasks. Its purpose is similar to the static word2vec embedding, but it is much more expressive and makes it possible to solve more complex tasks with greater accuracy.

We can illustrate how self-attention works with the following diagram, which shows the multi-head self-attention of the word *economy* (different colors represent different attention heads):

Figure 7.8 – Multi-head self-attention of the word "economy"
(generated by https://github.com/jessevig/bertviz)

We can see that the strongest link to *economy* comes from the word *market*, which makes sense because the two words form a phrase with a unique meaning. However, we can also see that different heads attend to different, further, parts of the input sequence.

As a conclusion to this section, let's outline the advantages of the attention mechanism compared to the way RNNs process sequences:

- **Direct access to the elements of the sequence**: An RNN encodes the information of the input elements in a single hidden (thought vector). In theory, it represents a distilled version of all sequence elements so far. In practice, it has limited representational power—it can only preserve meaningful information for a sequence with a maximum length of around 100 tokens before the newest tokens start erasing the information of the older ones.

  In contrast, the attention mechanism provides direct access to all input sequence elements. On one hand, this imposes a strict limit on the maximum sequence length. On the other hand, it makes it possible, as of the time of writing book, to have transformer-based LLMs, which can process sequences of more than 32,000 tokens.

- **Parallel processing of the input sequence**: An RNN processes the input sequence elements one by one, in the order of their arrival. Therefore, we cannot parallelize RNNs. Compare this with the attention mechanism—it consists exclusively of matrix multiplication operations, which are **embarrassingly parallel**. This makes it possible to train LLMs with billions of trainable parameters over large training datasets.

But these advantages come with one disadvantage—where an RNN preserves the order of the sequence elements, the attention mechanism, with its direct access, does not. However, we'll introduce a workaround to that limitation in the *Transformer encoder* section.

This concludes our theoretical introduction to TA. Next, let's implement it.

Implementing TA

In this section, we'll implement MHA, following the definitions from the *Transformer attention* section. The code in this section is part of the larger transformer implementation, which we'll discuss throughout the chapter. We won't include the full source code, but you can find it in the book's GitHub repo.

Note

This example is based on `https://github.com/harvardnlp/annotated-transformer`. Let's also note that PyTorch has native transformer modules (the documentation is available at `https://pytorch.org/docs/stable/generated/torch.nn.Transformer.html`). Still, in this section, we'll implement TA from scratch to understand it better.

We'll start with the implementation of regular scaled dot product attention. As a reminder, it implements the formula $\text{Attention}(\mathbf{Q}, \mathbf{K}, \mathbf{V}) = \text{softmax}\left(\mathbf{Q}\mathbf{K}^{\top}/\sqrt{d_k}\right)\mathbf{V}$, where $\mathbf{Q}$ = query, $\mathbf{K}$ = key, and $\mathbf{V}$ = value:

```python
def attention(query, key, value, mask=None, dropout=None):
    d_k = query.size(-1)
    # 1) and 2) Compute the alignment scores with scaling
    scores = (query @ key.transpose(-2, -1)) \
                / math.sqrt(d_k)
    if mask is not None:
        scores = scores.masked_fill(mask == 0, -1e9)
    # 3) Compute the attention scores (softmax)
    p_attn = scores.softmax(dim=-1)
    if dropout is not None:
        p_attn = dropout(p_attn)
    # 4) Apply the attention scores over the values
    return p_attn @ value, p_attn
```

The `attention` function includes `dropout`, as it is part of the full transformer implementation. Once again, we'll leave the `mask` parameter and its purpose for later. Let's also note a novel detail—the use of the `@` operator (`query @ key.transpose(-2, -1)` and `p_attn @ value`), which, as of Python 3.5, is reserved for matrix multiplication.

Next, let's continue with the MHA implementation. As a reminder, the implementation follows the formula: $\text{MultiHead}(\mathbf{Q}, \mathbf{K}, \mathbf{V}) = \text{Concat}(\mathbf{head}_1, \mathbf{head}_2 \ldots \mathbf{head}_h)\mathbf{W}^o$. Here, $\mathbf{head}_i = \text{Attention}(\mathbf{Q}\mathbf{W}_i^Q, \mathbf{K}\mathbf{W}_i^K, \mathbf{V}\mathbf{W}_i^V)$.

We'll implement it as a subclass of `torch.nn.Module`, called `MultiHeadedAttention`. We'll start with the constructor:

```python
class MultiHeadedAttention(torch.nn.Module):
    def __init__(self, h, d_model, dropout=0.1):
        """
        :param h: number of heads
        :param d_model: query/key/value vector length
        """
        super(MultiHeadedAttention, self).__init__()
        assert d_model % h == 0
        # We assume d_v always equals d_k
        self.d_k = d_model // h
        self.h = h
        # Create 4 fully connected layers
        # 3 for the query/key/value projections
        # 1 to concatenate the outputs of all heads
        self.fc_layers = clones(
            torch.nn.Linear(d_model, d_model), 4)
```

```
        self.attn = None
        self.dropout = torch.nn.Dropout(p=dropout)
```

Note that we use the `clones` function (implemented on GitHub) to create four identical FC `self.fc_layers` instances. We'll use three of them for the **Q/K/V** multi-head linear projections—$\mathbf{W}_i^Q$, $\mathbf{W}_i^K$, and $\mathbf{W}_i^V$. The fourth FC layer is to merge the concatenated results of the outputs of the different heads, $\mathbf{W}^O$. We'll store the current attention results in the `self.attn` property.

Next, let's implement the `MultiHeadedAttention.forward` method. Please bear in mind that the declaration should be indented, as it is a property of the `MultiHeadedAttention` class:

```
def forward(self, query, key, value, mask=None):
    if mask is not None:
        # Same mask applied to all h heads.
        mask = mask.unsqueeze(1)

    batch_samples = query.size(0)
    # 1) Do all the linear projections in batch from d_model => h x
d_k
    projections = [
        l(x).view(batch_samples, -1, self.h, self.d_k)
        .transpose(1, 2)
        for l, x in zip(self.fc_layers, (query, key, value))
    ]
    query, key, value = projections
    # 2) Apply attention on all the projected vectors in batch.
    x, self.attn = attention(
        query, key, value,
        mask=mask,
        dropout=self.dropout)
    # 3) "Concat" using a view and apply a final linear.
    x = x.transpose(1, 2).contiguous() \
        .view(batch_samples, -1, self.h * self.d_k)
    return self.fc_layers[-1](x)
```

We iterate over the `query`/`key`/`value` tensors and their reference projection, `self.fc_layers`, and produce `query`/`key`/`value` projections with the following snippet:

```
l(x).view(batch_samples, -1, self.h, self.d_k).transpose(1, 2)
```

Then, we apply regular attention over the projections using the attention function we first defined. Next, we concatenate the outputs of the multiple heads, and finally, we feed them to the last FC layer (`self.fc_layers[-1]`) and return the results.

Now that we've discussed the TA, let's continue with the transformer model itself.

Building transformers with attention

We've spent the better part of this chapter touting the advantages of the attention mechanism. It's time to reveal the full **transformer** architecture, which, unlike RNNs, relies solely on the attention mechanism (*Attention Is All You Need*, https://arxiv.org/abs/1706.03762). The following diagram shows two of the most popular transformer flavors, **post-ln** and **pre-ln** (or **post-normalization** and **pre-normalization**):

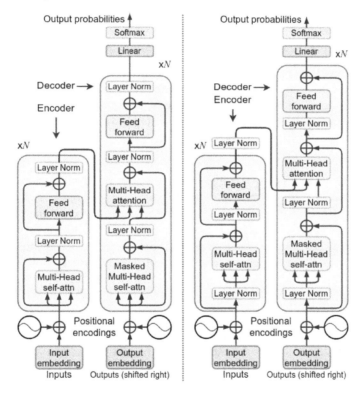

Figure 7.9 – Left: the original (post-normalization, post-ln) transformer; right: pre-normalization (pre-ln) transformer (inspired by https://arxiv.org/abs/1706.03762)

It looks scary, but fret not—it's easier than it seems. In this section, we'll discuss the transformer in the context of the seq2seq task, which we defined in the *Introducing seq2seq models* section. That is, it will take a sequence of tokens as input, and it will output another, different, token sequence. As with the seq2seq model, it has two components—an **encoder** and a **decoder**. We'll start with the encoder (the left-hand component of both sections of the preceding diagram).

Transformer encoder

The encoder begins with an input sequence of one-hot-encoded tokens. The most popular tokenization algorithms are **byte-pair encoding** (**BPE**), WordPiece, and Unigram (*Chapter 6*). The tokens are transformed into d_{model}-dimensional embedding vectors. The transformation works in the way we described in *Chapter 6*. We have a lookup table (matrix)—the index of the one-hot-encoded token indicates the matrix row, which represents the embedding vector. The embedding vectors are further multiplied by $\sqrt{d_{model}}$. They are initialized randomly and are trained with the whole model (this is opposed to initializing them with an algorithm such as word2vec).

The next step adds positional information to the existing embedding vector. This is necessary because the attention mechanism doesn't preserve the order of sequence elements. This step modifies the embedding vectors in a way that implicitly encodes that information within them.

The original transformer implementation uses **static positional encoding**, represented by special positional encoding vectors with the same size as the token embeddings. We add these vectors, using element-wise addition, to all embedding vectors of the sequence, depending on their position. The static encoding is unique for each position of the sequence but is constant with regard to the elements of the sequence. Because of this, we can precompute the positional encodings only once and use them subsequently.

An alternative way to encode positional information is with relative position representations (*Self-Attention with Relative Position Representations*, https://arxiv.org/abs/1803.02155). Here, the positional information is dynamically encoded in the key-value matrices, **K/V**, of the attention blocks. Each element of the input sequence has a different position in relation to the rest of the elements. Therefore, the relative position encoding is computed dynamically for each token. This encoding is applied to the **K** and **V** matrices as an additional part of the attention formula.

The rest of the encoder is composed of a stack of *N=6* identical blocks that come in two flavors: post-ln and pre-ln. Both types of blocks share the following sublayers:

- A multi-head self-attention mechanism, like the one we described in the *Transformer attention* section. Since the self-attention mechanism works across the whole input sequence, the encoder is **bidirectional** by design. That is, the context of the current token includes both the tokens that come before and the ones that come after it in the sequence. This is opposed to a regular RNN, which only has access to the tokens that came before the current one. Each position in an encoder block can attend to all positions in the previous encoder block.

  We feed the embedding of each token as a query, **q**, to the multi-head self-attention (we can feed the full input sequence in one pass as an input matrix, **Q**). At the same time, the embeddings of its context act as the key-value store **K/V**. The output vector of the multi-head self-attention operation serves as input for the rest of the model.

> **MHA and activation functions**
>
> The MHA produces h attention vectors for each of the h attention heads for each input token. Then, they are linearly projected with an FC layer that combines them. The whole attention block doesn't have an explicit activation function. But let's recall that the attention block ends with a non-linearity, softmax. The dot product of the key-value vectors is an additional non-linearity. In that strict sense, the attention block doesn't need additional activation.

- A simple FC FFN, which is defined by the following formula:

$$FFN(x) \;=\; ActivationFunc\big(\mathbf{W}_1 \mathbf{x} + b_1\big)\,\mathbf{W}_2 + b_2$$

  The network is applied to each sequence element, $\mathbf{x}$, separately. It uses the same set of parameters ($\mathbf{W}_1$, $\mathbf{W}_2$, b_1, and b_2) across different positions, but different parameters across the different encoder blocks. The original transformer uses **rectified linear unit** (**ReLU**) activations. However, more recent models use one of its variations, such as **sigmoid linear units** (**SiLUs**). The role of the FFN is to process the MHA output in a way that better fits the input for the next block.

The difference between pre-ln and post-ln blocks lies in the position of the normalization layer. Each post-ln sublayer (both the MHA and FFN) has a residual connection around itself and ends with normalization and dropout over the sum of that connection and its own output. The normalization layers in the post-ln transformer lie after the attention and the FFN, respectively. Therefore, the output of each post-ln sublayer is as follows:

$$LayerNorm\big(\mathbf{x} + SubLayer(\mathbf{x})\big)$$

In contrast, the pre-ln blocks (the right section of *Figure 7.9*), in the two encoder normalization layers lie before the attention and the FFN, respectively. Therefore, the output of each pre-ln sublayer is this:

$$\mathbf{x} + SubLayer(LayerNorm(\mathbf{x}))$$

The difference between the two flavors manifests itself during training. Without going into too many details, the aptly named paper *Understanding the Difficulty of Training Transformers* (https://arxiv.org/abs/2004.08249) suggests that the post-ln transformer's strong dependency on residual connections amplifies the fluctuation caused by parameter changes (for example, adaptive learning rate) and destabilizes the training. Because of this, the post-ln training starts with a warmup phase with a low learning rate, before ultimately increasing it. This is opposed to the usual learning rate schedule that starts with a large value, which only decreases with the progression of the training. The pre-ln blocks don't have such a problem and don't need a warmup phase. However, they could suffer from **representation collapse**, where the hidden representation in deeper blocks (those closer to the end of the NN) will be similar and thus contribute little to model capacity. In practice, both types of blocks are in use.

So far, so good with the encoder. Next, let's build upon our attention implementation by building the encoder as well.

Implementing the encoder

In this section, we'll implement the post-ln encoder, which is composed of several different submodules. Let's start with the main class, `Encoder`:

```
class Encoder(torch.nn.Module):
    def __init__(self, block: EncoderBlock, N: int):
        super(Encoder, self).__init__()
        self.blocks = clones(block, N)
        self.norm = torch.nn.LayerNorm(block.size)

    def forward(self, x, mask):
        """Iterate over all blocks and normalize"""
        for layer in self.blocks:
            x = layer(x, mask)

        return self.norm(x)
```

It stacks N instances of EncoderBlock (`self.blocks`), followed by a LayerNorm normalization, `self.norm`. Each instance serves as input to the next, as the definition of the `forward` method shows. In addition to the regular input, `x`, `forward` also takes as input a `mask` parameter. However, it is only relevant to the decoder part, so we won't focus on it here.

Next, let's see the implementation of the `EncoderBlock` class:

```
class EncoderBlock(torch.nn.Module):
    def __init__(self,
                 size: int,
                 self_attn: MultiHeadedAttention,
                 ffn: PositionwiseFFN,
                 dropout=0.1):
        super(EncoderBlock, self).__init__()
        self.self_attn = self_attn
        self.ffn = ffn
        # Create 2 sub-layer connections
        # 1 for the self-attention
        # 1 for the FFN
        self.sublayers = clones(SublayerConnection(size, dropout), 2)
        self.size = size
    def forward(self, x, mask):
        x = self.sublayers[0](x, lambda x: self.self_attn(x, x, x,
mask))
        return self.sublayers[1](x, self.ffn)
```

Each encoder block consists of multi-head self-attention (`self.self_attn`) and FFN (`self.ffn`) sublayers (`self.sublayers`). Each sublayer is wrapped by its residual connection, **layer normalization** (**LN**), and dropout, implemented by the `SublayerConnection` class and instantiated with the familiar `clone` function:

```
class SublayerConnection(torch.nn.Module):
    def __init__(self, size, dropout):
        super(SublayerConnection, self).__init__()
        self.norm = torch.nn.LayerNorm(size)
        self.dropout = torch.nn.Dropout(dropout)
    def forward(self, x, sublayer):
        return x + self.dropout(sublayer(self.norm(x)))
```

The `SublayerConnection.forward` method takes as input the data tensor, `x`, and `sublayer`, which is an instance of either `MultiHeadedAttention` or `PositionwiseFFN` (it matches the sublayer definition LayerNorm(x + SubLayer(x)) from the *Transformer encoder* section).

The only component we haven't defined yet is `PositionwiseFFN`, which implements the formula $\text{FFN}(x) = \text{ActivationFunc}(\mathbf{W}_1 \mathbf{x} + b_1) \mathbf{W}_2 + b_2$. We'll use SiLU activation. Let's add this missing piece:

```
class PositionwiseFFN(torch.nn.Module):
    def __init__(self, d_model: int, d_ff: int, dropout=0.1):
        super(PositionwiseFFN, self).__init__()
        self.w_1 = torch.nn.Linear(d_model, d_ff)
        self.w_2 = torch.nn.Linear(d_ff, d_model)
        self.dropout = torch.nn.Dropout(dropout)
    def forward(self, x):
        return self.w_2(
            self.dropout(
                torch.nn.functional.silu(
                    self.w_1(x)
        )))
```

This concludes our implementation of the encoder. Next, let's focus our attention on the decoder.

Transformer decoder

The decoder generates the output sequence, based on a combination of the encoder output and its own previously generated sequence of tokens (we can see the decoder on the right side of both sections in *Figure 7.9* at the beginning of the *Building transformers with attention* section). In the context of a seq2seq task, the full encoder-decoder transformer is an autoregressive model. First, we feed the initial sequence—for example, a sentence to translate or a question to answer—to the encoder. This can happen in a single pass, if the sequence is short enough to fit the maximum size of the query matrix, $\mathbf{Q}$. Once the encoder processes all sequence elements, the decoder will take the encoder output and start generating the output sequence one token at a time. It will append each generated token to the initial input sequence. We'll feed the new, extended sequence to the encoder once again. The new output of the encoder will initiate the next token generation step of the decoder, and so on. In effect, the target token sequence is the same as the input token sequence, shifted by one (similar to the seq2seq decoder).

The decoder uses the same embedding vectors and positional encoding as the encoder. It continues with a stack of $N=6$ identical decoder blocks. Each block consists of three sublayers and each sublayer employs residual connections, dropout, and normalization. As with the encoder, the blocks come in post-ln and pre-ln flavors. The sublayers are as follows:

- A masked multi-head self-attention mechanism. The encoder's self-attention is bidirectional—it can attend to all elements of the sequence, regardless of whether they come before or after the current element. However, the decoder only has a partially generated target sequence. Therefore, the decoder is **unidirectional**—the self-attention can only attend to the preceding sequence elements. During inference, we have no choice but to run the transformer in a sequential way so that it can produce each token of the output sequence one by one. However, during training, we can feed the whole target sequence simultaneously, as it's known in advance. To avoid illegal forward attention, we can **mask out** illegal connections by setting $-\infty$ on all such values in the input of the attention softmax. We can see the mask component in *Figure 7.6* of the *Transformer attention* section and the result of the mask operation here:

$$\text{mask}(\mathbf{Q}\,\mathbf{K}^{\mathsf{T}}) \;=\; \text{mask}\!\left(\begin{bmatrix} e_{11} & \cdots & e_{1n} \\ \vdots & \ddots & \vdots \\ e_{m1} & \cdots & e_{mn} \end{bmatrix} \right) \;=\; \begin{bmatrix} e_{11} & -\infty & -\infty & & -\infty \\ e_{21} & e_{22} & -\infty & \cdots & -\infty \\ e_{31} & e_{32} & e_{33} & & -\infty \\ & \vdots & & \ddots & \vdots \\ e_{m1} & e_{m2} & e_{m3} & \cdots & e_{mn} \end{bmatrix}$$

- A regular attention (not self-attention) mechanism, where the queries come from the previous decoder layer, and the keys and values come from the encoder output. This allows every position in the decoder to attend all positions in the original input sequence. This mimics the typical encoder-decoder attention mechanisms, which we discussed in the *Introducing seq2seq models* section.

- FFN, which is similar to the one in the encoder.

The decoder ends with an FC layer, followed by a softmax operation, which produces the most probable next word of the sentence.

We can train the full encoder-decoder model using the teacher-forcing process we defined in the *Introducing seq2seq models* section.

Next, let's implement the decoder.

Implementing the decoder

In this section, we'll implement the decoder in a similar pattern to the encoder. We'll start with the implementation of the main module, `Decoder`:

```
class Decoder(torch.nn.Module):
    def __init__(self, block: DecoderBlock, N: int, vocab_size: int):
        super(Decoder, self).__init__()
        self.blocks = clones(block, N)
        self.norm = torch.nn.LayerNorm(block.size)
        self.projection = torch.nn.Linear(block.size, vocab_size)
    def forward(self, x, encoder_states, source_mask, target_mask):
        for layer in self.blocks:
            x = layer(x, encoder_states, source_mask, target_mask)
        x = self.norm(x)
        return torch.nn.functional.log_softmax(self.projection(x),
 dim=-1)
```

It consists of N instances of `DecoderBlock` (`self.blocks`). As we can see in the `forward` method, the output of each `DecoderBlock` instance serves as input to the next. These are followed by the `self.norm` normalization (an instance of `LayerNorm`). The decoder ends with an FC layer (`self.projection`), followed by a softmax to produce the most probable next word. Note that the `Decoder.forward` method takes an additional parameter, `encoder_states`, which is passed to the `DecoderBlock` instances. `encoder_states` represents the encoder output and is the link between the encoder and the decoder. In addition, the `source_mask` parameter provides the mask of the decoder self-attention.

Next, let's implement the `DecoderBlock` class:

```
class DecoderBlock(torch.nn.Module):
    def __init__(self,
                 size: int,
                 self_attn: MultiHeadedAttention,
                 encoder_attn: MultiHeadedAttention,
                 ffn: PositionwiseFFN,
                 dropout=0.1):
        super(DecoderBlock, self).__init__()
        self.size = size
        self.self_attn = self_attn
        self.encoder_attn = encoder_attn
        self.ffn = ffn
        self.sublayers = clones(SublayerConnection(size,
                dropout), 3)
    def forward(self, x, encoder_states, source_mask, target_mask):
        x = self.sublayers[0](x, lambda x: \
```

```
                    self.self_attn(x, x, x, target_mask))
        x = self.sublayers[1](x, lambda x: \
                    self.encoder_attn(x, encoder_states,
                    encoder_states, source_mask))
        return self.sublayers[2](x, self.ffn)
```

This implementation follows the `EncoderBlock` pattern but is adapted to the decoder: in addition to self-attention (`self_attn`), we also have encoder attention (`encoder_attn`). Because of this, we instantiate three `sublayers` instances (instances of the familiar `SublayerConnection` class): for self-attention, encoder attention, and the FFN.

We can see the combination of multiple attention mechanisms in the `DecoderBlock.forward` method. `encoder_attn` takes as a query the output of the preceding decoder block (`x`) and key-value combination from the encoder output (`encoder_states`). In this way, regular attention establishes the link between the encoder and the decoder. On the other hand, `self_attn` uses `x` for the query, key, and value.

This concludes the decoder implementation. We'll proceed with building the full transformer model in the next section.

Putting it all together

We now have implementations of the encoder and the decoder. Let's combine them in the full `EncoderDecoder` class:

```
class EncoderDecoder(torch.nn.Module):
    def __init__(self,
                    encoder: Encoder,
                    decoder: Decoder,
                    source_embeddings: torch.nn.Sequential,
                    target_embeddings: torch.nn.Sequential):
        super(EncoderDecoder, self).__init__()
        self.encoder = encoder
        self.decoder = decoder
        self.source_embeddings = source_embeddings
        self.target_embeddings = target_embeddings

    def forward(self, source, target, source_mask, target_mask):
        encoder_output = self.encoder(
            x=self.source_embeddings(source),
            mask=source_mask)
        return self.decoder(
            x=self.target_embeddings(target),
            encoder_states=encoder_output,
```

```
        source_mask=source_mask,
        target_mask=target_mask)
```

It combines encoder, decoder, source_embeddings, and target_embeddings. The forward method takes the source sequence and feeds it to encoder. Then, decoder takes its input from the preceding output step (x=self.target_embeddings(target)), the encoder states (encoder_states=encoder_output), and the source and target masks. With these inputs, it produces the predicted next token (word) of the sequence, which is also the return value of the forward method.

Next, we'll implement the build_model function, which instantiates all the classes we implemented to produce a single transformer instance:

```
def build_model(source_vocabulary: int,
                target_vocabulary: int,
                N=6, d_model=512, d_ff=2048, h=8, dropout=0.1):
    c = copy.deepcopy
    attn = MultiHeadedAttention(h, d_model)
    ff = PositionwiseFFN(d_model, d_ff, dropout)
    position = PositionalEncoding(d_model, dropout)
    model = EncoderDecoder(
      encoder=Encoder(
        EncoderBlock(d_model, c(attn), c(ff), dropout), N),
        decoder=Decoder(
          DecoderBlock(d_model, c(attn), c(attn),
                  c(ff), dropout), N, target_vocabulary),
        source_embeddings=torch.nn.Sequential(
          Embeddings(d_model, source_vocabulary), c(position)),
        target_embeddings=torch.nn.Sequential(
          Embeddings(d_model, target_vocabulary), c(position)))

    # Initialize parameters with random weights
    for p in model.parameters():
        if p.dim() > 1:
            torch.nn.init.xavier_uniform_(p)

    return model
```

Besides the familiar MultiHeadedAttention and PositionwiseFFN, we also create a position variable (an instance of the PositionalEncoding class). This class implements the static positional encoding we described in the *Transformer encoder* section (we won't include the full implementation here).

Now, let's focus on the `EncoderDecoder` instantiation: we are already familiar with `encoder` and `decoder`, so there are no surprises there. But the embeddings are a tad more interesting. The following code instantiates the source embeddings (but this is also valid for the target ones):

```
source_embeddings=torch.nn.Sequential(Embeddings(d_model, source_
vocabulary), c(position))
```

We can see that they are a sequential list of two components:

- An instance of the `Embeddings` class, which is simply a combination of `torch.nn.Embedding` further multiplied by $\sqrt{d_{model}}$ (we'll omit the class definition here)

- Positional encoding `c(position)`, which adds the static positional data to the embedding vector

Once we have the input data preprocessed in this way, it can serve as input to the core part of the encoder-decoder.

In the next section, we'll discuss the major variants of the transformer architecture.

Decoder-only and encoder-only models

So far, we've discussed the full encoder-decoder variant of the transformer architecture. But in practice, we are going to mostly use two of its variations:

- **Encoder-only**: These models use only the encoder part of the full transformer. Encoder-only models are bidirectional, following the properties of encoder self-attention.

- **Decoder-only**: These models use only the decoder part of the transformer. Decoder-only models are unidirectional, following the properties of the decoder's masked self-attention.

I know that these dry definitions sound vague, but don't worry—in the next two sections we'll discuss one example of each type to make it clear.

Bidirectional Encoder Representations from Transformers

Bidirectional Encoder Representations from Transformers (**BERT**; see https://arxiv.org/abs/1810.04805), as the name gives away, is an encoder-only (hence bidirectional) model that learns representations. These representations serve as a base for solving various downstream tasks (the pure BERT model doesn't solve any specific problem). The following diagram shows generic pre-ln and post-ln encoder-only models with softmax outputs (which also apply to BERT):

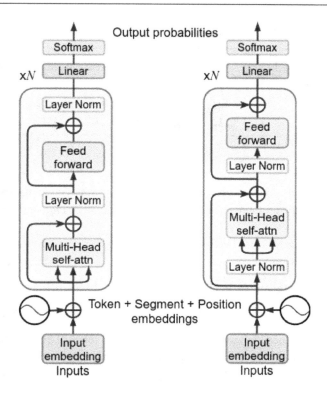

Figure 7.10 – Left: post-ln encoder-only model; right: pre-ln encoder-only model

BERT model sizes

BERT comes in two variations—BERT$_{BASE}$ and BERT$_{LARGE}$. BERT$_{BASE}$ has 12 encoder blocks, each with 12 attention heads, 768-dimensional attention vectors (the d_{model} parameter), and a total of 110M parameters. BERT$_{LARGE}$ has 24 encoder blocks, each with 16 attention heads, 1,024-dimensional attention vectors, and a total of 340M parameters. The models use WordPiece tokenization and have a 30,000-token vocabulary.

Let's start with the way BERT represents its input data, which is an important part of its architecture. We can see an input data representation in the following diagram:

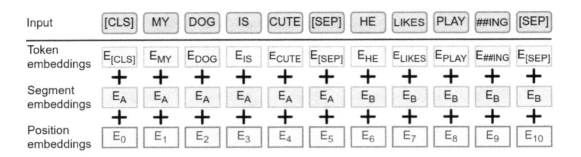

Figure 7.11 – BERT input embeddings as the sum of the token embeddings, the segmentation embeddings, and the position embeddings (source: https://arxiv.org/abs/1810.04805)

Because BERT is encoder-only, it has two special modes of input data representation so that it can handle a variety of downstream tasks:

- A single sequence (for example, in classification tasks, such as **sentiment analysis**, or **SA**)

- A pair of sequences (for example, machine translation or **question-answering** (**QA**) problems)

The first token of every sequence is always a special classification token, [CLS]. The encoder output, corresponding to this token, is used as the aggregate sequence representation for classification tasks. For example, if we want to apply SA over the sequence, the output corresponding to the [CLS] input token will represent the sentiment (positive/negative) output of the model (this example is relevant when the input data is a single sequence). This is necessary because the [CLS] token acts as a query, while all other elements of the input sequence act as the key/value store. In this way, all tokens of the sequence participate in the weighted attention vector, which serves as input to the rest of the model. Selecting another token besides [CLS] excludes this token from the attention formula, which introduces unfair bias against it and results in an incomplete sequence.

If the input data is a pair of sequences, we pack them together in a single sequence, separated by a special [SEP] token. On top of that, we have additional learned segmentation embedding for every token, which indicates whether it belongs to sequence *A* or sequence *B*. Therefore, the input embeddings are the sum of the token embeddings, the segmentation embeddings, and the position embeddings. Here, the token and position embeddings serve the same purpose as they do in the regular transformer.

Now that we are familiar with the input data representation, let's continue with the training.

BERT training

BERT training is a two-step process (this is also valid for other transformer-based models):

- **Pre-training**: Train the model with unlabeled data over different pre-training tasks

- **Fine-tuning**: A form of **transfer learning** (**TL**), where we initialize the model with the pre-trained parameters and fine-tune them over a labeled dataset of a specific downstream task

We can see the pre-training on the left side and the fine-tuning on the right side of the following diagram:

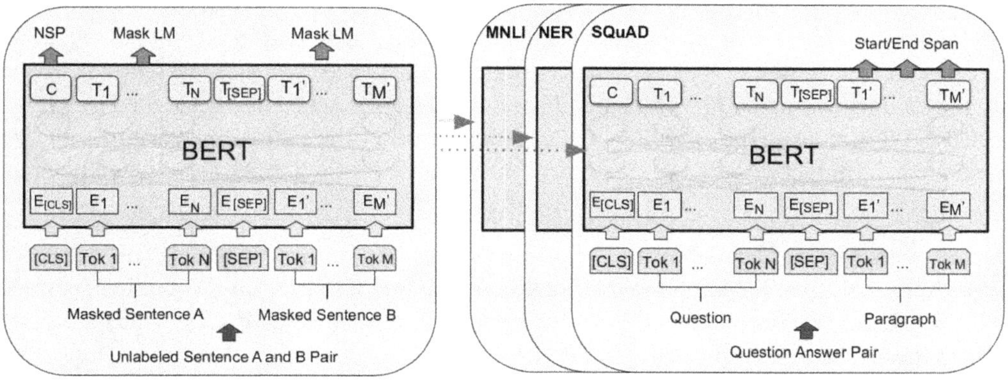

Figure 7.12 – Left: pre-training; right: fine-tuning (source: https://arxiv.org/abs/1810.04805)

Here, **Tok N** represents the one-hot-encoded input tokens, **E** represents the token embeddings, and **T** represents the model output vector. The topmost labels represent the different tasks we can use the model for in each of the training modes.

The authors of the paper pre-trained the model using two unsupervised training tasks: **masked language modeling** (**MLM**) and **next sentence prediction** (**NSP**).

We'll start with MLM, where the model is presented with an input sequence and its goal is to predict a missing word in that sequence. MLM is similar in nature to the **continuous bag-of-words** (**CBOW**) objective of the word2vec model (see *Chapter 6*). To solve this task, the BERT encoder output is extended with an FC layer with softmax activation, which outputs the most probable word given the input sequence. Each input sequence is modified by randomly masking 15% (according to the paper) of the WordPiece tokens. Within these 15%, we can replace the target token with either a special [MASK] token (80% of the time), a random word (10% of the time), or leave the word as is (10% of the time). This is necessary because the vocabulary of the downstream tasks doesn't have the [MASK] token. On the other hand, the pre-trained model might expect it, which could lead to unpredictable behavior.

Next, let's continue with NSP. The authors argue that many important downstream tasks, such as question answering or **natural language inference** (**NLI**), are based on understanding the relationship between two sentences, which is not directly captured by language modeling.

NLI

NLI determines whether a sentence, which represents a **hypothesis**, is either true (**entailment**), false (**contradiction**), or undetermined (**neutral**) given another sentence, called a **premise**. For example, given the premise *I am running*, we have the following hypothesis: *I am sleeping* is false; *I am listening to music* is undetermined; *I am training* is true.

The authors of BERT propose a simple and elegant unsupervised solution to pre-train the model to understand sentence relationships (displayed on the left side of *Figure 7.12*). We'll train the model on binary classification, where each input sample starts with a `[CLS]` token and consists of two sequences (let's use sentences for simplicity), *A* and *B*, separated by a `[SEP]` token. We'll extract sentences *A* and *B* from the training corpus. In 50% of the training samples, *B* is the actual next sentence that follows *A* (labeled as `is_next`). In the other 50%, *B* is a random sentence from the corpus (`not_next`). As we mentioned, the model outputs `is_next`/`not_next` labels on the `[CLS]` corresponding input.

Next, let's focus on the fine-tuning task, which follows the pre-training task (the right side of *Figure 7.12*). The two steps are very similar, but instead of creating a masked sequence, we simply feed the BERT model with the task-specific unmodified input and output and fine-tune all the parameters in an end-to-end fashion. Therefore, the model that we use in the fine-tuning phase is the same model that we'll use in the actual production environment.

Let's continue with some of the downstream tasks we can solve with BERT.

BERT downstream tasks

The following diagram shows how to solve several different types of tasks with BERT:

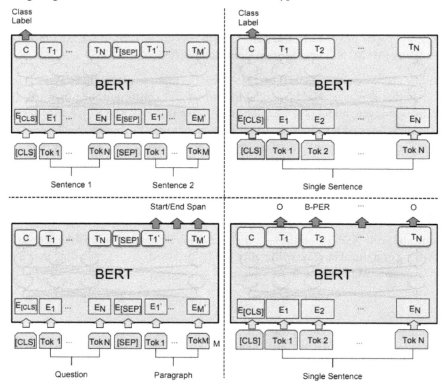

Figure 7.13 – BERT applications for different tasks (source: https://
arxiv.org/abs/1810.04805)

Let's discuss them:

- The top-left scenario illustrates how to use BERT for sentence-pair classification tasks, such as NLI. In short, we feed the model with two concatenated sentences and only look at the [CLS] token output classification, which will output the model result. For example, in an NLI task, the goal is to predict whether the second sentence is an entailment, a contradiction, or neutral with respect to the first one.

- The top-right scenario illustrates how to use BERT for single-sentence classification tasks, such as SA. This is very similar to sentence-pair classification. In both cases, we'll extend the encoder with an FC layer and a binary softmax, with N possible classes (N is the number of classes for each task).

- The bottom-left scenario illustrates how to use BERT on a QA dataset. Given that sequence A is a question and sequence B is a passage from *Wikipedia*, which contains the answer, the goal is to predict the text span (start and end) of the answer within this passage. The model outputs the probability for each token of sequence B to be either the start or the end of the answer.

- The bottom-right scenario illustrates how to use BERT for **named entity recognition** (**NER**), where each input token is classified as some type of entity.

This concludes our section dedicated to the BERT model. Next, let's focus on decoder-only models.

Generative Pre-trained Transformer

In this section, we'll discuss a decoder-only model, known as **Generative Pre-trained Transformer** (**GPT**; see *Improving Language Understanding by Generative Pre-Training*, https://cdn.openai.com/research-covers/language-unsupervised/language_understanding_paper.pdf). This is the first of a series of GPT models, released by OpenAI, which led to the now-famous GPT-3 and GPT-4.

> **GPT model size**
>
> GPT has 12 decoder layers, each with 12 attention heads, and 768-dimensional attention vectors. The FFN is 3,072-dimensional. The model has a total of 117 million parameters (weights). GPT uses BPE tokenization and has a token vocabulary size of 40,000.

We can see the GPT decoder-only architecture in the following diagram:

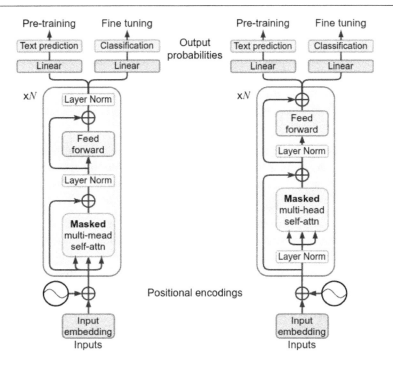

Figure 7.14 – Left: post-ln decoder-only model; right: pre-ln decoder-only model;
different outputs for the pre-training and fine-tuning training steps

> **Note**
>
> We discuss the decoder-only architecture in the context of the original GPT paper, but it applies to the broad class of decoder-only models.

It is derived from the decoder we discussed in the *Transformer decoder* section. The model takes as input token embeddings and adds static positional encoding. This is followed by a stack of N decoder blocks. Each block has two sublayers:

- **Masked multi-head self-attention**: Let's put emphasis on the masked part. It determines the main properties of the decoder-only model—it is unidirectional and autoregressive. This is opposed to bidirectional encoder-only models.

- **FFN**: This sublayer has the same purpose as in the encoder-decoder model.

The sublayers contain the relevant residual links, normalization, and dropout. The decoder comes in pre-ln and post-ln flavors.

The model ends with an FC layer, followed by a softmax operation, which can be adapted to suit the specific task at hand.

The main difference between this decoder and the one in the full encoder-decoder transformer is the lack of an attention sublayer, which links the encoder and decoder blocks in the full model. Since the current architecture doesn't have an encoder part, the sublayer is obsolete. This makes the decoder very similar to the encoder, except for masked self-attention. Hence, the main difference between the encoder-only and decoder-only models is that they are bidirectional and unidirectional, respectively.

As with BERT, the training of GPT is a two-step process, which consists of unsupervised pre-training and supervised fine-tuning. Let's start with the pre-training, which resembles the seq2seq training algorithm (the decoder part) we described in the *Introducing seq2seq models* section. As a reminder, we train the original seq2seq model to transform an input sequence of tokens into another, different output sequence of tokens. Examples of such tasks include machine translation and question answering. The original seq2seq training is supervised because matching the input and output sequences counts as labeling. Once the full input sequence is fed into the seq2seq encoder, the decoder starts generating the output sequence one token at a time. In effect, the seq2seq decoder learns to predict the next word in the sequence (as opposed to predicting any masked word in the full sequence, as with BERT). Here, we have a similar algorithm, but the output sequence is the same as the input sequence. From a language modeling point of view, the pre-training learns to approximate the conditional probability of the next token, t_i, given an input sequence of tokens, $t_{i-k} \dots t_{i-1}$, and the model parameters, θ: $P\left(t_i \middle| t_{i-k} \dots t_{i-1}, \theta\right)$.

Let's illustrate the pre-training with an example. We'll assume that our input sequence is [[START], t_1, t_2, ..., t_{N-1}] and that we'll denote the training pairs as {input: label}. Our training pairs are going to be {[[START]]: t_1}, {[[START], t_1]: t_2}, and {[[START], t_1, ..., t_{N-1}]: t_N}. We can see the same scenario displayed in the following diagram:

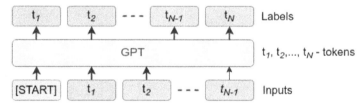

Figure 7.15 – GPT pre-training to predict the next word of the same input/output sequence

Next, let's discuss the supervised fine-tuning step, which is similar to BERT fine-tuning. The following diagram illustrates how the tasks of sequence classification and NLI work in GPT:

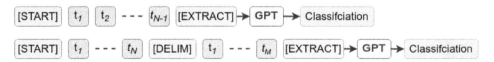

Figure 7.16 – GPT fine-tuning; top: text classification; bottom: NLI

In both cases, we have special [START] and [EXTRACT] tokens. The [EXTRACT] token plays the same role as [CLS] in BERT—we take the output of that token as the result of the classification. But here, it's at the end of the sequence, rather than the start. Again, the reason for this is that the decoder is unidirectional and only has full access to the input sequence at its end. The NLI task concatenates the premise and the entailment, separated by a special [DELIM] token.

This concludes our introduction to GPT—the prototypical example of a decoder-only model. With this, we've introduced the three major transformer architectures—encoder-decoder, encoder-only, and decoder-only. This is also a good place to conclude the chapter.

Summary

Our focus in this chapter was the attention mechanism and transformers. We started with the seq2seq model, and we discussed Bahdanau and Luong attention in its context. Next, we gradually introduced the TA mechanism, before discussing the full encoder-decoder transformer architecture. Finally, we focused on encoder-only and decoder-only transformer variants.

In the next chapter, we'll focus on LLMs, and we'll explore the Hugging Face transformers library.

8

Exploring Large Language Models in Depth

In recent years, interest in transformers has skyrocketed in the academic world, industry, and even the general public. The state-of-the-art transformer-based architectures today are called **large language models** (**LLMs**). The most captivating feature is their text-generation capabilities, and the most popular example is ChatGPT (`https://chat.openai.com/`). But in their core lies the humble transformer we introduced in *Chapter 7*. Luckily, we already have a solid foundation of transformers. One remarkable aspect of this architecture is that it has changed little in the years since it was introduced. Instead, the capabilities of LLMs have grown with their size (the name gives it away), lending credibility to the phrase *quantitative change leads to qualitative change.*

The success of LLMs has further fueled the research in the area (or is it the other way around?). On the one hand, large industrial labs (such as Google, Meta, Microsoft, or OpenAI) invest heavily to push the boundaries for even larger LLMs. On the other hand, the agile, open source community finds creative ways to achieve a lot with limited resources.

In this chapter, we'll explore the current LLM landscape from theoretical and practical perspectives. We will survey many of the latest LLMs, their properties, and their training. Furthermore, we'll see how to apply them for our purposes with the help of the Hugging Face Transformers library.

In this chapter, we're going to cover the following main topics:

- Introducing LLMs
- LLM architecture
- Training LLMs
- Emergent abilities of LLMs
- Introducing Hugging Face Transformers

Technical requirements

We'll implement the example in this chapter using Python, PyTorch, and the Hugging Face Transformers library (`https://github.com/huggingface/transformers`). If you don't have an environment with these tools, fret not—the example is available as a Jupyter notebook on Google Colab. The code examples are in the book's GitHub repository: `https://github.com/PacktPublishing/Python-Deep-Learning-Third-Edition/tree/main/Chapter08`.

Introducing LLMs

In this section, we'll take a more systematic approach and dive deeper into transformer-based architectures. As we mentioned in the introduction, the transformer block has changed remarkably little since its introduction in 2017. Instead, the main advances have come in terms of larger models and larger training sets. For example, the original GPT model (GPT-1) has 117M parameters, while GPT-3 (*Language Models are Few-Shot Learners*, `https://arxiv.org/abs/2005.14165`) has 175B, a thousandfold increase. We can distinguish two informal transformer model categories based on size:

- **Pre-trained language models** (**PLMs**): Transformers with fewer parameters, such as **Bidirectional Encoder Representations from Transformers** (**BERT**) and **generative pre-trained transformers** (**GPT**), fall into this category. Starting with BERT, these transformers introduced the two-step pre-training/FT paradigm. The combination of the attention mechanism and unsupervised pre-training (**masked language modeling** (**MLM**) or **next-word prediction** (**NWP**) creates effective general-purpose semantic features, which we can use for a number of downstream tasks. Because of this, PLMs perform better than other **natural language processing** (**NLP**) algorithms, such as **recurrent neural networks** (**RNNs**). Combined with their highly parallelizable architecture, this has inspired a lot of follow-up work on transformers, which produced improved models and eventually led to the next category.

- **LLMs**: These are transformer models with billions of parameters. LLMs differ qualitatively from PLMs in the following ways:

  - **Emergent capabilities**: They can solve a series of complex tasks, which we will discuss in the *Emergent abilities of LLMs* section

  - **Prompting interface**: LLMs can interact with humans with natural language instead of special APIs

  - **Fusion of research and engineering**: The scale of LLMs requires researchers to have strong engineering skills in large-scale data processing and parallel training

Today, LLMs are almost exclusively decoder-only models because the main applications of the current LLMs revolve around text generation (for example, chatbots such as ChatGPT). This has happened at the expense of encoder-only and encoder-decoder architectures. To better understand why, let's see how a chatbot works. It starts with a user-generated message (known as a **prompt**). A prompt is the initial input sequence to the decoder-based model, which generates a response one token at a time. The response is added back to the input sequence. A special token separates the prompts and the responses. Once the LLM generates a response, the user may make another prompt. In this case, we concatenate the new prompt to the existing sequence and task the LLM to create a new response based on the extended sequence. The LLM has no mechanism for memorizing the existing chat session other than including it as part of the input sequence. This process can continue indefinitely. However, once it reaches the maximum length of the context window, it will start truncating the initial parts of the sequence (we can think of this as a sliding window).

> **Note**
>
> Parts of this chapter are based on the paper *A Survey of Large Language Models* (`https://arxiv.org/abs/2303.18223`). We'll refer to it simply as *the survey*.

LLM architecture

In *Chapter 7*, we introduced the **multi-head attention** (**MHA**) mechanism and the three major transformer variants—encoder-decoder, encoder-only, and decoder-only (we used BERT and GPT as prototypical encoder and decoder models). In this section, we'll discuss various bits and pieces of the LLM architecture. Let's start by focusing our attention (yes—it's the same old joke) on the attention mechanism.

LLM attention variants

The attention we discussed so far is known as **global attention**. The following diagram displays the **connectivity matrix** of a bidirectional global self-attention mechanism (context window with size *n=8*):

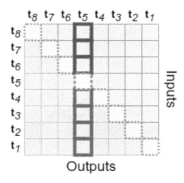

Figure 8.1 – Global self-attention with a context window with size n=8

Each row and column represent the full input token sequence, $[t_1...t_8]$. The dotted colored diagonal cells represent the current input token (query), t_i. The uninterrupted colored cells of each column represent all tokens (keys) that t_i can attend to. For example, t_5 attends to all preceding tokens, $[t_1...t_4]$, and all succeeding tokens, $[t_6...t_8]$. The term *global* implies that t_i attends to all tokens. Hence all cells are colored. As we'll see in the *Sparse attention* section, there are attention variants where not all tokens participate. We'll denote these tokens with transparent cells. The diagram depicts bidirectional self-attention, as the query can attend to both preceding (down) and succeeding (up) elements. The query will only attend to the elements below the current input token in the unidirectional case. For example, t_5 will only attend to $[t_1...t_4]$.

As we'll see, one of the main challenges of the attention mechanism is its time and space complexity.

Attention complexity

Despite its advantages, the attention mechanism (particularly global attention) has some drawbacks. One of them is that space and time complexity increase quadratically with the increase of the context window. That's because the mechanism is implemented with the help of matrices and matrix multiplication.

> **Matrix multiplication time complexity**
>
> The time complexity of the multiplication of two $n \times n$ matrices is $O(n^3)$ because the classic implementation uses three nested loops. In practice, the algorithm is optimized and is less complex. For the purposes of this section, we'll use the complexity of the classic implementation.

For example, a context window with size $n=4$ results in $n \times n = 4 \times 4$ **Q** and **V** matrices with 16 total cells each. But a context window of $n=8$ results in $n \times n = 8 \times 8$ **Q** and **V** matrices with 64 total cells each. Therefore, a two-times-larger context window requires four times more memory. Since the time complexity of matrix multiplication is $O(n^3)$, increasing the context window from $n=4$ to $n=8$ would increase the number of operations from $4^3 = 64$ to $8^3 = 512$.

Next, let's focus on the transformer block, where we have a **feed-forward network** (**FFN**), multi-head self-attention, and four linear projections (**fully connected** (**FC**) layers)—three for the **Q/K/V** pre-attention split and one that combines the attention heads' outputs. We'll discuss each component's relative weight in the block's computational load. Let's denote the embedding size with d_{model}, the key dimension with d_k, the value dimension with d_v ($d_k = d_v = d_{model}/h = d$), the context window size with n, the number of heads with h, and the size of the hidden layer in the FFN with *ffn* (the usual convention is *ffn*=4*d). The time complexity of the different components is shown here:

- $O(h \times 4 \times n \times d^2) = O(n \times d^2)$: The three input linear projections for all heads
- $O(h \times n^2 \times d) = O(n^2 \times d)$: The h self-attention heads
- $O(h \times n \times d^2) = O(n \times d^2)$: The fourth output linear projection after the self-attention heads
- $O(n \times d \times \textit{ffn} + n \times d \times \textit{ffn}) = O(8n \times d^2) = O(n \times d^2)$: The FFN module

The full combined complexity of the block is $O(n \times d^2 + n^2 \times d)$. We can see that it depends on the ratio between the length of the context window, n, and the embedding size, d. If $d>>n$, then the computational time of the linear projections will overshadow the time of the attention heads and vice versa. In practice, $d>>n$ is the most common scenario. But in either case, the attention mechanism has at least quadratic space and time complexity. Let's see some solutions to this challenge.

Multi-query and grouped-query attention

MHA branches the input data to multiple heads using three linear projections per head. The following diagram shows two optimizations of this configuration:

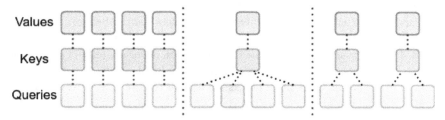

Figure 8.2 – Left: MHA; center: multi-query attention (MQA); right: grouped-query attention (GQA) (inspired by https://arxiv.org/abs/2305.13245)

Let's discuss them (apart from MHA, which we introduced in *Chapter 7*):

- **MQA** (*Fast transformer decoding: One write-head is all you need*, https://arxiv.org/abs/1911.02150): The different heads share key and value projections, as opposed to unique projections in MHA. Since the input sequence is the same as well, all heads share the same key-value store and only differ in their queries. This optimization reduces both the memory and computational requirements, with little performance penalty.

- **GQA** (*GQA: Training Generalized Multi-Query Transformer Models from Multi-Head Checkpoints*, https://arxiv.org/abs/2305.13245): A hybrid between MHA and MQA, which shares single key and value heads for a *subgroup* of query heads. The authors show that GQA is almost as fast as MQA and achieves quality close to MHA.

In the next section, we'll discuss attention optimization, which takes into account the specifics of GPU memory management.

FlashAttention

In this section, we'll introduce FlashAttention (*FlashAttention: Fast and Memory-Efficient Exact Attention with IO-Awareness*, https://arxiv.org/abs/2205.14135; *FlashAttention-2: Faster Attention with Better Parallelism and Work Partitioning*, https://arxiv.org/abs/2307.08691). This is not a new attention mechanism but an implementation of global attention, which considers the specifics of the GPU hardware. A GPU has a large number of computational cores that can perform relatively simple but highly parallelizable operations (such as matrix multiplication). It has two memory levels: small but fast cache (L1 and L2) and large but relatively slow **high bandwidth memory** (**HBM**). To perform an operation, it transfers the necessary data from the HBM to the cache. The cores use the cache for their calculations. Once the operation is done, the result is stored back in the HBM. The main bottleneck in this pipeline is the data transfers rather than the actual computation (the fewer data transfers, the better).

Next, let's focus on the attention block, which has five operations: 1) matrix multiplication ($\mathbf{Q}\mathbf{K}^\top$), 2) mask, 3) softmax, 4) dropout, and 5) matrix multiplication (**V**). The standard implementation performs the operations sequentially, starting with the first matrix multiplication. Once it's done, it proceeds with the mask, and so on. Each operation involves two-way data transfer between the HBM and the cache. These transfers are unnecessary because the results of operation i are transferred from the cache to the HBM just to be sent back from the HBM to the cache for operation $i+1$. FlashAttention proposes a special **fused kernel** to solve the inefficiency. It splits the **Q/K/V** matrices into smaller blocks that can fit in the cache. Once these blocks are transferred there, the fused kernel performs all five operations without intermediate data transfers. Only the final result is sent back to the HBM. Splitting the matrices into blocks is possible because matrix multiplication is embarrassingly parallel. But the other innovation of FlashAttention is the ability to split the softmax operation, which isn't as trivial (we won't go into details about how it's implemented). The operation is done once all matrix blocks pass through this pipeline.

Splitting matrix multiplication

Let's say we want to multiply the matrices **A** and **B**. Because of the way matrix multiplication works, we can split **B** by column into two matrices, $\mathbf{B}_1$ and $\mathbf{B}_2$. Then, we perform two matrix multiplications on each device: $\mathbf{A}\mathbf{B}_1$ and $\mathbf{A}\mathbf{B}_2$. Finally, we concatenate the output of the two operations in a single matrix, equivalent to the matrix produced by the original multiplication, **AB**.

In the next section, we'll discuss solving the performance issue with new attention mechanisms.

Sparse attention

Sparse attention is a class of methods where the output vector attends to a subset of all key vectors instead of the entire context window. For example, if we can attend to four vectors of interest from the entire eight-vector context, we could reduce the necessary computations twice.

The following diagram displays three bidirectional sparse attention mechanisms:

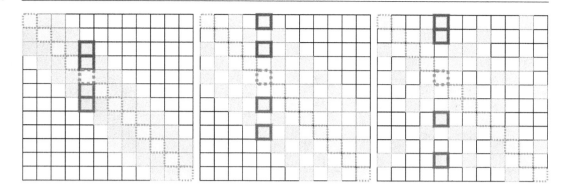

Figure 8.3 – Left: local attention; center: dilated local attention;
right: random attention; context window size n=12

The mechanisms follow the same notation as the ones in *Figure 8.2*, with one addition—the transparent cells represent tokens (keys), which the query doesn't attend to.

On the left, we have bidirectional **local attention** (or **sliding window attention**), first introduced in *Image Transformer*, https://arxiv.org/abs/1802.05751). The query attends to a limited context window of the nearest w keys around the current token ($\frac{1}{2}w$ to the left and $\frac{1}{2}w$ to the right). The self-attention block still takes the full n-sized sequence as input, but each token attends to a limited w-sized local context. This way, the memory footprint is the same as global attention, but the time complexity is reduced to $O(n \times w \times d)$ instead of $O(n \times n \times d)$.

To understand why local attention works, let's return to **convolutional neural networks** (**CNNs**). Recall that the earlier layers of a CNN have small receptive fields and capture smaller, simpler features. Conversely, the deeper CNN layers have large receptive fields that capture larger and more complex features. We can apply the same principle to transformers. Research has shown that the initial transformer blocks learn simple token features and local syntax, while the deeper layers learn more complex context-dependent aspects of token semantics. Because of this, we can apply local attention to the earlier transformer blocks and reserve global attention for the deeper ones without sacrificing performance.

Dilated attention (*Figure 8.3*, center) is a modification of local attention, which works in a similar way to the dilated convolutions we introduced in *Chapter 4*. Unlike local attention, here, the context window is not continuous. Instead, there is a gap of g cells (which could be more than one) between each context token. This makes it possible to attend to a wider context with the same n of computations.

Next, we have bidirectional **random attention** (*Figure 8.3*, right), where the current query (token) attends to a subset of r keys (tokens) from the full context window. The time complexity is reduced to $O(n \times r \times d)$ instead of $O(n \times n \times d)$. The attention pattern can be viewed as a directed graph. In the case of random attention, this graph is also random. That is, the information can flow rapidly between any pair of nodes without considering the actual structure of the data, which might be biased.

It is also possible to combine global and local attention. One such example is **Longformer** (*Longformer: The Long-Document Transformer*, https://arxiv.org/abs/2004.05150), displayed in the following diagram:

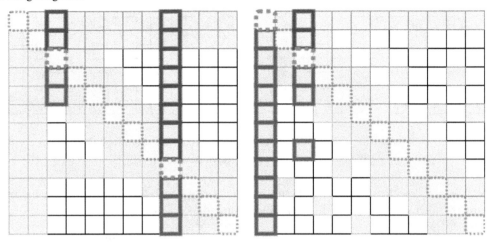

Figure 8.4 – Combined local and global attention; left: Longformer block; right: Big Bird block

It introduces a drop-in replacement self-attention block in an otherwise unmodified transformer model. The block represents a combination of global and local (or dilated) attention. It applies local attention to most input tokens, but a few can use global attention. The left section of *Figure 8.4* shows the combined self-attention block and one example of input tokens that apply local and global attention. More specifically, the authors use the Longformer block in a unidirectional BERT-style model to solve MLM and **question-answering** (**QA**) tasks (*Chapter 7*). They only apply global attention to special tokens such as [CLS] in MLM tasks. As the diagram shows, global attention works in both directions. The special token can attend to all other tokens, but the other tokens can also attend to the special token in addition to their local attention context. In the case of autoregressive language modeling (unidirectional model), they apply only dilated local attention, as there are no tokens with special significance. The full Longformer model uses dilated attention with a larger context window and *g* in the deeper layers, leaving the earlier ones with only local attention.

Big Bird (*Figure 8.4*, right; *Big Bird: Transformers for Longer Sequences*, https://arxiv.org/abs/2007.14062) is similar to Longformer but adds random attention.

Next, let's discuss the **sparse transformer** attention developed by OpenAI (*Generating Long Sequences with Sparse Transformers*, https://arxiv.org/abs/1904.10509). A sparse transformer introduces unidirectional strided and fixed attention schemes, displayed in the following diagram:

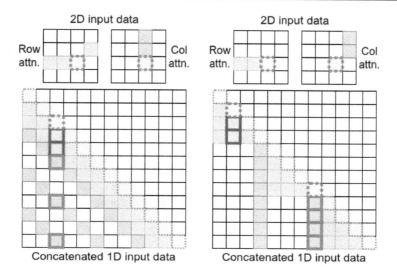

Figure 8.5 – Left: strided sparse attention with l=4; right: fixed sparse attention; input image size 4×4; sequence length n=12 (inspired by `https://arxiv.org/abs/1904.10509`)

To understand how they work, let's discuss the context of the paper. It proposes a unified decoder-only model to generate new images, text, or audio. Depending on the use case, the input and output data can be a two-dimensional image tensor (we'll omit the color dimension for simplicity). However, the transformer accepts as input a one-dimensional sequence. We can solve this by concatenating the rows of the image in a single one-dimensional tensor. Once done, we can treat the image like a regular sequence and feed it to the model. *Figure 8.5* displays a strided (left) and fixed attention (right) connectivity matrix for a two-dimensional image (top) and its equivalent concatenated one-dimensional sequence (bottom). Let's note that the bottom expanded sequence doesn't match the dimensions of the top image—it should be with length *n=16*, which reflects the 4×4 image, instead of *n=12* as it is now. Since this is a generative decoder-only model, it uses unidirectional attention, even though the concept of direction doesn't exist in the same way in images as in text.

Next, let's discuss the two attention schemes. We'll start with strided attention, where the current token attends to the preceding row and column of the input image. These are two separate mechanisms split between different attention heads:

- **Row head**: Equivalent to unidirectional local attention, which attends to the previous $l \approx \sqrt{n}$ tokens, where $\sqrt{n}$ is the length of one entire row of the 2D input image. Let's denote the index of the current input token with i and the tokens it attends to with j. We can summarize the row mechanism in the following way:

$$(i - j) < l$$

- **Column head**: Equivalent to unidirectional dilated attention with a stride (gap) of $l \approx \sqrt{n}$ (the same as the row head). Assuming that the input image is square, the column head jumps the equivalent of one row ($\sqrt{n}$) and attends to a location representing the previous cell in a virtual column of the one-dimensional sequence. We can summarize column strided attention in the following way:

$$(i - j) \bmod l = 0$$

This scheme performs best for 2D input data, such as images, because the row/column split reflects the underlying data structure. The time complexity of this scheme is $O(n \times l \times d) \approx O(n \times \sqrt{n} \times d)$.

Next, we have **fixed attention**, which attends to a fixed column and the elements after the latest column element. It performs better on non-periodic data, such as text. Once again, this is a combination of two separate mechanisms split between different heads:

- **Column head**: Attends to a fixed column, which doesn't necessarily match the column of the current input token, $\mathbf{t}_i$. Multiple input tokens attend to the same column, which makes it possible to attend to the entire length of the sequence. We can summarize the column mechanism in the following way:

$$l - c \leq j \bmod l \leq l$$

Here, c is a parameter (8, 16, or 32). For example, if $l=64$ and $c=16$, then all positions greater than 64 can attend to positions 48-64, all positions greater than 128 can attend to 112-128, and so on.

- **Row head**: The first head is similar to the row head in strided attention. But instead of attending to the length of one entire row, it only attends to the location of the current column head. The row head provides local context. We can summarize it in the following way:

$$floor\left(\frac{j}{l}\right) = floor\left(\frac{i}{l}\right)$$

Here, *floor* rounds down the result of the division to the nearest whole number.

Next, let's focus our attention (I can't stop myself) on a special case of decoder-only architecture and various aspects of LLM architecture.

Prefix decoder

In this section, we'll introduce **prefix** (or **non-causal**) **decoder** (*Unified Language Model Pre-training for Natural Language Understanding and Generation*, https://arxiv.org/abs/1905.03197). This is a decoder-only model that introduces a new type of attention pattern, displayed in the following diagram:

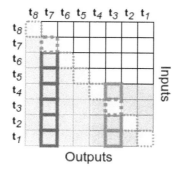

Figure 8.6 – Prefix decoder self-attention pattern (inspired by https://arxiv.org/abs/1905.03197)

We split the input sequence into two segments—t_1 through t_4 (**source** or **prefix**), and t_5 through t_8 (**target**). The tokens of the source segment have bidirectional access to all other tokens of that segment. However, the target segment tokens have unidirectional access to the preceding tokens of the whole (source and target) input sequence. For example, t_3 is part of the source segment and can attend to t_1, t_2, and t_4. Conversely, t_7 is part of the target and can only attend to tokens t_1 through t_6 (but not t_8). The prefix decoder is a hybrid between encoder-decoder and decoder models. The source segment acts as an encoder, and the target acts as a decoder, yet the underlying architecture is decoder-based.

We can use the prefix decoder for **sequence-to-sequence** (**seq2seq**) tasks, such as machine translation or text summarization, for which we would normally use a full encoder-decoder. To do so, we concatenate the input and output sequence using special start-of-sequence ([SOS]) and end-of-sequence ([EOS]) tokens. For example, let's take the text summarization task. We represent the text sequence to summarize (S1) and its summarization (S2) as a single sequence: [[SOS] , S1, [EOS] , S2, [EOS]]. The source sequence, [[SOS] , S1, [EOS]], falls within the bidirectional part of the attention pattern, and the target sequence, [S2, [EOS]], falls within the unidirectional one. We pre-train the model with the help of MLM, where we mask random tokens from the full sequence. We fine-tune the model by randomly masking some tokens in the target sequence and learning to recover the masked words. Let's note that the [EOS] token can also participate in the masking. In this way, the model learns when to generate [EOS] tokens and terminate the generation of the target sequence.

Next, let's get into more details about various aspects of the LLM architecture.

Transformer nuts and bolts

The following table provides a detailed summary of the main transformer network configurations and their variants:

Configuration	Method	Equation
Normalization position	Post Norm	Norm(**x**+Sulayerb(**x**))
	Pre Norm	**x** + Sublayer(Norm(**x**))
	Sandwich Norm	**x** + Norm(Sublayer(Norm(**x**)))
Normalization method	LayerNorm	$\frac{\mathbf{x}-\mu}{\sqrt{\sigma}} \cdot \gamma + \beta, \quad \mu = \frac{1}{d}\sum_{i=1}^{d} x_i, \quad \sigma = \sqrt{\frac{1}{d}\sum_{i=1}^{d}(x_i - \mu))^2}$
	RMSNorm	$\frac{\mathbf{x}}{\text{RMS}(\mathbf{x})} \cdot \gamma, \quad \text{RMS}(\mathbf{x}) = \sqrt{\frac{1}{d}\sum_{i=1}^{d} x_i^2}$
	DeepNorm	LayerNorm($\alpha \cdot \mathbf{x}$ + Sublayer(**x**))
Activation function	ReLU	ReLU(**x**) = max(**x**, **0**)
	GeLU	GeLU(**x**) = $0.5\mathbf{x} \otimes [1 + \text{erf}(\mathbf{x}/\sqrt{2})], \quad \text{erf}(x) = \frac{2}{\sqrt{\pi}}\int_0^x e^{-t^2} dt$
	Swish	Swish(**x**) = **x** $\otimes$ sigmoid(**x**)
	SwiGLU	SwiGLU($\mathbf{x_1}, \mathbf{x_2}$) = Swish($\mathbf{x_1}$) $\otimes$ $\mathbf{x_2}$
	GeGLU	GeGLU($\mathbf{x_1}, \mathbf{x_2}$) = GeLU($\mathbf{x_1}$) $\otimes$ $\mathbf{x_2}$
Position embedding	Absolute	$\mathbf{x}_i = \mathbf{x}_i + \mathbf{p}_i$
	Relative	$A_{ij} = \mathbf{W}_q\mathbf{x}_i\mathbf{x}_j^T\mathbf{W}_k^T + r_{i-j}$
	RoPE	$A_{ij} = \mathbf{W}_q\mathbf{x}_i\mathbf{R}_{\theta,i-j}\mathbf{x}_j^T\mathbf{W}_k^T$
	Alibi	$A_{ij} = \mathbf{W}_q\mathbf{x}_i\mathbf{R}_{\theta,i-j}\mathbf{x}_j^T\mathbf{W}_k^T \, A_{ij} = \mathbf{W}_q\mathbf{x}_i\mathbf{x}_j^T\mathbf{W}_k^T - m(i-j)$

Figure 8.7 – Different transformer configurations (source: https://arxiv.org/abs/2303.18223)

We're already familiar with many of these—we introduced the three different normalization positions in *Chapter 7*. We also introduced two of the three normalization methods in *Chapter 3*. By default, most transformers use **layer normalization (LN)**. However, some models use **RMSNorm** because of its superior training speed and performance. Last but not least, **DeepNorm** (*DeepNet: Scaling Transformers to 1,000 Layers*, https://arxiv.org/abs/2203.00555) is new to us. As the paper's name suggests, this normalization helped build a 1,000-layer transformer. The authors argue that in pre-**layer normalization (pre-ln)** architectures, the gradients at the bottom layers tend to be larger than the ones at the top layers, degrading the performance compared to **post-layer normalization (post-ln)** models. On the other hand, post-ln models are unstable due to exploding gradients. To overcome this, they propose a simple yet effective normalization of the residual connections:

$$\text{LayerNorm}(\alpha\mathbf{x} + \text{SubLayer(x)})$$

Here, α is a constant applied at the output of the residual connection. Its value depends on the transformer type (encoder or decoder) and the model depth (number of blocks). The theoretical justification of DeepNorm is that it bounds the model update by that constant.

Next, let's discuss the activation functions. More specifically, we'll discuss the activation function (ActivationFunc) of the first layer of the **feed-forward network (FFN)** sublayer, as this is the only explicit activation in the transformer block. As a reminder, we can define the original FFN as follows:

$$\text{FFN}(x) = \text{ActivationFunc}\big(\mathbf{W}_1\mathbf{x} + b_1\big)\mathbf{W}_2 + b_2$$

We discussed most activations in *Chapter 3*, except for **SwiGLU** and **GeGLU** (*GLU Variants Improve Transformer*, https://arxiv.org/abs/2002.05202). They are variations of **Gated Linear Unit** (**GLU**, *Language Modeling with Gated Convolutional Networks*, https://arxiv.org/abs/1612.08083), which is more of a fusion between layer and activation function rather than pure activation. We can define GLU as follows:

$$ActivationGLU(\text{x}) = ActivationFunc(\mathbf{W}\text{x} + b) \otimes (\mathbf{V}\text{x} + c)$$

Here, *ActivationFunc* is a specific activation function (Swish for *Swi*GLU and *GeLU* for *Ge*GLU), $\otimes$ is the element-wise product of two vectors, and **W** and **V** are weight matrices, which represent linear projections (that is, FC layers). GLU introduces an additional linear projection, **V**, parallel to the original path of the network, **W**. Thanks to the element-wise product, the path with activation, **W**, acts as a gate to the signal coming from the **V** path. This is like **Long Short-Term Memory** (**LSTM**) gates. We can now define the FFN with GLU activation:

$$FFN_{ActivationGLU}(\mathbf{x}) = \left(ActivationFunc(\mathbf{W}_1\text{x}) \otimes \mathbf{V}\text{x} \right) \mathbf{W}_2$$

Let's note that the authors have excluded the bias from the modified FFN. This is also a good place to mention that different LLMs have different bias configurations, listed next:

- Use bias in both the linear projections and the attention blocks themselves
- Use bias in the linear projections but not in the attention blocks
- Don't use bias in either the linear projections or the attention blocks

According to some experiments, the lack of biases stabilizes the training.

Next, let's focus on the various types of positional embeddings we haven't mentioned so far. Unfortunately (or fortunately), discussing them in detail goes beyond the scope of this book. But the important thing to remember is that we have either absolute (static) or relative (dynamic) positional encodings. In the first case, we modify the input token embedding vectors. In the second case, we modify the **K/V** attention matrices relative to their position of the current input token.

The survey summarizes the suggestions from existing literature for detailed transformer configuration. For stronger generalization, it suggests using pre-RMSNorm normalization, and SwiGLU or GeGLU activation functions. In addition, using LN immediately after embedding layers is likely to incur performance degradation. As for position embeddings, **Rotary Positional Embedding** (**RoPE**) or **Attention with Linear Biases** (**AliBi**) perform better on long sequences than other methods.

Now that we're familiar with the architecture properties of LLMs, let's discuss specific model instances.

Models

The following table represents a summary of some of the popular recent LLMs:

Model	Category	Size	Normalization	PE	Activation	Bias	#L	#H	d_{model}	MCL
GPT3	Causal decoder	175B	Pre LayerNorm	Learned	GeLU	✓	96	96	12288	2048
PanGU-α	Causal decoder	207B	Pre LayerNorm	Learned	GeLU	✓	64	128	16384	1024
OPT	Causal decoder	175B	Pre LayerNorm	Learned	ReLU	✓	96	96	12288	2048
PaLM	Causal decoder	540B	Pre LayerNorm	RoPE	SwiGLU	×	118	48	18432	2048
BLOOM	Causal decoder	176B	Pre LayerNorm	ALiBi	GeLU	✓	70	112	14336	2048
MT-NLG	Causal decoder	530B	-	-	-	-	105	128	20480	2048
Gopher	Causal decoder	280B	Pre RMSNorm	Relative	-	-	80	128	16384	2048
Chinchilla	Causal decoder	70B	Pre RMSNorm	Relative	-	-	80	64	8192	-
Galactica	Causal decoder	120B	Pre LayerNorm	Learned	GeLU	×	96	80	10240	2048
LaMDA	Causal decoder	137B	-	Relative	GeGLU	-	64	128	8192	-
Jurassic-1	Causal decoder	178B	Pre LayerNorm	Learned	GeLU	✓	76	96	13824	2048
LLaMA	Causal decoder	65B	Pre RMSNorm	RoPE	SwiGLU	×	80	64	8192	2048
LLaMA 2	Causal decoder	70B	Pre RMSNorm	RoPE	SwiGLU	×	80	64	8192	4096
GLM-130B	Prefix decoder	130B	Post DeepNorm	RoPE	GeGLU	✓	70	96	12288	2048
T5	Encoder-decoder	11B	Pre RMSNorm	Relative	ReLU	×	24	128	1024	512

Figure 8.8 – Model cards of recent LLMs with public configuration details (modified from https://arxiv.org/abs/2303.18223p)

Here, **PE** denotes position embedding, **#L** denotes the number of transformer layers, **#H** denotes the number of attention heads per layer, d_{model} denotes the size of hidden states, and **MCL** denotes the maximum context length during training.

We'll start with the GPT series of models (developed by OpenAI), which is outlined in the following diagram:

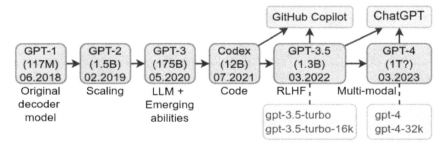

Figure 8.9 – The evolution of the GPT series of models (inspired by https://arxiv.org/abs/2303.18223)

We're already familiar with GPT-1, so let's move on to **GPT-2** (*Language Models are Unsupervised Multitask Learners*, https://d4mucfpksywv.cloudfront.net/better-language-models/language-models.pdf). As a recurring theme of this chapter, it is similar to GPT-1 (except that it uses pre-normalization), but it's much larger (1.5B versus 117M parameters). It has a larger token vocabulary and requires a larger training dataset size. GPT-1 and GPT-2 are the only open source models of the series. Next, we have the **GPT-3** collection of eight variants (closed source), ranging from 125M to 175B parameters (GPT-3 refers to the largest model). It uses the same architecture as GPT-2 but adds alternating sparse and MHA layers. This is the first model of the series that falls into the LLM category. Thanks to its size and sophisticated training, it exhibits so-called **emergent abilities** (we will discuss them in their namesake section) and serves as a base for the next models. The first derivative model is **Codex** (*Evaluating Large Language Models Trained on Code*, `https://arxiv.org/abs/2107.03374`), which can generate source code from Python docstrings prompts. To do this, the model is fine-tuned on the publicly available GitHub source code data. Initially, GitHub Copilot was based on Codex. Next, we have the GPT-3.5 collection of models. It is not public and has no official paper, so we can only speculate about its properties, but we'll assume it's similar to GPT-3. As with Codex, it has the ability to generate code based on natural language task descriptions. It also uses **reinforcement learning with human feedback** (**RLHF**) for FT (again, we'll discuss it in its namesake section), which improves the model's responses. GPT-3.5 is available through OpenAI's API in two subvariants—`gpt-3.5-turbo` with a context length of 4,096 tokens and `gpt-3.5-turbo-16k` with 16,384 tokens. The current version of Copilot is based on GPT-3.5. The newest model, GPT-4, accepts multimodal inputs (images and text) but outputs text only. It is also closed, but it might have more than 1T parameters. According to Sam Altman, CEO of OpenAI, training GPT-4 has cost more than $100 million (`https://www.wired.com/story/openai-ceo-sam-altman-the-age-of-giant-ai-models-is-already-over/`). GPT-4 is also available through OpenAI's API with two subvariants—`gpt-4` with a context length of 8,192 tokens and `gpt-4-32k` with 32,768 tokens.

Next, let's discuss the **LlaMa** series of pre-trained (and not fine-tuned) models released by Meta. The first version (*LLaMA: Open and Efficient Foundation Language Models*, `https://arxiv.org/abs/2302.13971`) has four variants, ranging from 6B to 65B parameters. This is one of the most popular LLMs in the open source community because Meta has also released its weights (although they are not licensed for commercial use). This way, the company has done the heavy lifting of pre-training the model. The open source community uses it as a **foundation model** because it can be fine-tuned with relatively little compute. Recently, Meta released **Llama 2**—an updated version of Llama (*Llama 2: Open Foundation and Fine-Tuned Chat Models*, `https://ai.meta.com/research/publications/llama-2-open-foundation-and-fine-tuned-chat-models`). It has three variants with 7B, 13B, and 70B parameters. Llama 2 uses GQA and 40% more pre-training data than Llama 1. In addition, each variant also has a version fine-tuned using RLHF. The model's license allows commercial use (with some limitations).

This concludes our survey on the architecture of LLMs. Next, let's discuss their training.

Training LLMs

Since most LLMs are decoder-only, the most common LLM pre-training task is NWP. The large number of model parameters (up to hundreds of billions) requires comparatively large training datasets to prevent overfitting and realize the full capabilities of the models. This requirement poses two significant challenges: ensuring training data quality and the ability to process large volumes of data. In the following sections, we'll discuss various aspects of the LLM training pipeline, starting from the training datasets.

Training datasets

We can categorize the training data into two broad categories:

- **General**: Examples include web pages, books, or conversational text. LLMs almost always train on general data because it's widely available and diverse, improving the language modeling and generalization capabilities of LLMs.

- **Specialized**: Code, scientific articles, textbooks, or multilingual data for providing LLMs with task-specific capabilities.

The following table lists the most popular language modeling datasets:

Corpora	Size	Source	Latest Update
BookCorpus	5GB	Books	Dec-2015
Gutenberg	-	Books	Dec-2021
C4	800GB	CommonCrawl	Apr-2019
CC-Stories-R	31GB	CommonCrawl	Sep-2019
CC-NEWS	78GB	CommonCrawl	Feb-2019
REALNEWs	120GB	CommonCrawl	Apr-2019
OpenWebText	38GB	Reddit	Continuous
Pushift.io	2TB	Reddit	Continuous
Wikipedia	21GB	Wikipedia	Continuous
The Pile	800GB	-	Dec-2020
ROOTS	1.6TB	-	Jun-2022
StackExchange	82GB	-	Continuous
ArXiv	3.8GB	Scientific	Continuous
GitHub	-	Code	Continuous

Figure 8.10 – Language modeling datasets (modified from https://arxiv.org/abs/2303.18223)

Let's discuss them:

- **Books**: We'll focus on two datasets:

  - **BookCorpus** (*Aligning Books and Movies: Towards Story-like Visual Explanations by Watching Movies and Reading Books*, `https://arxiv.org/abs/1506.06724`): Includes 11,000 fictional books with close to 1B words (released in 2015).

  - **Project Gutenberg** (`https://www.gutenberg.org/`): Includes 70,000 fictional books.

- **Common Crawl** (`https://commoncrawl.org/`): Petabyte-sized web crawling database. The data is split by the date obtained, starting from 2008. The latest archive contains 3.1B web pages (390 TiB of uncompressed content), scraped from 44 million hosts or 35 million registered domains. It contains a lot of low-quality data, but there are multiple subsets with higher-quality data:

  - **Colossal, cleaned version of Common Crawl** (**C4**): An 800 GiB dataset developed by Google. The original dataset is unavailable for download, but Google has published the tools to recreate it from the Common Crawl database. In 2019, the **Allen Institute for AI** (**AI2**, `https://allenai.org/`) released a recreation, available at `https://huggingface.co/datasets/allenai/c4`. Its most popular sub-variant is the *en* version, which removes all documents that contain words from the so-called *badwords filter* (a list of bad words is available at `https://github.com/LDNOOBW/List-of-Dirty-Naughty-Obscene-and-Otherwise-Bad-Words`).

  - **CC-News**: Articles from news sites all over the world.

  - **RealNews**: News articles extracted from the 5,000 news domains indexed by *Google News*.

  - **CC-Stories-R**: A dataset for common sense reasoning and language modeling. It consists of Common Crawl documents with the most overlapping n-grams with the questions in common sense reasoning tasks. The new training corpus represents the top 1.0% of the highest-ranked documents.

- **Reddit links**: One way to overcome the low signal-to-noise ratio of Common Crawl is to rely on human-curated content. Enter Reddit, where users can post textual content or links, and other users can upvote these submissions (the upvotes are known as *karma*). We'll mention two Reddit-based datasets:

  - **WebText** (released alongside the GPT-2 model): Contains a subset of 45 million Reddit-submitted links with a karma of three or more. The documents behind these links form the LLM training data. WebText is not publicly available, but there is an open source version called **OpenWebText** (`https://github.com/jcpeterson/openwebtext`).

  - **Pushshift** (`https://arxiv.org/abs/2001.08435`): Contains all link submissions and comments posted on Reddit.

Reddit API pricing controversy

The rise of LLMs has made Reddit data much more valuable than before. Because of this, the company has decided to introduce fees for access to its previously free API. This measure mainly targets AI companies that plan to train their LLMs using the data. However, the proposal has led many of the site's voluntary moderators (Reddit relies on them) to announce a strike by temporarily closing the previously open communities they moderate. At the time of writing, the disagreement is still ongoing.

- **The Pile** (*An 800GB Dataset of Diverse Text for Language Modeling*, `https://arxiv.org/abs/2101.00027`): Composed of 22 diverse and high-quality datasets derived from various sources, including PubMed, arXiv, GitHub, Stack Exchange, Hacker News, YouTube, and others. The Pile also introduces the OpenWebText2 and BookCorpus2 extensions of the original OpenWebText and BookCorpus datasets.

- **ROOTS** (*The BigScience ROOTS Corpus: A 1.6TB Composite Multilingual Dataset*, `https://arxiv.org/abs/2303.03915`): A web-scale curated dataset covering 46 natural languages and 13 programming languages.

- **Wikimedia** (`https://dumps.wikimedia.org/`): Because of its high-quality content, this is an excellent source of training data.

- **Stack Exchange** (`https://archive.org/details/stackexchange`): A network of QA topic sites with a rating system. The most popular representative is **Stack Overflow**. It releases a tri-monthly anonymized data dump with all user-contributed content.

- **arXiv** (`https://www.kaggle.com/datasets/Cornell-University/arxiv`): The primary scientific data source, which contains more than 2.2B scientific articles.

- **GitHub**: The GH Archive project (`https://www.gharchive.org/`) records, archives, and provides access to the public GitHub timeline.

In practice, the LLM pre-training step uses a mix of several datasets. The following screenshot shows the distribution of the sources of pre-training data for several representative LLMs:

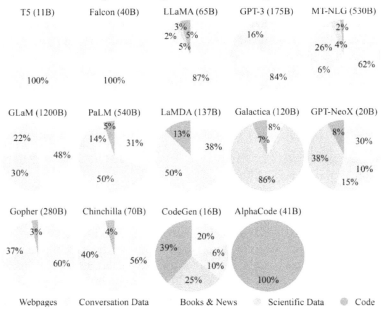

Figure 8.11 – Ratios of various data sources in the pre-training data for existing LLMs (source: https://arxiv.org/abs/2303.18223)

Mixing datasets is not a trivial process and requires several processing steps. Let's discuss them:

- **Remove low-quality or irrelevant data**: For example, web pages contain large amounts of HTML tags, JavaScript, or **cascading style sheets** (**CSS**). Yet, we're only interested in human-readable text (except when we want to train the model explicitly to understand HTML). In this case, we'll have to remove the HTML and JavaScript and only leave the text.

- **Remove personally identifiable information** (**PII**): Data is often extracted from web pages, which might contain personal information. This step aims to remove such data from the training set.

- **Tokenization**: We discussed tokenization in depth in *Chapter 6*, and we won't discuss it here.

Finally, let's introduce a practical transformer scaling law (*Scaling Laws for Neural Language Models*, `https://arxiv.org/abs/2001.08361`). Because of their scale, training LLMs can be expensive. Therefore, it is important not to train the model more (or less) than necessary. Based on empirical experiments, the scaling law proposes an optimal ratio between the amount of training compute (expressed in **floating-point operations per second**, or **FLOPS**), C, the model size (number of parameters), N, and the training dataset size (number of tokens), D:

$$C \approx 6ND$$

Now that we know the steps to build the training set, let's focus on the actual pre-training.

Pre-training properties

Similar to other **neural networks** (**NNs**), pre-training of LLMs works with gradient descent and backpropagation. But because of their size, the training has specific properties, which we'll discuss in this section.

Adam optimizer

Most LLMs use Adam (*Adam: A Method for Stochastic Optimization*, `https://arxiv.org/abs/1412.6980`) or one of its modifications. Although we've used it in many examples so far, we haven't discussed it in detail. Time to remedy this omission.

A reminder of the weight update formula

In *Chapter 2*, we learned that we use backpropagation to compute the gradient (first derivative) of the loss function, $J(\theta)$, with respect to every parameter, θ_j: $\partial J(\theta)/\partial \theta_j$. Once we have the gradient, we can perform the weight update with the formula $\theta_j \leftarrow \theta_j - \eta \partial J(\theta)/\partial \theta_j$, where η is the learning rate. We can add momentum (or velocity) to that formula. To do so, we'll assume we are at step t of the training process. Then, we can calculate the momentum of the current update based on the momentum of the update at step $t\text{-}1$: $v_t \leftarrow \mu v_{t-1} - \eta \partial J(\theta)/\partial \theta_j$, where μ is a momentum rate in the [0:1] range. In addition, we can add L2 regularization (or weight decay; see *Chapter 3*): $v_t \leftarrow \mu v_{t-1} - \eta\left(\left(\partial J(\theta)/\partial \theta_j\right) + \lambda \theta_j\right)$, where λ is the weight decay coefficient. Finally, we can perform the weight update: $\theta_j \leftarrow \theta_j + v_t$.

Adam calculates individual and adaptive learning rates for every weight based on previous weight updates (momentum). Let's see how that works:

1. Compute the first moment (or mean) and the second moment (or variance) of the gradient:

$$m_t \leftarrow \beta_1 m_{t-1} + (1 - \beta_1) \frac{\partial J(\theta)}{\partial \theta_j}$$

$$v_t \leftarrow \beta_2 v_{t-1} + (1 - \beta_2) \left(\frac{\partial J(\theta)}{\partial \theta_j} \right)^2$$

Here, β_1 and β_2 are hyperparameters with default values of 0.9 and 0.95, respectively. The two formulas are very similar to the momentum one. The relationship between m_t (m_{t-1}) and v_t (v_{t-1}) acts as a simulation of a moving average. But instead of averaging across multiple previous values, we take the latest previous value, m_{t-1} (v_{t-1}), and assign it a weight coefficient, β_1 (β_2).

2. The initial values of m_t and v_t are 0, so they will have a bias toward 0 in the initial phase of the training. To understand why this could be a problem, let's assume that at $t=1$, $\beta_1 = 0.9$ and $\partial J(\theta)/\theta_j = 5$. Then, $m_1 = 0.9*0 + (1 - 0.9)*5 = 0.5$ is much less than the actual gradient of 5. We can compensate for this bias with bias-corrected versions of m_t and v_t:

$$\widehat{m}_t \leftarrow \frac{m_t}{1 - \beta_1^t}$$

$$\widehat{v}_t \leftarrow \frac{v_t}{1 - \beta_2^t}$$

3. Perform the weight update with the following formula:

$$\theta_j \leftarrow \theta_j - \eta \frac{\widehat{m}_t}{\sqrt{\widehat{v}_t} + \varepsilon}$$

Here, ε is some small value to prevent division by 0.

AdamW (*Decoupled Weight Decay Regularization*, https://arxiv.org/abs/1711.05101) improves Adam with decoupled weight decay:

$$\theta_j \leftarrow \theta_j - \left(\eta \frac{\widehat{m}_t}{\sqrt{\widehat{v}_t} + \varepsilon} + \lambda \theta_j \right)$$

Recall that the L2 regularization participates in the loss function and then, through the derivative process, is transferred (as weight decay) to the weight update formula. In this case, the regularization will pass through all the transformations of the cost function and will be subject to them. As the name suggests, decoupled weight decay bypasses all these transformations and participates directly in the preceding formula.

One issue with Adam and AdamW is the increased memory consumption—the optimizer stores at least two additional values (m_t and v_t) for every model parameter.

Parallel processing

The scale of LLMs necessitates special steps for efficient training. First, we'll discuss how to train LLMs across multiple devices. More specifically, we'll discuss a combination of three different types of parallelism (also referred to as 3D parallelism):

- **Data parallelism**: It works when the model is small enough to fit on a single device:

    I.   Create identical copies of the entire model and its optimizer states (including the random seeds) across all devices.

    II. Split each batch of the training set into unique subsets (shards) and distribute them across all devices.

    III. Each device computes its gradient based on its unique subset of the input batch.

    IV. Aggregate the gradients of all devices into a single gradient update.

    V.   Distribute the aggregated updates across the devices and perform weight updates on each device. This way, we start and end each training step with identical models.

- **Model (or pipeline)**: Split the model across multiple devices on an operation (layer) level. For example, if our model has 9 layers, we can send layers 1 through 6 to one device and layers 7 through 9 to another. In this way, we can train models that don't fit in the memory of a single device. Not only that, but we can apply this method even on a single device. In this case, we'll load the first set of operations (1-6) and compute their output. Then, we'll unload them and load the following subset (7-9). The output of the first set will serve as input for the second. Backpropagation works in the same way but in the opposite direction. One issue with model parallelism is that if we use multiple devices, the second one will idle until the first produces output.

- **Tensor (or horizontal)**: Split the model across different devices on the tensor level, which solves the idling problem of model parallelism. To understand how this works, let's recall that matrix multiplication is the most computationally intensive operation of contemporary NNs. But, as we discussed in the *FlashAttention* section, it is also embarrassingly parallel. Therefore, we can split it across devices.

Zero redundancy optimizer

Zero redundancy optimizer (**ZeRO**) (*ZeRO: Memory Optimizations Toward Training Trillion Parameter Models*, https://arxiv.org/abs/1910.02054) is a hybrid between data and model parallelism. The following diagram illustrates the three stages of ZeRO:

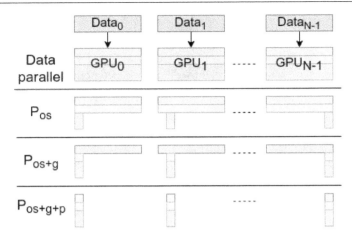

Figure 8.12 – ZeRO (inspired by https://arxiv.org/abs/1910.02054)

The first line represents a case of a data-parallel system. Each GPU receives a unique shard of the input mini-batch. It also holds an identical copy of the model parameters (first colored rectangle of the GPUi block), gradients (second rectangle), and optimizer states (third rectangle). The size of the optimizer states that they take up most of the memory during training (for example, Adam stores multiple values per model parameter). The following three lines represent the three stages of ZeRO:

1. **Optimizer state partitioning** (P_{os}): Each GPU holds an identical copy of the entire model parameters and its gradients, but the optimizer states are partitioned across the GPUs, and each holds only a portion.

2. **Add gradient partitioning** (P_{os+g}): Each GPU holds an identical copy of the entire model parameters, but the gradients and the optimizer states are partitioned.

3. **Add model parameters** (P_{os+g+p}): Each GPU holds a portion of all components.

To understand how the algorithm works, we'll assume we use P_{os+g+p}, a model with N layers and N GPUs. Each layer is distributed on one GPU—the first layer is on GPU_0, the second layer is on GPU_1, and so on. Let's start with the forward phase. First, GPU_0 receives $Data_0$. Since it holds the first layer of the model, it can feed it the input and calculate its activations independently. At the same time, GPU_0 broadcasts the parameters of the first layer to all other GPUs. Each GPU now holds the first layer parameters in addition to its own portion of the model parameters. In this way, GPU_i can process its own input, $Data_i$, through the first layer, just as GPU_0 did. Once a GPU computes the activations of the first layer, it deletes its parameters from its memory (except GPU_0, which preserves them). We repeat the same steps, this time with the second layer. GPU_1 broadcasts its parameters so that all GPUs can continue with the forward phase. After this, all but GPU_1 delete the second layer parameters. This process continues until all GPUs produce model output. Then, the loss function is aggregated across all GPUs. Next, the backward phase starts, which works in the same way as the forward one, but this time the GPUs broadcast both the gradients and the optimizer states.

Mixed precision training

Mixed precision training (https://arxiv.org/abs/1710.03740) is the idea that not all values have to be stored with 32-bit (double or full) floating-point precision (**FP32** or **Float32** data format). Research has shown that storing some values into lower 16-bit (single or half) floating-point precision (**FP16** of **Float16**) doesn't degrade the model performance. The weights, activations, gradients, and optimizer states are stored as FP16. In addition, the model retains an FP32 master copy of the weights. The forward and backward passes use FP16 weights, but the results are optimal when the weight update operation uses the FP32 master copy. One possible explanation is that the weight update formula uses the gradients multiplied by the learning rate, and the result might become too small to be represented in FP16. Another explanation is that the ratio of the weight value to the weight update is very large, which could lead to the weight update becoming zero.

> Bfloat16 and TensorFloat32
>
> The Google Brain division developed the **brain floating-point** format (hence the name, **bfloat**) for **machine learning** (**ML**) applications. The standard FP16 format has one sign bit, five exponent bits, and ten mantissa bits. In contrast, bfloat16 has eight exponent and seven mantissa bits. The exponent bits are the same as FP32. Bfloat16 comes close to FP32 in terms of performance on ML tasks. We also have **TensorFloat-32** (**TF32**)—a 19-bit format developed by NVIDIA for ML purposes with 8 exponent and 10 mantissa bits.

Pre-training peculiarities and summary

In this section, we'll discuss some LLM pre-training peculiarities. Let's start with the mini-batch size. Ideally, we would compute the gradients over the entire training dataset and only then perform one weight update. However, large datasets and models make this computationally infeasible. The opposite extreme is to perform one weight update per training sample. But then, the training would be susceptible to outlier samples, which might steer the loss function to suboptimal local minima. Mini-batch training is a compromise that makes it possible to fit within the available computational resources and avoid the influence of outlier samples. But in theory, the larger the mini-batch size, the better. LLM training is distributed across multiple devices, which makes it possible (and even desirable) to use large batch sizes. The batch size can vary between 32K to 8.25M tokens depending on the model. In addition, it can be dynamic and increase as the training progresses. Empirical experiments have demonstrated that this technique stabilizes the training.

Next, let's focus on the learning rate, η. Although Adam implements an adaptive learning rate, most LLMs start with a **warmup phase** to stabilize the training. More specifically, during the first 0.1% to 0.5% of the training steps, the learning rate gradually increases from around $\eta = 1*10^{-6}$ to $\eta = 1*10^{-5}$. Then, the learning rate gradually decreases to around 10% of its maximum value following a cosine (or linear) decay strategy.

LLM training also uses **gradient clipping**—a technique that prevents the exploding gradients problem. One way to implement it is to clip by value:

$$\text{If } |\mathbf{g}| \geq max\_threshold \text{ or } |\mathbf{g}| \leq min\_threshold \text{ then } \mathbf{g} \leftarrow relevant\_threshold$$

Here, **g** is a vector with all gradient values ($|\mathbf{g}|$ is the norm or the vector or its absolute value). First, we select *min_threshold* and *max_threshold* values. Then, we clip the value of the gradient to the threshold in the weight update formula if it exceeds these boundaries.

Another option is to clip by norm:

$$\text{If } |\mathbf{g}| \geq threshold, \text{ then } \mathbf{g} \leftarrow threshold * \mathbf{g}/|\mathbf{g}|$$

Here, $\mathbf{g}/|\mathbf{g}|$ is a unit vector. It has the same direction as the original, but its length is 1. The value of every element in the unit vector is in the [0:1] range. By multiplying it by the *threshold*, every element lies within the [0: threshold] range. In this way, norm clipping scales the gradients within the pre-defined threshold.

The following table summarizes the training properties of some popular LLMs:

Model	Batch Size (#tokens)	Learning Rate	Warmup	Decay Method	Optimizer	Precision Type	Weight Decay	Grad Clip	Dropout
GPT3 (175B)	32K→3.2M	6×10^{-5}	yes	cosine decay to 10%	Adam	FP16	0.1	1.0	-
PanGu-α (200B)	-	2×10^{-5}	-	-	Adam	-	0.1	-	-
OPT (175B)	2M	1.2×10^{-4}	yes	manual decay	AdamW	FP16	0.1	-	0.1
PaLM (540B)	1M→4M	1×10^{-2}	no	inverse square root	Adafactor	BF16	lr^2	1.0	0.1
BLOOM (176B)	4M	6×10^{-5}	yes	cosine decay to 10%	Adam	BF16	0.1	1.0	0.0
MT-NLG (530B)	64 K→3.75M	5×10^{-5}	yes	cosine decay to 10%	Adam	BF16	0.1	1.0	-
Gopher (280B)	3M→6M	4×10^{-5}	yes	cosine decay to 10%	Adam	BF16	-	1.0	-
Chinchilla (70B)	1.5M→3M	1×10^{-4}	yes	cosine decay to 10%	AdamW	BF16	-	-	-
Galactica (120B)	2M	7×10^{-6}	yes	linear decay to 10%	AdamW	-	0.1	1.0	0.1
LaMDA (137B)	256K	-	-	-	-	BF16	-	-	-
Jurassic-1 (178B)	32 K→3.2M	6×10^{-5}	yes	-	-	-	-	-	-
LLaMA (65B)	4M	1.5×10^{-4}	yes	cosine decay to 10%	AdamW	-	0.1	1.0	-
LLaMA 2 (70B)	4M	1.5×10^{-6}	yes	cosine decay to 10%	AdamW	-	0.1	1.0	-
GLM (130B)	0.4M→8.25M	8×10^{-5}	yes	cosine decay to 10%	AdamW	FP16	0.1	1.0	0.1
T5 (11B)	64K	1×10^{-2}	no	inverse square root	AdaFactor	-	-	-	0.1
ERNIE 3.0 Titan (260B)	-	1×10^{-4}	-	-	Adam	FP16	0.1	1.0	-
PanGu-Σ (1.085T)	0.5M	2×10^{-5}	yes	-	Adam	FP16	-	-	-

Figure 8.13 – LLM training properties (modified from https://arxiv.org/abs/2303.18223)

This concludes our introduction to pre-training LLMs. Next, let's focus on the FT phase.

FT with RLHF

So far, we have focused on the pre-training phase of LLMs. The pre-training objective of an LLM is to predict the next token based on (primarily) a web page training dataset. However, pre-trained models can express undesirable behaviors. They can often make up facts, generate biased or toxic text, or simply not follow user instructions. Yet, their purpose is to interact with humans in a *helpful*, *honest*, and *harmless* way. In this section, we'll discuss the technique of RLHF, which makes it possible to fine-tune the LLM for better alignment with human values (also known as **alignment tuning**). More specifically, we'll focus on the technique described in *Training language models to follow instructions with human feedback* (`https://arxiv.org/abs/2203.02155`) by OpenAI. They apply RLHF on a GPT-3 model to produce the GPT-3.5 family of models. This is part of the secret sauce that makes ChatGPT so good at interacting with users.

The FT starts where the pre-training ends—with the pre-trained LLM. The following diagram shows the three steps of the RLHF process:

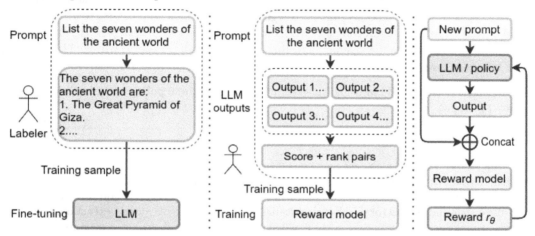

Figure 8.14 – Left: supervised FT; middle: reward model training; right:
LLM RLHF (inspired by https://arxiv.org/abs/2203.02155)

First is **supervised FT** (**SFT**, *Figure 8.14*—left). It uses human labelers to create a dataset of [`prompt:` `response`] samples, where `prompt` and `response` are source and target token sequences, respectively. This dataset serves to fine-tune the LLM using the same target as pre-training—to predict the next token of the response, given the prompt. The fine-tuned LLM serves as a base for the next two steps.

> **On the need for pre-training and three-step FT**
>
> The SFT step implicitly answers an unasked question—why do we need pre-training and three-step FT to train our model? The reason is that generating a human-labeled training dataset is not scalable and represents a major bottleneck. For example, the pre-training dataset can have over a trillion tokens; generating prompts and their respective responses of such magnitude with human labelers is infeasible. Therefore, we need pre-training to provide the LLM with a solid foundation, which we can fine-tune with a much smaller dataset.

The second step is **reward model** (**RM**) training (*Figure 8.14*—center). We start with a dataset of prompts, and we use the fine-tuned LLM to generate several responses for each prompt. Then, a human labeler assigns scalar scores to the responses according to their suitability to the prompt, using their own judgment. This step aims to train the RM to assign a score, r_θ, to a response like a human labeler. But to do so, the labeler also compares the responses and ranks (orders) them from best to worst, regardless of their scalar score. This is necessary because different labelers can give inconsistent scores of the same response, but ranking the responses is much more consistent. The prompts and the ranked responses form a new dataset, where each pair of winner and loser responses creates one training sample. For example, the responses A > B > C will form the pairs [(A, B), (A, C), (B, C)]. This dataset trains the RM, which is based on the fine-tuned LLM. Its output next-token classifier is replaced with a randomly initialized regression layer, which outputs the predicted scalar score of a given response. The RM computes the scores of both responses of each pair. The difference between them participates in the loss function. This is an example of **transfer learning** (**TL**), which aims to train the new regression layer on top of the original LLM.

The third step is to train the LLM using RL with the RM and **proximal policy optimization** (**PPO**) (*Figure 8.14—*right).

A recap of RL

To understand the third step, let's recap our introduction to RL from *Chapter 1*. We have a system of environment and an agent. The agent can take one of a number of actions that change the state of the environment. The environment reacts to the agent's actions and provides its modified state and reward (or penalty) signals that help the agent to decide its next action. The decision-making algorithm of the agent is called a policy. The agent's goal is to maximize the total rewards received throughout the training episodes.

In this scenario, the policy of the agent is represented by the fine-tuned LLM. The token vocabulary represents the actions it can take—that is, the agent's action is to select the next token in the sequence. The environment presents the LLM with a random prompt, and the agent (LLM) generates a response. Then, the RM, part of the environment, scores the generated response. The RM score is the reward sent to the agent and serves to update its parameters.

In the next section, we'll discuss what makes LLMs different from other models.

Emergent abilities of LLMs

In this section, we'll discuss the phenomenon of **emergent abilities** of LLMs, first summarized in `https://arxiv.org/abs/2206.07682`. The paper defines emergent abilities as follows:

> *An ability is emergent if it is not present in smaller models but is present in larger models.*

These abilities represent a qualitative difference between large and small language models, which cannot be predicted by extrapolation.

We'll start with the ability known as **few-shot prompting** (or **in-context learning**), popularized by GPT-3. Here, the initial user prompt is an instruction the LLM has to follow through its response without any additional training. The prompt itself may describe with natural text one or more training examples (hence, the term *few-shot*). This is the only context that the LLM can use for training before generating its response. The following diagram shows an example of a few-shot prompt:

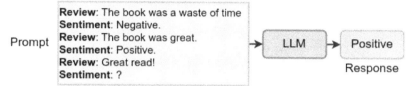

Figure 8.15 – An example of a few-shot prompt (inspired by https://arxiv.org/abs/2206.07682)

Next, let's discuss the ability of LLMs to solve complex multi-step reasoning tasks with the help of a **chain-of-thought** (**CoT**) prompting strategy (*Chain-of-Thought Prompting Elicits Reasoning in Large Language Models*, https://arxiv.org/abs/2201.11903). This type of prompt presents the LLM with a series of intermediate steps that can guide the model to reach the final task answer. The following diagram shows a comparison between regular and CoT prompts:

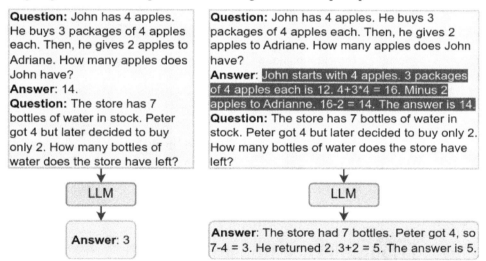

Figure 8.16 – Left: regular one-shot prompt; right: CoT one-shot
prompt (inspired by https://arxiv.org/abs/2201.11903)

It is speculated that this ability is obtained by including source code in the training data.

Let's also note that the alignment tuning we discussed in the *FT with RLHF* section is also an emergent ability, as it only improves the performance of large models.

The following diagram shows how the performance on various tasks significantly improves with the scale of the model:

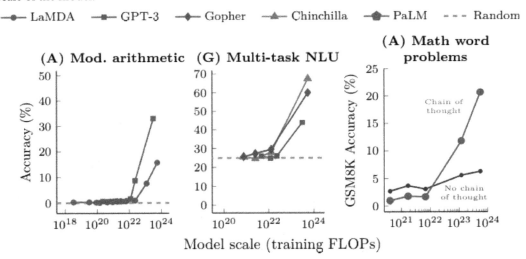

Figure 8.17 – Emergent abilities are only present in large-scale models
(source: https://arxiv.org/abs/2206.07682)

The x axis shows the training computational time for each model (measured in FLOPS), and the y axis shows the model accuracy. The graphs show the model accuracy on three different benchmarks:

- An arithmetic benchmark that tests multiplication of 2-digit numbers, as well as the addition and subtraction of 3-digit numbers

- 57 college-level tests covering a range of topics, including math, history, law, and more

- Chain-of-thought versus regular prompt on math word problems, like the one described in *Figure 8.16*

This concludes our theoretical introduction to LLMs. Next, let's see how to use them in practice.

Introducing Hugging Face Transformers

So far, we have discussed in depth the architecture and training properties of LLMs. But the sad truth is that these models are so large it is unlikely that you or I would build one from scratch. Instead, we'll probably use a pre-trained model. In this section, we'll see how to do that with the Hugging Face Transformers library (https://github.com/huggingface/transformers). As the name suggests, its focus is the transformer architecture. It supports three different backends—PyTorch, TensorFlow, and JAX (as usual, we'll focus on PyTorch). It is open source and available for commercial use. The company behind it, Hugging Face, also develops the Hugging Face Hub—a complementary service to the library cloud-based platform. It supports hosting and/or running Git repositories (such as GitHub), transformer models, datasets, and web applications (intended for **proof-of-concept** (**POC**) demos of ML applications). With that, let's proceed with our first example.

We'll start with a basic use case—we'll load a pre-trained Llama 2 chat 7B model and use it to generate a response to the user's prompt:

1. First, we add in the `import` statements:

```
import torch
from transformers import AutoTokenizer, pipeline
```

2. Then, we define the model's name in a variable:

```
model = "meta-llama/Llama-2-7b-chat-hf"
```

Every transformer model has a unique identifier, which works for the Hugging Face model hub. The Hub hosts all models, and the library can automatically download the model weights behind the scenes. In this case, we use the smallest Llama 2 7B RLHF-optimized model to preserve computational resources.

3. Next, let's load the model tokenizer:

```
tokenizer = AutoTokenizer.from_pretrained(model)
```

The various LLM models use different tokenizers. The `AutoTokenizer` instance can select the right one based on the model identifier.

4. Let's see the `tokenizer` properties by printing it with `print(tokenizer)`:

```
LlamaTokenizerFast(name_or_path='meta-llama/Llama-2-7b-chat-hf',
vocab_size=32000, model_max_length=10000000000000000198846248386
56, is_fast=True, padding_side='left', truncation_side='right',
special_tokens={'bos_token': AddedToken("<s>", rstrip=False,
lstrip=False, single_word=False, normalized=False), 'eos_
token': AddedToken("</s>", rstrip=False, lstrip=False, single_
word=False, normalized=False), 'unk_token': AddedToken("<unk>",
rstrip=False, lstrip=False, single_word=False,
normalized=False)}, clean_up_tokenization_spaces=False)
```

The tokenizer is based on a byte-level **byte-pair encoding** (**BPE**). This output gives us useful information about the token vocabulary size, special tokens, and other properties.

5. Then, we create a `pipeline` instance:

```
text_gen_pipeline = pipeline(
    task='text-generation',
    model=model,
    tokenizer=tokenizer,
    torch_dtype=torch.bfloat16,
    device_map='auto',
)
```

The pipeline abstraction makes it easy to use the models for inference. The `task` parameter determines the type of task to solve. The library supports multiple tasks, also covering images and audio. `pipeline` will return different objects, depending on the task. It also takes care of downloading and initializing the model. In addition, we set the datatype to `torch.bfloat16` to reduce the memory footprint. The `device_map='auto'` parameter allows the Accelerate library (`https://github.com/huggingface/accelerate`) to run the model across any distributed configuration automatically.

6.   We can see the model definition with the following command: `print(text_gen_pipeline.model)`. For example, the command output for the largest 70B Llama 2 model, `Llama-2-70b-hf`, is this:

```
LlamaForCausalLM(
    (model): LlamaModel(
        (embed_tokens): Embedding(32000, 8192)
        (layers): ModuleList(
            (0-79): 80 x LlamaDecoderLayer(
                (self_attn): LlamaAttention(
                    (q_proj): Linear(in=8192, out=8192)
                    (k_proj): Linear(in=8192, out=1024)
                    (v_proj): Linear(in=8192, out=1024)
                    (o_proj): Linear(in=8192, out=8192)
                    (rotary_emb): LlamaRotaryEmbedding()
                )
                (mlp): LlamaMLP(
                    (gate_proj): Linear(in=8192, out=28672)
                    (up_proj): Linear(in=8192, out=28672)
                    (down_proj): Linear(in=28672, out=8192)
                    (act_fn): SiLUActivation()
                )
                (input_layernorm): LlamaRMSNorm()
                (post_attention_layernorm): LlamaRMSNorm()
            )
        )
        (norm): LlamaRMSNorm()
    )
    (lm_head): Linear(in=8192, out=32000)
)
```

To fit the page line length, I have modified the original output: `in` stands for `in_features`, `out` stands for `out_features`, and all linear layers have an additional `bias=False` parameter. The token vocabulary size is 32,000, and the embedding size (d_{model}) is 8,192. The model has 80 identical decoder blocks (`LlamaDecoderLayer`). Each block contains a self-attention sublayer (`*_proj` are the projections), a FFN with a single hidden layer

(LlamaMLP), rotary embeddings (LlamaRotaryEmbedding), **root mean square** (**RMS**) normalization (LlamaRMSNorm), and SiLU activation (SiLUActivation). Let's note that the activation differs from the SwiGLU activation defined in the paper.

7. Then, we run the inference:

```
sequences = text_gen_pipeline(
    text_inputs='What is the answer to the ultimate question of
life, the universe, and everything?',
    max_new_tokens=200,
    num_beams=2,
    top_k=10,
    top_p=0.9,
    do_sample=True,
    num_return_sequences=2,
)
```

Here, text_inputs is the user prompt, which serves as the initial input sequence. The num_return_sequences=2 parameter indicates that model will generate two separate responses (more on that later). Here's the first response:

```
Answer: The answer to the ultimate question of life, the
universe, and everything is 42.
Explanation:
The answer 42 is a humorous and satirical response to the idea
that there is a single, definitive answer to the ultimate
questions of life, the universe, and everything. The answer was
first proposed by Douglas Adams in his science fiction series
"The Hitchhiker's Guide to the Galaxy," where
```

As we can see, the response is factually correct, but it was interrupted because of the limit set by the max_new_tokens=200 parameter.

Let's analyze the rest of the arguments of the text_gen_pipeline call, as they all relate to the strategy of generating new tokens. The LLM ends with a softmax operation, which outputs a probability distribution over all tokens of the vocabulary. The simplest way to select the next token is a greedy strategy, which always takes the one with the highest probability. However, this is often suboptimal because it can hide high-probability words behind low-probability ones. To clarify, a token might be assigned a low probability at the current state of the generated sequence, and another token would be selected in its place. This means that the potential sequence, which includes the current low-probability token, will not exist. Therefore, even if it had high-probability tokens down the line, we would never know because the low-probability token blocks it from further exploration. One way to solve this is with a **beam search** strategy by setting do_sample=True. In this case, the algorithm takes the probability of the entire current sequence rather than just the probability of the latest token. Therefore, the new token will be the one that maximizes the overall probability of the sequence instead of its local probability. The num_beams=2 parameter indicates that the algorithm always keeps the two sequences (beams) with the highest probability. Since we can have more than one output sequence, the `num_return_sequences=2` parameter indicates the number of sequences to return. For example, if `num_beams=5` and `num_return_sequences=3`, the algorithm will return the three highest-probability sequences out of all five available (`num_return_sequences > num_beams` are invalid arguments). The `early_stopping=True` parameter indicates that the generation is finished when all beam hypotheses reach the end-of-sequence (`[EOS]`) token. The `top_k=10` parameter indicates that the algorithm will only sample the top 10 highest-probability tokens, regardless of their sequence probabilities. `top_p=0.9` is like `top_k`, but instead of sampling only from the most likely *k* tokens, it selects from the smallest possible set of tokens whose combined probability exceeds the probability *p*.

This concludes our introduction to the Transformers library and the whole chapter.

Summary

LLMs are very large transformers with various modifications to accommodate the large size. In this chapter, we discussed these modifications, as well as the qualitative differences between LLMs and regular transformers. First, we focused on their architecture, including more efficient attention mechanisms such as sparse attention and prefix decoders. We also discussed the nuts and bolts of the LLM architecture. Next, we surveyed the latest LLM architectures with special attention given to the GPT and LlaMa series of models. Then, we discussed LLM training, including training datasets, the Adam optimization algorithm, and various performance improvements. We also discussed the RLHF technique and the emergent abilities of LLMs. Finally, we introduced the Hugging Face Transformers library.

In the next chapter, we'll discuss transformers for **computer vision** (**CV**), multimodal transformers, and we'll continue our introduction to the Transformers library.

9

Advanced Applications of Large Language Models

In the previous two chapters, we introduced the transformer architecture and learned about its latest large-scale incarnations, known as **large language models** (**LLMs**). We discussed them in the context of **natural language processing** (**NLP**) tasks. NLP was the original transformer application and is still the field at the forefront of LLM development today. However, the success of the architecture has led the research community to explore the application of transformers in other areas, such as computer vision.

In this chapter, we'll focus on these areas. We'll discuss transformers as replacements for convolutional networks (CNNs, *Chapter 4*) for tasks such as image classification and object detection. We'll also learn how to use them as generative models for images instead of text, as we have done until now. We'll also implement a model fine-tuning example – something we failed to do in *Chapter 8*. And finally, we'll implement a novel LLM-driven application.

In this chapter, we will cover the following main topics:

- Classifying images with Vision Transformer
- Detection transformer
- Generating images with stable diffusion
- Fine-tuning transformers
- Harnessing the power of LLMs with LangChain

Technical requirements

We'll implement the example in this chapter using Python, PyTorch, the Hugging Face Transformers library (`https://github.com/huggingface/transformers`), and the LangChain framework (`https://www.langchain.com/`, `https://github.com/langchain-ai/langchain`). If you don't have an environment with these tools, fret not – the example is available as a Jupyter Notebook on Google Colab. The code examples can be found in this book's GitHub repository: `https://github.com/PacktPublishing/Python-Deep-Learning-Third-Edition/tree/main/Chapter09`.

Classifying images with Vision Transformer

Vision Transformer (**ViT**, *An Image is Worth 16x16 Words: Transformers for Image Recognition at Scale*, `https://arxiv.org/abs/2010.11929`) proves the adaptability of the attention mechanism by introducing a clever technique for processing images. One way to use transformers for image inputs is to encode each pixel with four variables – pixel intensity, row, column, and channel location. Each pixel encoding is an input to a simple **neural network** (**NN**), which outputs a d_{model}-dimensional embedding vector. We can represent the three-dimensional image as a one-dimensional sequence of these embedding vectors. It acts as an input to the model in the same way as a token embedding sequence does. Each pixel will attend to every other pixel in the attention blocks.

This approach has some disadvantages related to the length of the input sequence (context window). Unlike a one-dimensional text sequence, an image has a two-dimensional structure (the color channel doesn't increase the number of pixels). Therefore, the input sequence length increases quadratically as the image size increases. Even a small 64×64 image would result in an input sequence with a length of 64*64=4,096. On one hand, this makes the model computationally intensive. On the other hand, as each pixel attends to the entire long sequence, it will be hard for the model to learn the structure of the image. CNNs approach this problem by using filters, which restrict the input size of a unit only to its immediate surrounding area (receptive field). To understand how ViT solves this problem, let's start with the following figure:

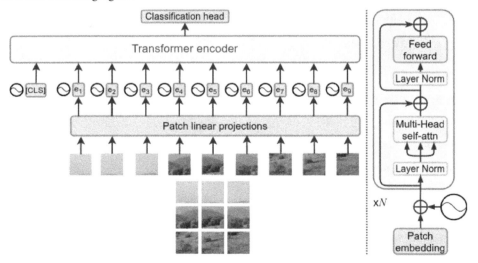

Figure 9.1 – Vision Transformer. Inspired by `https://arxiv.org/abs/2010.11929`

Let's denote the input image resolution with *(H, W)* and the number of channels with *C*. Then, we can represent the input image as a tensor, $\mathbf{x} \in \mathbb{R}^{H \times W \times C}$. ViT splits the image into a sequence of two-dimension square patches, $\mathbf{x}_p \in \mathbb{R}^{N \times P^2 \times C}$ *(Figure 9.1)*. Here, *(P, P)* is the resolution of each image patch *(P=16)* and $N = (H \times W)/P^2$ is the number of patches (which is also the input sequence length). The sequence of patches serves as input to the model, in the same way as a token sequence does.

Next, the input patches, $\mathbf{x}_p$, serve as input to a linear projection, which outputs a d_{model}-dimensional **patch embedding** vector for each patch. The patch embeddings form the input sequence, $\mathbf{z}_0$. We can summarize the patch-to-embedding process with the following formula:

$$\mathbf{z}_o = \left[\mathbf{x}_{cls}; \mathbf{x}_p^{(1)} E; \mathbf{x}_p^{(2)} E; \ldots \mathbf{x}_p^{(N)} E \right] + \mathbf{E}_{pos}$$

Here, $\mathbf{E} \in \mathbb{R}^{(P^2 \cdot C) \times d_{model}}$ is the linear projection and $\mathbf{E}_{pos} \in \mathbb{R}^{(N+1) \times d_{model}}$ is the static positional encoding (the same as in the original transformer).

Once we have the embedding sequence, ViT processes it with a standard encoder-only pre-normalization transformer, similar to BERT *(Chapter 7)*. It comes in three variants, displayed as follows:

Variant	Layers	d_{model}	FFN size	Heads	Params
ViT-Base	12	768	3072	12	86M
ViT-Large	24	1024	4096	16	307M
ViT-Huge	32	1280	5120	16	632M

Figure 9.2 – ViT variants. Based on https://arxiv.org/abs/2010.11929

The encoder architecture uses unmasked self-attention, which allows a token to attend to the full sequence rather than the preceding tokens only. This makes sense because the preceding or the next element doesn't carry the same meaning in the relationship between pixels of an image as the order of the elements in a text sequence. The similarities between the two models don't end here. Like BERT, the input sequence starts with a special [CLS] ($\mathbf{x}_{cls}$) token (for classification tasks). The model output for the $\mathbf{x}_{cls}$ token is the output for the full image. In this way, the $\mathbf{x}_{cls}$ token attends to the entire input sequence (that is, the entire image). Alternatively, if we take the model output for any other patch, we will introduce an imbalance between the selected patch and the others of the sequence.

Following the example of BERT, ViT has pre-training and fine-tuning phases. Pre-training uses large general-purpose image datasets (such as ImageNet) while fine-tuning trains the model on smaller task-specific datasets.

The model ends with a **classification head**, which contains one hidden layer during pre-training and no hidden layers during fine-tuning.

One issue with ViT is that it performs best when pre-training with very large datasets, such as JFT-300M with 300M labeled images (*Revisiting Unreasonable Effectiveness of Data in Deep Learning Era*, https://arxiv.org/abs/1707.02968). This makes the training a lot more computationally intensive versus a comparable CNN. Many further variants of ViT try to solve this challenge and propose other improvements to the original model. You can find out more in *A Survey on Visual Transformer* (`https://arxiv.org/abs/2012.12556`), which is regularly updated with the latest advancements in the field.

With that, let's see how to use ViT in practice.

Using ViT with Hugging Face Transformers

In this section, we'll implement a basic example of ViT image classification with the help of Hugging Face Transformers and its `pipeline` abstraction, which we introduced in *Chapter 8*. Let's start:

1. Import the `pipeline` abstraction:

    ```
    from transformers import pipeline
    ```

2. Create an image classification pipeline instance. The pipeline uses the ViT-Base model:

    ```
    img_classification_pipeline = pipeline(
        task="image-classification",
        model="google/vit-base-patch16-224")
    ```

3. Run the instance with an image of a bicycle from Wikipedia:

    ```
    img_classification_pipeline("https://upload.wikimedia.org/
    wikipedia/commons/thumb/4/41/Left_side_of_Flying_Pigeon.
    jpg/640px-Left_side_of_Flying_Pigeon.jpg")
    ```

    This code outputs the following top-5 class probability distribution (only the first class is displayed):

    ```
    [{'score': 0.4616938531398773, 'label': 'tricycle,
        trike, velocipede'}]
    ```

This example is simple enough, but let's dive in and analyze the ViT model itself. We can do this with the `print(img_classification_pipeline.model)` command, which outputs the following:

```
ViTForImageClassification(
  (vit): ViTModel(
    (embeddings): ViTEmbeddings(
      (patch_embeddings): ViTPatchEmbeddings(
        (projection): Conv2d(3, 768,
                      kernel_size=(16, 16),
                      stride=(16, 16))
      )
      (dropout): Dropout(p=0.0)
    )
```

```
        (encoder): ViTEncoder(
            (layer): ModuleList(
                (0-11): 12 x ViTLayer(
                    (attention): ViTAttention(
                        (attention): ViTSelfAttention(
                            (query): Linear(in_f=768,
                                    out_f=768)
                            (key): Linear(in_f=768, out_f=768)
                            (value): Linear(in_f=768,
                                    out_f=768)
                            (dropout): Dropout(p=0.0)
                        )
                        (output): ViTSelfOutput(
                            (dense): Linear(in_f=768,
                                    out_f=768)
                            (dropout): Dropout(p=0.0)
                        )
                    )
                    (intermediate): ViTIntermediate(
                        (dense): Linear(in_f=768, out_f=3072)
                        (intermediate_act_fn):GELUActivation()
                    )
                    (output): ViTOutput(
                        (dense): Linear(in_f=3072, out_f=768)
                        (dropout): Dropout(p=0.0)
                    )
                    (layernorm_before): LayerNorm((768,))
                    (layernorm_after): LayerNorm((768,))
                )
            )
        )
        (layernorm): LayerNorm((768,))
    )
    (classifier): Linear(in_f=768, out_f=1000)
)
```

The model works with 224×224 input images. Here, `in_f` and `out_f` are shortened for `in_features` and `out_features`, respectively. Unlike other models, ViT uses bias in all `Linear` layers (the `bias=True` input parameter is not displayed). Let's discuss the components of the model in the order that they appear:

- `ViTEmbeddings`: The patch embedding block. It contains a 2D convolution with a 16×16 filter size, stride of 16, three input channels (one for each color), and 768 output channels (d_{model} = 768). Applying the convolutional filter at each location produces one 768-dimensional patch embedding per location of the input image. Since the patches form a two-dimensional grid (the same as the input), the output is flattened to a one-dimensional sequence. This block also adds positional encoding information, which is not reflected in its string representation. The dropout probability of all dropout instances is 0 because the model runs in inference rather than training mode.

- `ViTEncoder`: The main encoder model contains 12 `ViTLayer` pre-ln (`LayerNorm`) encoder block instances. Each contains the following:

  - `ViTAttention` attention block: `ViTSelfAttention` multi-head attention and its output linear projection, `ViTSelfOutput`. All **fully connected** (**FC**) layers in the block have a size equal to d_{model} = 768, despite the multiple attention heads. This is possible because each of the N attention heads has a size of d_{model}/N.

  - **Feed-forward network** (**FFN**): A combination of two linear layers, `ViTIntermediate` plus `GELUActivation` and `ViTOutput`.

  - Classification head (`classifier`): In inference mode, the classification head has only one `Linear` layer with 1,000 outputs (because the model was fine-tuned on the ImageNet dataset).

Next, let's see how object detection with transformers works.

Understanding the DEtection TRansformer

DEtection TRansformer (**DETR**, *End-to-End Object Detection with Transformers*, `https://arxiv.org/abs/2005.12872`) introduces a novel transformer-based object detection algorithm.

A quick recap of the YOLO object detection algorithm

We first introduced YOLO in *Chapter 5*. It has three main components. The first is the backbone – that is, a CNN model that extracts features from the input image. Next is the neck – an intermediate part of the model that connects the backbone to the head. Finally, the head outputs the detected objects using a multi-step algorithm. More specifically, it splits the image into a grid of cells. Each cell contains several pre-defined anchor boxes with different shapes. The model predicts whether any of the anchor boxes contains an object and the coordinates of the object's bounding box. Many of the boxes will overlap and predict the same object. The model filters the overlapping objects with the help of intersection-over-union and non-maximum suppression.

Like YOLO, DetR starts with a CNN backbone. However, it replaces the neck and the head with a full post-normalization transformer encoder-decoder. This negates the need for hand-designed components such as the non-maximum suppression procedure or anchor boxes. Instead, the model outputs a set of bounding boxes and class labels for the detected objects. To understand how it works, we'll start with the following figure, which displays the components of DetR:

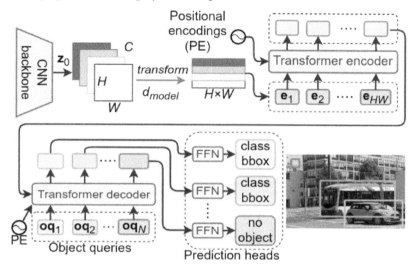

Figure 9.3 – DetR architecture. Inspired by https://arxiv.org/abs/2005.12872

First, the backbone CNN extracts the features from the input image, the same as in YOLO. Its outputs are the feature maps of the last convolutional layer. The original input image is a tensor with a shape of $\mathbf{x}_{img} \in \mathbb{R}^{C_0 \times H_0 \times W_0}$, where $C_0 = 3$ is the number of color channels and H_0/W_0 are the image dimensions. The last convolution output is a tensor with a shape of $\mathbf{z}_0 \in \mathbb{R}^{C \times H \times W}$. Typically, the number of output feature maps is $C=2048$, and their height and width are $H = H_0/32$ and $W = W_0/32$, respectively.

However, the three-dimensional (excluding the batch dimension) backbone output is incompatible with the expected input tensor of the encoder, which should be a one-dimensional input sequence of d_{model}-sized embedding tensors ($d_{model} < C$). To solve this, the model applies 1×1 bottleneck convolution, which downsamples the number of channels from C to d_{model}, followed by a flattening operation. The transformed tensor becomes $\mathbf{z}_0 \in \mathbb{R}^{d_{model} \times H \cdot W}$, which we can use as a transformer input sequence.

Next, let's focus on the actual transformer, which is displayed in detail in the following figure:

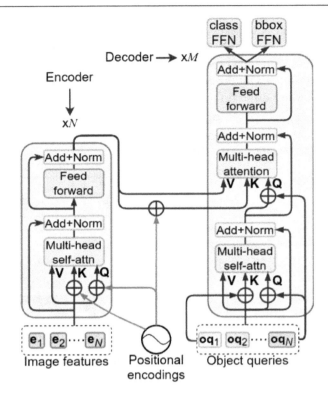

Figure 9.4 – DetR transformer in detail. Inspired by https://arxiv.org/abs/2005.12872

The encoder maps the input sequence to a sequence of continuous representations, just like the original encoder (*Chapter 7*). One difference is that the model adds fixed absolute positional encodings to each **Q/K** tensor of all attention layers of the encoder, as opposed to static positional encodings added only to the initial input tensor of the original transformer.

The decoder is where it gets more interesting. First, let's note that the fixed positional encodings also participate in the decoder's encoder-decoder attention block. Since they participate in all self-attention blocks of the encoder, we propagate them to the encoder-decoder attention to level the playing field.

Next, the encoder takes as input a sequence of N **object queries**, represented by tensors, $\mathbf{oq}_i \in \mathbb{R}^{N \times d_{model}}$. We can think of them as slots, which the model uses to detect objects. The model output for each input object query represents the properties (bounding box and class) of one detected object. Having N object queries means that the model can detect N objects at most. Because of this, the paper's authors propose to use N, which is significantly larger than the typical number of objects in an image. Unlike the original transformer, the decoder's attention here isn't masked, so it can detect all objects in parallel rather than sequentially.

At the start of the training process, the object query tensors are initialized randomly. The training itself updates both the model weights and the query tensors – that is, the model learns the object queries alongside the model weights. They act as learned positional encodings of the detected objects and serve the same purpose as the initial fixed input positional encodings. Because of this, we add the object queries to the encoder-decoder attention and the self-attention layers of the decoder blocks in the same way we add the input positional encodings to the encoder. This architecture has a sort of *bug* – the very first self-attention layer of the first decoder block will take as input the same object query twice, making it useless. Empirical experiments show that this doesn't degrade the model performance. For the sake of simplicity, the implementation doesn't have a unique first decoder block without self-attention but uses the standard decoder block instead.

Encoding configurations

The model can work with multiple configurations of the fixed and learned encodings:

- add both types of encodings only to input data;
- add the fixed encodings to the input data and the learned encodings to the input and all decoder attention layers;
- add the fixed encodings to the data and all encoder attention layers and the learned encodings only to the decoder input;
- add both types of encodings to the input data and every attention layer of the encoder and the decoder.

The model works best in the fourth configuration, but for the sake of simplicity, it can be implemented in the first.

The object queries make it possible to not impose prior geometric limitations such as grid cells and anchor boxes in YOLO. Instead, we specify only the maximum number of objects to detect and let the model do its magic. The learned queries tend to specialize over different regions of the image. However, this is a result of the training and the properties of the training dataset, as opposed to manually crafted features.

The model ends with a combination of two heads: a three-layer perceptron with ReLU activations and a separate FC layer. The perceptron is called an FFN, which differs from the FFNs in the transformer blocks. It predicts the detected object bounding box height, width, and normalized center coordinates concerning the input image. The FC has softmax activation and predicts the class of the object. Like YOLO, it includes an additional special background class, which indicates that no object is detected within the slot. Having this class is even more necessary because some slots will inevitably be empty, as N is much larger than the number of objects in the image.

Predicting a set of unrestricted bounding boxes poses a challenge to the training because it is not trivial to match the predicted boxes with the ground-truth ones. The first step is to pad the ground-truth boxes for each image with dummy entries, so the number of ground-truth boxes becomes equal to the number of predicted ones, N. Next, the training uses one-to-one **bipartite matching** between the predicted and ground-truth boxes. Finally, the algorithm supervises each predicted box to be closer to the ground-truth box it was matched to. You can check out the paper for more details on the training.

> **DetR for image segmentation**
>
> The authors of DetR extend the model for image segmentation. The relationship between DetR for detection and segmentation is similar to the one between Faster R-CNN and Mask R-CNN (*Chapter 5*). DetR for segmentation adds a third head that's implemented with upsampling convolutions. It produces binary segmentation masks for each detected object in parallel. The final result merges all masks using pixel-wise argmax.

Using DetR with Hugging Face Transformers

In this section, we'll implement a basic example of DetR object detection with the help of Hugging Face Transformers and its `pipeline` abstraction, which we introduced in *Chapter 8*. This example follows the ViT pattern, so we'll include the full code without any comments. Here it is:

```
from transformers import pipeline
obj_detection_pipeline = pipeline(
    task="object-detection",
    model="facebook/detr-resnet-50")
obj_detection_pipeline("https://upload.wikimedia.org/wikipedia/
commons/thumb/4/41/Left_side_of_Flying_Pigeon.jpg/640px-Left_side_of_
Flying_Pigeon.jpg")
```

The last call returns a list of detected objects in the following form:

```
{'score': 0.997983455657959,
   'label': 'bicycle',
   'box': {'xmin': 16, 'ymin': 14, 'xmax': 623, 'ymax': 406}}
```

Next, we can see the model definition with the `print(obj_detection_pipeline.model)` command. Here, `in_f` and `out_f` are shortened for `in_features` and `out_features`, respectively. DetR uses bias in all `Linear` layers (the `bias=True` input parameter is not displayed). We'll omit the backbone definition.

Let's discuss the model elements in the order that they appear, starting with the 1×1 bottleneck convolution (we have $d_{model} = 256$):

```
(input_projection): Conv2d(2048, 256,
                    kernel_size=(1, 1),
                    stride=(1, 1))
```

Next, we have the object query embedding (*N=100*). As we mentioned, the object queries are learned alongside the weight updates during training:

```
(query_position_embeddings): Embedding(100, 256)
```

The following is the encoder with six post-ln encoder blocks, ReLU activation, and FFN with one 2,048-dimensional hidden layer. Note that the positional encodings are not displayed (the same applies to the decoder):

```
(encoder): DetrEncoder(
  (layers): ModuleList(
    (0-5): 6 x DetrEncoderLayer(
      (self_attn): DetrAttention(
        (k_proj): Linear(in_f=256, out_f=256)
        (v_proj): Linear(in_f=256, out_f=256)
        (q_proj): Linear(in_f=256, out_f=256)
        (out_proj): Linear(in_f=256, out_f=256)
      )
      (self_attn_layer_norm): LayerNorm((256,))
      (activation_fn): ReLU()
      (fc1): Linear(in_f=256, out_f=2048)
      (fc2): Linear(in_f=2048, out_f=256)
      (final_layer_norm): LayerNorm((256,))
    )
  )
)
```

Then, we have the decoder with six post-ln decoder blocks and the same properties as the encoder:

```
(decoder): DetrDecoder(
  (layers): ModuleList(
    (0-5): 6 x DetrDecoderLayer(
      (self_attn): DetrAttention(
        (k_proj): Linear(in_f=256, out_f=256)
        (v_proj): Linear(in_f=256, out_f=256)
        (q_proj): Linear(in_f=256, out_f=256)
        (out_proj): Linear(in_f=256, out_f=256)
      )
      (activation_fn): ReLU()
      (self_attn_layer_norm): LayerNorm((256,))
      (encoder_attn): DetrAttention(
        (k_proj): Linear(in_f=256, out_f=256)
        (v_proj): Linear(in_f=256, out_f=256)
        (q_proj): Linear(in_f=256, out_f=256)
        (out_proj): Linear(in_f=256, out_f=256)
      )
      (encoder_attn_layer_norm): LayerNorm((256,))
      (fc1): Linear(in_f=256, out_f=2048)
      (fc2): Linear(in_f=2048, out_f=256)
```

```
        (final_layer_norm): LayerNorm((256,))
    )
  )
  (layernorm): LayerNorm((256,))
)
```

Finally, we have the output FFN and linear layer. The FFN outputs four values (the bounding box coordinates), and the linear layer can detect 91 classes and the background:

```
(class_labels_classifier): Linear(in_f=256, out_f=92)
(bbox_predictor): DetrMLPPredictionHead(
  (layers): ModuleList(
    (0-1): 2 x Linear(in_f=256, out_f=256)
    (2): Linear(in_f=256, out_f=4)
  )
)
```

Next, let's see how we can generate new images with transformers.

Generating images with stable diffusion

In this section, we'll introduce **stable diffusion** (**SD**, *High-Resolution Image Synthesis with Latent Diffusion Models*, https://arxiv.org/abs/2112.10752, https://github.com/ Stability-AI/stablediffusion). This is a generative model that can synthesize images based on text prompts or other types of data (in this section, we'll focus on the text-to-image scenario). To understand how it works, let's start with the following figure:

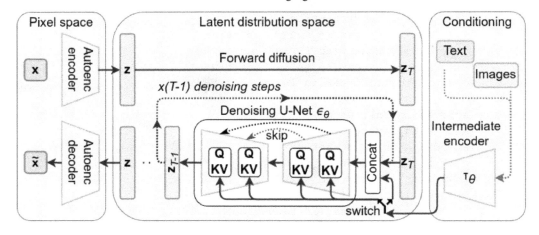

Figure 9.5 – Stable diffusion model and training. Inspired by https://arxiv.org/abs/2112.10752

SD combines an autoencoder (**AE**, the *Pixel space* section of *Figure 9.5*), denoising diffusion probabilistic models (**DDPM** or simply **DM**, the *Latent distribution space* section of *Figure 9.5* and *Chapter 5*), and transformers (the *Conditioning* section of *Figure 9.5*). Before we dive into each of these components, let's outline their role in the training and inference pipelines of SD. Training involves all of them – AE encoder, forward diffusion, reverse diffusion (**U-Net**, *Chapter 5*), AE decoder, and conditioning. Inference (generating images from text) only involves reverse diffusion, AE decoder, and conditioning. Don't worry if you don't understand everything you just read, as we'll go into more detail in the following sections. We'll start with the AE, continue with the conditioning transformer, and combine it when we discuss the diffusion process.

Autoencoder

Although we mentioned AEs briefly in *Chapter 1*, we'll introduce this architecture in more detail here, starting with the following figure:

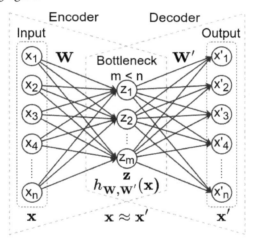

Figure 9.6 – An AE

An AE is a feed-forward neural network that tries to reproduce its input. In other words, an AE's target value (label), **y**, equals the input data, **x**. We can formally say that it tries to learn an identity function, $h_{w,w'}(\mathbf{x}) = \mathbf{x}$ (a function that repeats its input). In its most basic form, an AE consists of hidden (or bottleneck) and output layers (**W** and **W'** are the weight matrices of these layers). Like U-Net, we can think of the autoencoder as a virtual composition of two components:

- **Encoder**: This maps the input data to the network's internal latent representation. For the sake of simplicity, in this example, the encoder is a single FC bottleneck layer. The internal state is just its activation tensor, **z**. The encoder can have multiple hidden layers, including convolutional ones (as in SD). In this case, **z** is the activation of the last layer.

- **Decoder**: This tries to reconstruct the input from the network's internal state, **z**. The decoder can also have a complex structure that typically mirrors the encoder. While U-Net tries to translate the input image into a target image of some other domain (for example, a segmentation map), the autoencoder simply tries to reconstruct its input.

We can train the autoencoder by minimizing a loss function, known as the **reconstruction error**. It measures the distance between the original input and its reconstruction.

The latent tensor, **z**, is the focus of the entire AE. The key is that the bottleneck layer has fewer units than the input/output ones. Because the model tries to reconstruct its input from a smaller feature space, we force it to learn only the most important features of the data. Think of the compact data representation as a form of compression (but not lossless). We can use only the encoder part of the model to generate latent tensors for downstream tasks. Alternatively, we can use only the decoder to synthesize new images from generated latent tensors.

During training, the encoder maps the input sample to the latent space, where each latent attribute has a discrete value. An input sample can have only one latent representation. Therefore, the decoder can reconstruct the input in only one possible way. In other words, we can generate a single reconstruction of one input sample. However, we want to generate new images conditioned on text prompts rather than recreating the original ones. One possible solution to this task is **variational autoencoders** (**VAEs**). A VAE can describe the latent representation in probabilistic terms. Instead of discrete values, we'll have a probability distribution for each latent attribute, making the latent space continuous. We can modify the latent tensor to influence the probability distribution (that is, the properties) of the generated image. In SD, the DM component, combined with the conditioning text prompts, acts as this modifier.

With this short detour completed, let's discuss the role of the convolutional encoder in SD (the *Pixel space* section of *Figure 9.5*). During training, the AE encoder creates a compressed initial latent representation tensor, $z \in \mathbb{R}^{h \times w \times c}$, of the input image, $x \in \mathbb{R}^{H \times W \times 3}$. More specifically, the encoder downsamples the image by a factor, $f = H/h = W/w$, where $f = 2^m$ (m is an integer selected by empirical experiments). Then, the entire diffusion process (forward and reverse) works with the compressed **z** rather than the original image, **x**. Only when the reverse diffusion ends does the AE decoder upsample the newly generated representation, **z**, into the final generated image, $\tilde{x}$. In this way, the smaller **z** allows the use of a smaller and more computationally efficient U-Net, which benefits both the training and the inference. The paper's authors refer to this combination of AEs and diffusion models as **latent diffusion models**.

The AE training is separate from the U-Net training. Because of this, we can train the AE once and then use it for multiple downstream tasks with different U-Net configurations.

Conditioning transformer

The conditioning transformer, τ_θ (*Figure 9.5*), produces a latent representation of the text description of the desired image. SD provides this representation to the U-Net so that it can influence its output. For this to work, the text latent representation has to live in the same semantic (not just dimensional) space as the image latent representation of the U-Net. To achieve this, the latest version of SD, 2.1, uses the OpenCLIP open source model as a conditioning transformer (*Reproducible scaling laws for contrastive language-image learning*, https://arxiv.org/abs/2212.07143). **CLIP** stands for **contrastive language-image pre-training**. This technique was introduced by OpenAI (*Learning Transferable Visual Models From Natural Language Supervision*, https://arxiv.org/abs/2103.00020). Let's discuss it in more detail, starting with the following figure:

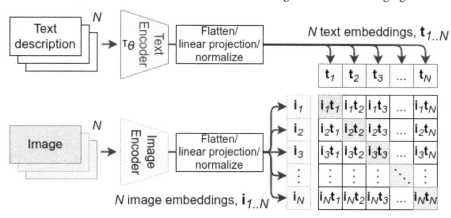

Figure 9.7 – CLIP. Inspired by https://arxiv.org/abs/2103.00020

It has two main components:

- **Text encoder**: This is a transformer that takes a token-encoded text sequence as input and outputs an embedding vector, **t**. The output is simply the activation of the last layer of the last transformer block rather than a task-specific head. The text encoder follows the principles we described in the last two chapters. For example, OpenAI CLIP uses a transformer-decoder with **byte-pair encoding** (**BPE**) tokenization. The input token sequence could include the special [EOS] token. The model output at this token serves as an embedding vector of the entire sequence. In the context of SD, we're only interested in the text encoder, and all other components of the CLIP system are only necessary for its training.

- **Image encoder**: This is either a ViT or a CNN (most often ResNet). It takes an image as input and outputs its embedding vector, **i**. Like the text encoder, this is the activation of the highest layer of the model and not a task-specific head.

For CLIP to work, the embedding vectors of the two encoders must have the same size, d_{model}. When necessary (for example, in the case of the CNN image encoder), the encoder's output tensor is flattened to a one-dimensional vector. If the dimensions of the two encoders still differ, we can add linear projections (FC layers) to equalize them.

Next, let's focus on the actual pre-training algorithm. The training set contains N text-image pairs, where the text of each pair describes the content of its corresponding image. We feed all text representations to the text encoder and the images to the image encoder to produce the $\mathbf{t}_{1..N}$ and $\mathbf{i}_{1..N}$ embeddings, respectively. Then, we compute a cosine similarity between every two embedding vectors (a total of $N{\times}N$ similarity measurements). Within these measurements, we have N correctly matching text-image pairs (the table diagonal of *Figure 9.5*) and $N{\times}N$-N incorrect pairs (all pairs outside the table diagonal). The training updates the weights of the two encoders so that the similarity scores for the correct pairs are maximized and the incorrect ones are minimized. Should the training prove successful, we'll have similar embeddings for text prompts that correctly describe what's on the image and dissimilar embeddings in all other cases. During SD training, we optimize the text encoder alongside the U-Net (but not the full CLIP system).

Now that we know how to produce semantically correct text embeddings, we can proceed with the actual diffusion model.

Diffusion model

DM is a type of generative model that has forward and reverse phases. Forward diffusion starts with the latent vector, $\mathbf{z}$, produced by the AE encoder (which takes an image, $\mathbf{x}$, as input). Then, it gradually adds random Gaussian noise to $\mathbf{z}$ through a series of T steps until the final (latent) representation, $\mathbf{z}_T$, is pure noise. Forward diffusion uses an accelerated algorithm, which produces $\mathbf{z}_T$ in a single step instead of T steps (*Chapter 5*).

Reverse diffusion does the opposite and starts with pure noise. It gradually tries to restore the original latent tensor, $\mathbf{z}$, by removing small amounts of noise in a series of T denoising steps. In practice, we're interested in reverse diffusion to generate images based on latent representations (forward diffusion only participates in the training). It is usually implemented with a U-Net type of CNN (*Figure 9.5*, ϵ_θ). It takes the noise tensor, $\mathbf{z}_t$, at step t as input and outputs an approximation of the noise added to the original latent tensor, $\mathbf{z}$ (that is, only the noise and not the tensor itself). Then, we subtract the predicted noise from the current U-Net input, $\mathbf{z}_t$, and feed the result as new input to the U-Net, $\mathbf{z}_{t+1}$. During training, the cost function measures the difference between the predicted and actual noise and accordingly updates the U-Net weights after each denoising step. This process continues until (hopefully) only the original tensor, $\mathbf{z}$, remains. Then, the AE decoder uses it to produce the final image.

The pure form of DM has no way to influence the properties of the generated image (this is known as conditioning) because we start with random noise, which results in random images. SD allows us to do just that – a way to condition the U-Net to generate images based on specific text prompts or other data types. To do this, we need to integrate the output embedding of the conditioning transformer with the denoising U-Net. Let's assume that we have a text prompt, $\mathbf{y}$. We feed it to the conditioning transformer, τ_θ, which produces an output, $\tau_\theta(\mathbf{y}) \in \mathbb{R}^{M{\times}d_\tau}$, where d_τ is the embedding size of the conditioning transformer, and M is the number of tokens of the input sequence. Usually, $M{=}1$ because we only take the output at the [EOS] token. Then, we map its output to the intermediate layers of the

U-Net via a **cross-attention** layer. In this layer, the key and value tensors represent the conditioning transformer outputs, and the query tensors represent the intermediate U-Net layers (*Figure 9.5*):

$$\mathbf{Q} = \mathbf{W}_Q^{(i)} \cdot \varphi_i(\mathbf{z}_t)$$

$$\mathbf{K} = \mathbf{W}_K^{(i)} \cdot \tau_\theta(\mathbf{y})$$

$$\mathbf{V} = \mathbf{W}_V^{(i)} \cdot \tau_\theta(\mathbf{y})$$

Here, i is the i-th intermediate U-Net layer, $\varphi_i(\mathbf{z}_t) \in \mathbb{R}^{N \times d_{i\varepsilon}}$ is the flattened activation of that layer, and $d_{i,\varepsilon}$ is the flattened activation tensor size. $\mathbf{W}_Q^{(i)} \in \mathbb{R}^{d \times d_{i\varepsilon}}$, $\mathbf{W}_K^{(i)} \in \mathbb{R}^{d \times d_\tau}$, and $\mathbf{W}_V^{(i)} \in \mathbb{R}^{d \times d_\tau}$ are learnable projection matrices, where d is the chosen size of the actual cross-attention embedding. We have a unique set of three matrices for each of the i intermediate U-Net layers with cross-attention. In its simplest form, we can add one or more cross-attention blocks after the output of an intermediate U-Net layer. The blocks can have a residual connection, which preserves the unmodified intermediate layer output and augments it with the attention vector. Note that the output of the intermediate convolutional layers has four dimensions: [batch, channel, height, width]. However, the standard attention blocks use two-dimensional input: [batch, dim]. One solution is to flatten the convolutional output before feeding it to the attention block. Alternatively, we can preserve the channel dimension and only flatten the height and width: [batch, channel, height*width]. In this case, we can assign one attention head to the output of each convolutional channel.

> **Note**
> *Figure 9.5* has a *switch* component, which allows us to concatenate the text prompt representation and the U-Net input rather than using cross-attention in the intermediate layers. This use case is for tasks other than text-to-image, which is the focus of this section.

With that, let's see how to use SD in practice.

Using stable diffusion with Hugging Face Transformers

In this section, we'll use SD to generate an image conditioned on a text prompt. In addition to the Transformers library, we'll also need **Diffusers** (https://github.com/huggingface/diffusers) – a library for pre-trained diffusion models for generating images and audio. Please note that the diffusers SD implementation requires the presence of a GPU. You can run this example in the Google Colab notebook with GPU enabled. Let's start:

1. Do the necessary imports:

    ```
    import torch
    from diffusers import StableDiffusionPipeline
    ```

2. Instantiate an SD pipeline (`sd_pipe`) with SD version 2.1. We don't use the main transformers `pipeline` abstraction, which we used in the preceding examples. Instead, we use `StableDiffusionPipeline`, which comes from the `diffusers` library. We'll also move the model to a `cuda` device (an NVIDIA GPU) if it is available:

```
sd_pipe = StableDiffusionPipeline.from_pretrained(
    "stabilityai/stable-diffusion-2-1",
    torch_dtype=torch.float16)
sd_pipe.to('cuda')
```

3. Let's run `sd_pipe` for 100 denoising steps with the following text prompt:

```
prompt = \
    "High quality photo of a racing car on a track"
image = sd_pipe(
    prompt,
    num_inference_steps=100).images[0]
```

The generated `image` is as follows:

Figure 9.8 – SD-generated image

Unfortunately, the AE, U-Net, and conditioning transformer descriptions are large, and it would be impractical to include them here. Still, they are available in the Jupyter Notebook. Nevertheless, we can see a shortened summary of the entire SD pipeline with the `print(sd_pipe)` command:

```
StableDiffusionPipeline {
  "safety_checker": [null, null],
```

```
    "tokenizer": ["transformers", "CLIPTokenizer"],
    "text_encoder": ["transformers", "CLIPTextModel"],
    "unet": ["diffusers", "UNet2DConditionModel"],
    "vae": ["diffusers", "AutoencoderKL"],
    "scheduler": ["diffusers", "DDIMScheduler"]
}
```

Here, `transformers` and `diffusers` refer to the package of origin for the given component.

The first component is an optional `safety_checker` (not initialized), which can identify **not-safe-for-work** (**NSFW**) images.

Next, we have a BPE-based `CLIPTokenizer` tokenizer, with a token vocabulary size of around 50,000 tokens. It tokenizes the text prompt and feeds it to `text_encoder` of CLIPTextModel. The Hugging Face `CLIPTextModel` duplicates the OpenAI CLIP transformer-decoder (the model card is available at `https://huggingface.co/openai/clip-vit-large-patch14`).

Then, we have `UNet2DConditionModel`. The convolutional portions of the U-Net use residual blocks (*Chapter 4*). It has four downsampling blocks with a downsampling factor of 2 (implemented with convolutions with stride 2). The first three include `text_encoder` cross-attention layers. Then, we have a single mid-block, which preserves input size and contains one residual and one cross-attention sublayer. The model ends with four skip-connected upsampling blocks, symmetrical to the downsampling sequence. The last three blocks also include cross-attention layers. The model uses **sigmoid linear unit** (**SiLU**, *Chapter 3*) activations.

Next, we have the convolutional autoencoder, `AutoencoderKL`, with four downsampling residual blocks, one residual mid-block (the same as the one in U-Net), four upsampling residual blocks (symmetrical to the downsampling sequence), and SiLU activations.

Finally, let's focus on `scheduler` of DDIMScheduler, which is part of the `diffusers` library. It is one of multiple available schedulers. During training, a scheduler adds noise to a sample to train the DM. It defines how to update the latent tensor based on the U-Net output during inference.

Stable Diffusion XL

Recently, Stability AI released Stable Diffusion XL (*SDXL: Improving Latent Diffusion Models for High-Resolution Image Synthesis*, `https://arxiv.org/abs/2307.01952`). SDXL uses a three times larger U-Net. The larger size is due to more attention blocks and a larger attention context (the new version uses the concatenated outputs of two different text encoders). It also utilizes an optional **refinement model** (**refiner**) – a second U-Net in the same latent space as the first, specializing in high-quality, high-resolution data. It takes the output latent representation, **z**, of the first U-Net as input and uses the same conditioning text prompt.

With that, we've concluded our introduction to SD and the larger topic of transformers for computer vision. Next, let's see how to fine-tune a transformer-based model.

Exploring fine-tuning transformers

In this section, we'll use PyTorch to fine-tune a pre-trained transformer. More specifically, we'll fine-tune a **DistilBERT** transformer-encoder (*DistilBERT, a distilled version of BERT: smaller, faster, cheaper, and lighter*, `https://arxiv.org/abs/1910.01108`) to classify whether a movie review is positive or negative. We'll use the Rotten Tomatoes dataset (`https://huggingface.co/datasets/rotten_tomatoes`), which contains around 10,000 reviews, split equally between positive and negative (`https://huggingface.co/datasets/rotten_tomatoes`), licensed under Apache 2.0, derived from `https://huggingface.co/datasets/rotten_tomatoes/blob/main/rotten_tomatoes.py`. We'll implement the example with the help of the Transformers library's `Trainer` class (`https://huggingface.co/docs/transformers/main_classes/trainer`), which implements the basic training loop, model evaluation, distributed training on multiple GPUs/TPUs, mixed precision, and other training features. This is opposed to implementing the training from scratch, as we've been doing until now in our PyTorch examples. We'll also need the **Datasets** (`https://github.com/huggingface/datasets`) and **Evaluate** (`https://github.com/huggingfahttps://github.com/huggingface/evaluate`) packages. Let's start:

1. Load the dataset, which is split into `train`, `validation`, and `test` portions:

    ```
    from datasets import load_dataset
    dataset = load_dataset('rotten_tomatoes')
    ```

2. Load the DistilBERT WordPiece subword `tokenizer`:

    ```
    from transformers import AutoTokenizer
    tokenizer = AutoTokenizer.from_pretrained('distilbert-base-
    uncased')
    ```

3. Use `tokenizer` to tokenize the dataset. In addition, it'll pad or truncate each sample to the maximum length accepted by the model. The `batched=True` mapping speeds up processing by combining the data in batches (as opposed to single samples). The `Tokenizers` library works faster with batches because it parallelizes the tokenization of all the examples in a batch:

    ```
    tok_dataset = dataset.map(
        lambda x: tokenizer(
            text=x['text'],
            padding='max_length',
            truncation=True),
        batched=True)
    ```

4. Load the transformer `model`:

    ```
    from transformers import AutoModelForSequenceClassification
    model = AutoModelForSequenceClassification.from_pretrained(
        'distilbert-base-uncased')
    ```

The AutoModelForSequenceClassification class loads DistilBERT configuration for binary classification – the model head has a hidden layer and an output layer with two units. This configuration works for our task because we have to classify the movie reviews into two categories.

5. Initialize TrainingArguments of the Trainer instance. We'll specify output_dir for the location of the model predictions and checkpoints. We'll also run the evaluation once per epoch:

```
from transformers import TrainingArguments
training_args = TrainingArguments(
    output_dir='test_trainer',
    evaluation_strategy='epoch')
```

6. Initialize the accuracy evaluation metric:

```
import evaluate
accuracy = evaluate.load('accuracy')
```

7. Initialize trainer with all the necessary components for training and evaluation:

```
from transformers import Trainer
import numpy as np
trainer = Trainer(
    model=model,
    train_dataset=tok_dataset['train'],
    eval_dataset=tok_dataset['test'],
    args=training_args,
    compute_metrics=
        lambda x: accuracy.compute(
            predictions=x[0],
            references=x[1]),
    preprocess_logits_for_metrics=
        lambda x, _: np.argmax(x.cpu(), axis=-1)
)
```

It accepts the model, train, and evaluation datasets and the training_args instance. The compute_metrics function will compute the validation accuracy after each epoch. preprocess_logits_for_metrics will convert the one-hot encoded model output (x[0]) to indexed labels so that it can match the format of the ground-truth labels (x[1]) in the compute_metrics function.

8. Finally, we can run the training:

```
trainer.train()
```

The model will achieve around 85% accuracy in three epochs.

Next, let's see how to harness the power of LLMs with the LangChain framework.

Harnessing the power of LLMs with LangChain

LLMs are powerful tools, yet they have some limitations. One of them is the context window length. For example, the maximum input sequence of Llama 2 is 4,096 tokens and even less in terms of words. As a reference, most of the chapters in this book hover around 10,000 words. Many tasks wouldn't fit this length. Another LLM limitation is that its entire knowledge is stored within the model weights at training time. It has no direct way to interact with external data sources, such as databases or service APIs. Therefore, the knowledge can be outdated or insufficient. The **LangChain** framework can help us alleviate these issues. It does so with the following modules:

- **Model I/O**: The framework differentiates between classic LLMs and chat models. In the first case, we can prompt the model with a single prompt, and it will generate a response. The second case is more interactive – it presumes a back-and-forth communication between the human and the model in a chat form. Internally, both are LLMs; the difference comes from using different APIs. Regardless of the model type, a token sequence is the only way to feed it with input data. The I/O module provides helper prompt templates for different use cases. For example, the chat template maintains an explicit list of all messages instead of concatenating them in a single sequence. We also have a few-shot template, which provides an interface to include one or more instructive input/output examples within the input query.

  The module can also parse the model output (a token sequence converted into words). For example, if the output is a JSON string, a JSON parser can convert it into an actual JSON object.

- **Retrieval**: This retrieves external data and feeds it to the model input sequence. Its most basic function is to parse file formats such as CSV and JSON. It can also split larger documents into chunks if they don't fit within the context window size.

Vector databases

The primary output of an LLM and other neural networks (before any task-specific heads) are embedding vectors, which we use for downstream tasks such as classification or text generation. The universal nature of this data format has led to the creation of vector-specific databases (or stores). As the name suggests, these stores only work with vectors and support fast vector operations, such as different similarity measures over the whole database. We can query an input embedding vector against all other database vectors and find the most similar ones. This concept is similar to the **Q/K/V** attention mechanism but in an external database form, which allows it to work with a larger dataset than in-memory attention.

This retrieval module has integrations with multiple vector databases. This way, we can use the LLM to generate and store document embeddings (the document acts as an input sequence). Later, we can query the LLM to generate a new embedding for a given query and compare this query against the database to find the nearest matches. In this scenario, the role of the LLM is limited to generating vector embeddings.

- **Chains**: These are mechanisms that combine multiple LangChain components to create a single application. For example, we can create a chain that takes user input, formats it with a special prompt template, feeds it to an LLM, and parses the LLM output to JSON. We can branch chains or combine multiple chains.

- **Memory**: This maintains the input token sequence throughout a chain of steps or model interactions with the outside world, which can modify and extend the sequence dynamically. It can also use the emerging LLM abilities to create a shortened summary of the current historical sequence. The shortened version replaces the original in the input token sequence for future inputs. This compression allows us to use the context window more efficiently and store more information.

- **Agents**: An agent is an entity that can take actions that interact with the environment. In the current context, an LLM acts as the agent's reasoning engine to determine which actions the agent is to take and in which order. To help with this task, the agent/LLM can use special functions called **tools**. These can be generic utilities (for example, API calls), other chains, or even agents.

- **Callbacks**: We can use callbacks to plug into various points of the LLM application, which is useful for logging, monitoring, or streaming.

Next, let's solidify our understanding of LangChain with an example.

Using LangChain in practice

In this section, we'll use LangChain, LangChain Experimental (https://github.com/langchain-ai/langchain/tree/master/libs/experimental), and OpenAI's gpt-3.5-turbo model to answer the question: *What are the sum of the elevations of the deepest section of the ocean and the highest peak on Earth? Use metric units only.* To make things more interesting, we won't let an LLM generate the output one word at a time. Instead, we'll ask it to break up the solution into steps and use data lookup and calculations to find the right answer.

> **Note**
> This example is partially based on https://python.langchain.com/docs/modules/agents/agent_types/plan_and_execute. It requires access to the OpenAI API (https://platform.openai.com/) and SerpAPI (https://serpapi.com/).

Let's start:

1. Initialize LangChain's API wrapper for the gpt-3.5-turbo model. The temperature parameter (in the [0,1] range) determines how the model selects the next token. For example, if temperature=0, it will always output the highest probability token. The closer temperature is to 1, the more likely it is that the model selects a token with a lower probability:

```
from langchain.chat_models import ChatOpenAI
model = ChatOpenAI(temperature=0)
```

2. Define the tools that will help us solve this task. First is the search tool, which uses SerpAPI to perform Google searches. This allows the LLM to query Google for the elevations of the deepest part of the ocean and the highest mountain in our question:

```
# Tools
from langchain.agents.tools import Tool
# Search tool
from langchain import SerpAPIWrapper
search = Tool(
    name='Search',
    func=SerpAPIWrapper().run,
    description='Google search tool')
```

3. Next is the calculator tool, which will allow the LLM to compute the sum of the elevations. This tool uses a special few-shot learning LangChain `PromptTemplate` to query the LLM to calculate mathematical equations:

```
from langchain import LLMMathChain
llm_math_chain = LLMMathChain.from_llm(
    llm=model,
    verbose=True)
calculator = Tool(
    name='Calculator',
    func=llm_math_chain.run,
    description='Calculator tool')
```

4. Initialize a special `PlanAndExecute` agent. It accepts the LLM, the tools we just defined, as well as `planner` and `executor` agents, as arguments:

```
from langchain_experimental.plan_and_execute import
PlanAndExecute, load_agent_executor, load_chat_planner
agent = PlanAndExecute(
    planner=load_chat_planner(
        llm=model),
    executor=load_agent_executor(
        llm=model,
        tools=[search, calculator],
        verbose=True),
    verbose=True)
```

`planner` uses a special LangChain text prompt template that queries the LLM model to break up the solution of the task into subtasks (steps). The model generates a list-friendly formatted string, which the planner parses and returns as the list of steps that `executor` (itself an `agent`) executes.

5.   Finally, we can run `agent`:

```
agent.run('What is the sum of the elevations of the deepest
section of the ocean and the highest peak on Earth? Use metric
units only.')
```

The final answer is `The depth of the deepest section of the ocean in metric units is 19,783 meters`. Although the text description is off the mark, the computation seems correct.

Let's analyze part of the steps `planner` and `executor` take to reach the result. First, `planner` takes our initial query and asks the LLM to break it up into the following list of steps:

```
1. 'Find the depth of the deepest section of the ocean in metric
units.'
2. 'Find the elevation of the highest peak on Earth in metric units.'
3. 'Add the depth of the deepest section of the ocean to the elevation
of the highest peak on Earth.'
4. 'Round the sum to an appropriate number of decimal places.'
5. "Given the above steps taken, respond to the user's original
question. \n"
```

Next, `agent` iterates over each step and tasks `executor` to perform it. `executor` has an internal LLM planner, which can also break up the current step into subtasks. In addition to the step description, `executor` uses a special text prompt, instructing its LLM `model` to identify the list of `tools` (a tool has a name) it can use for each step. For example, `executor` returns the following result as output for the augmented version of the first step:

```
'Action: {
    "action": "Search",
    "action_input": "depth of the deepest section of the ocean in
metric units"
}'
```

The `Search` agent's `action` represents a new intermediate step to be executed after the current one. It will use the search tool to query Google with `action_input`. In that sense, the chain is dynamic, as the output of one step can lead to additional steps added to the chain. We add the result of each step to the input sequence of the future steps, and the LLM, via different prompt templates, ultimately determines the next actions.

This concludes our introduction to LangChain – a glimpse of what is possible with LLMs.

Summary

In this chapter, we discussed a variety of topics. We started with LLMs in the computer vision domain: ViT for image classification, DetR for object detection, and SD for text-to-image generation. Next, we learned how to fine-tune an LLM with the Transformers library. Finally, we used LangChain to implement a novel LLM-driven application.

In the next chapter, we'll depart from our traditional topics and dive into the practical field of MLOps.

Part 4: Developing and Deploying Deep Neural Networks

In this single-chapter part, we'll discuss some techniques and tools that will help us develop and deploy neural network models.

This part has the following chapter:

- *Chapter 10, Machine Learning Operations (MLOps)*

10

Machine Learning
Operations (MLOps)

So far in this book, we have focused on the theory of **neural networks** (**NNs**), various NN architectures, and the tasks we can solve with them. This chapter is a little different because we'll focus on some of the practical aspects of NN development. We'll delve into this topic because the development and production deployment of ML models (and NNs in particular) have some unique challenges. We can split this process into three steps:

1. **Training dataset creation**: Data collection, cleanup, storage, transformations, and feature engineering.

2. **Model development**: Experiment with different models and training algorithms and evaluate them.

3. **Deployment**: Deploy trained models in the production environment and monitor their performance in computational and accuracy terms.

This multi-step complex pipeline presupposes some of the challenges when solving ML tasks:

- **Diverse software toolkit**: Each step has multiple competing tools.

- **Model development is hard**: Each training instance has a large number of variables. These could be modifications in the NN architecture, variations in the training hyperparameters (such as learning rate or momentum), or different training data distributions. On top of that, NNs have sources of randomness, such as weight initialization or data augmentation. Therefore, if we cannot reproduce earlier results, it won't be easy to pinpoint the reason. Even if we have a bug in the code, it might not result in an easily detectable runtime exception. Instead, it might just deteriorate the model accuracy slightly. So that we don't lose track of all the experiments, we need a robust tracking and monitoring system.

- **Complex deployment and monitoring**: NNs require GPUs and batch-organized data for optimal performance. These requirements might collide with the real-world requirements of processing data in streams or sample-wise. In addition, the nature of the user data might change with time, which could cause **model drift**.

In this chapter, we will cover the following main topics:

- Understanding model development
- Exploring model deployment

Technical requirements

We'll implement the examples in this chapter using Python, PyTorch, **TensorFlow** (**TF**), and **Hugging Face** (**HF**) Transformers, among others. If you don't have an environment set up with these tools, fret not – the examples are available as Jupyter Notebooks on Google Colab. You can find the code examples in this book's GitHub repository: `https://github.com/PacktPublishing/Python-Deep-Learning-Third-Edition/tree/main/Chapter10`.

Understanding model development

In this section, we'll discuss various tools that will help us manage the model development phase of the ML solution life cycle. Let's start with the most important question – which NN framework should we choose?

Choosing an NN framework

So far in this book, we've mostly used PyTorch and TensorFlow. We can refer to them as **foundational** frameworks as these are the most important components of the entire NN software stack. They serve as a base for other components in the ML NN ecosystem, such as Keras or HF Transformers, which can use either of them as a backend (multi-backend support will come with Keras 3.0). In addition to TF, Google has also released JAX (`https://github.com/google/jax`), a foundational library that supports GPU-accelerated NumPy operations and Autograd. Other popular libraries such as NumPy, pandas, and scikit-learn (`https://scikit-learn.org`) go beyond the scope of this book as they are not strictly related to NNs. Because of the importance of foundational libraries, they are the first and most important choice in our toolkit. But which one should we choose if we start a project from scratch?

PyTorch versus TensorFlow versus JAX

Let's check the level of community adoption for these libraries. Our first stop is **Papers with Code** (`https://paperswithcode.com/`), which indexes ML papers, code, datasets, and results. The site also maintains the trend of paper implementations grouped by framework (`https://paperswithcode.com/trends`). As of September 2023, 57% of the new papers are using PyTorch. TF and JAX are distant second and third with 3% and 2%, respectively. This trend isn't new – PyTorch was released in 2016, but it has already surpassed TF in 2019. This particular data point indicates that PyTorch dominates cutting-edge research, which is what the most recent papers are. Therefore, if you want to always use the latest and greatest in the field, it's a good idea to stick to PyTorch. Next, let's look at the ML models, hosted on the HF platform (`https://huggingface.co/models`), where we can also filter by project framework. Out of ~335,000 total models hosted, ~131,000 use PyTorch, ~10,000 use TF, and ~9,000 use JAX. Again, this is a strong result in favor of PyTorch. However, this is not the full picture, as these results are for public and open source projects. They are not necessarily indicative of what companies use in production. More representative of this could be PyPI Stats (`https://pypistats.org/`), which provides aggregate download information on Python packages available from the **Python Package Index** (**PyPi**, `https://pypi.org/`). The picture here is a bit more nuanced – PyTorch has 11,348,753 downloads for the last month (August-September 2023) versus 16,253,288 for TF and 3,041,747 for JAX. However, we should be cautious with PyPi Stats because many automated processes (such as continuous integration) can inflate the PyPI download count, without indicating real-world use. In addition, the PyTorch download page advises installing the library through Conda (`https://conda.io/`). The monthly statistics show 759,291 PyTorch downloads versus 154,504 for TF and 6,260 for JAX. Therefore, PyTorch leads here as well. Overall, my conclusion is that PyTorch is more popular than TF, but both libraries are used in production environments.

My advice, which you can take with as many pinches of salt as you wish, would be to select PyTorch if you start a project now. This is why this book has put more emphasis on PyTorch compared to TF. One exception to this rule is if your project runs on mobile or edge devices (`https://en.wikipedia.org/wiki/Edge_device`) with limited computational power. TF has better support for such devices through the TF Lite library (`https://www.tensorflow.org/lite`).

But ultimately, you can work with your preferred software stack and then convert your models into other libraries for deployment. We'll see how this is possible in the next section.

Open Neural Network Exchange

Open Neural Network Exchange (**ONNX**, `https://onnx.ai/`) provides an open source format for NN-based and traditional ML models (we'll focus on NNs here). It defines an extensible computation graph model, built-in operators, and standard data types. In other words, ONNX provides a universal NN representation format, which allows us to convert models implemented with one library (for example, PyTorch) into others (such as TF), provided that both the source and target libraries support ONNX. In this way, you can train your model with one library and then convert it into another when deploying to production. This also makes sense because ONNX focuses on inference mode and not training (representing the training process using ONNX in experimental mode).

ONNX represents an NN as a computational **graph** of operations (as we discussed in *Chapter 2*), where the nodes are the operations (or **operators** in ONNX terms) and the edges are the input/output connections (or ONNX **variables**) between them. In addition, it implements a Python runtime for evaluating ONNX models and Ops. Its purpose is to clarify the semantics of ONNX and to help understand and debug ONNX tools and converters, but it is not intended for production, and it is not optimized for performance. We'll use it for its intended purpose and illustrate ONNX with a code example for a simple linear regression, defined as $Y = f(\mathbf{X}, \mathbf{A}, \mathbf{B}) = \mathbf{XA} + \mathbf{B}$. To do this, we'll use the onnx (`!pip install onnx`) Python package. Let's start:

1. Define the graph representation's input (X, A, B) and output (Y) variables:

```
import numpy as np
from onnx import TensorProto, numpy_helper
from onnx.helper import make_tensor_value_info

X = make_tensor_value_info(
    name='X',
    elem_type=TensorProto.FLOAT,
    shape=[None, None])
Y = make_tensor_value_info(
    'Y', TensorProto.FLOAT, [None])
A = numpy_helper.from_array(
    np.array([0.5, -0.6], dtype=np.float32),
    name='A')
B = numpy_helper.from_array(
    np.array([0.4], dtype=np.float32),
    name='B')
```

Here, `make_tensor_value_info` declares a named graph I/O variables (X and Y) with a type (`elem_type`) and shape. `shape=[None]` means any shape, and `shape=[None, None]` means a two-dimensional tensor without specific dimension sizes. On the other hand, A and B are the function parameters (weights), and we initialize them with pre-defined values from NumPy arrays.

2. Define the graph operations:

```
from onnx.helper import make_node
mat_mul = make_node(
    op_type='MatMul',
    inputs=['X', 'A'],
    outputs=['XA'])
addition = make_node('Add', ['XA', 'B'], ['Y'])
```

`mat_mul` represents matrix multiplication (`MatMul`) between the X and A input matrices into the XA output variable. `addition` sums the output of `mat_mul`, XA, with the bias, B.

ONNX operators

This example introduces the `MatMul` and `Add` ONNX operators. The full list of supported operators (available at `https://onnx.ai/onnx/operators/`) includes many other NN building blocks, such as activation functions, convolutions, pooling, and tensor operators (for example, `concat`, `pad`, `reshape`, and `flatten`). In addition, it supports the so-called **control flow** operators, which can create dynamic computational graphs. For example, the `if` operator executes one subgraph or another, depending on a Boolean value. ONNX itself doesn't implement the operators. Instead, the libraries that support it (such as PyTorch) have their own implementations. Conversely, the ONNX conversion will fail if your library model has operators that aren't supported by ONNX.

3. We now have the ingredients to define the computational `graph`:

```
from onnx.helper import make_graph
graph = make_graph(
    nodes=[mat_mul, addition],
    name='Linear regression',
    inputs=[X],
    outputs=[Y],
    initializer=[A, B])
```

We can see our computational graph in the following figure:

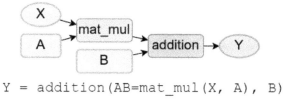

$$Y = addition(AB=mat\_mul(X, A), B)$$

Figure 10.1 – Linear regression ONNX computational graph

4. Use `graph` to create an `onnx_model` instance. The model allows you to add additional metadata to the graph, such as docstring, version, author, and license, among others:

```
from onnx.helper import make_model
onnx_model = make_model(graph)
onnx_model.doc_string = 'Test model'
onnx_model.model_version = 1
```

5. Check the model for consistency. This verifies that the input type or shapes match between the model components:

```
from onnx.checker import check_model
check_model(onnx_model)
print(onnx_model)
```

6. Finally, we can compute the output of the model for two random input samples with an instance of `ReferenceEvaluator`:

```
from onnx.reference import ReferenceEvaluator
sess = ReferenceEvaluator(onnx_model)
print(sess.run(
    output_names=None,
    feed_inputs={'X': np.random.randn(2, 2).astype(np.
float32)}))
```

The result of the computation is a NumPy array:

[array([[-0.7511951,   1.0294889], dtype=float32)]

7. ONNX allows us to serialize and deserialize both the model structure and its weights with **Protocol Buffers (protobuf**, https://protobuf.dev/). Here's how to do this:

```
with open('model.onnx', 'wb') as f:
    f.write(onnx_model.SerializeToString())

from onnx import load
with open('model.onnx', 'rb') as f:
    onnx_model = load(f)
```

Now that we have introduced ONNX, let's see how we can use it in practice by exporting PyTorch and TF models to ONNX.

In addition to `torch` and `tensorflow`, we'll also need the `torchvision`, `onnx`, and `tf2onnx` (https://github.com/onnx/tensorflow-onnx, `!pip install tf2onnx`) packages. Let's start with PyTorch:

1. Load a pre-trained model (`MobileNetV3`, refer to *Chapter 5*):

```
import torch
from torchvision.models import mobilenet_v3_small, MobileNet_V3_
Small_Weights
torch_model = mobilenet_v3_small(
  weights=MobileNet_V3_Small_Weights.DEFAULT)
```

2. Then, export the model:

```
torch.onnx.export(
    model=torch_model,
    args=torch.randn(1, 3, 224, 224),
    f="torch_model.onnx",
    export_params=True)
```

Most parameters speak for themselves. `args=torch.randn(1, 3, 224, 224)` specifies a dummy tensor. This is necessary because the serializer might invoke the model once to infer the graph structure and tensor sizes. The dummy tensor will serve as input for this invocation. However, this exposes one of the limitations of the conversion process: if the model includes a dynamic computational graph, the converter will only convert the path of the current invocation. `export_params` tells the exporter to include the model weights, besides the model structure.

3. Use ONNX to load the exported model and check it for consistency (spoiler: it works):

```
import onnx
torch_model_onnx = onnx.load('torch_model.onnx')
onnx.checker.check_model(torch_model_onnx)
```

Next, let's do the same but with TF. Unlike PyTorch, TF doesn't have out-of-the-box ONNX serialization support. Instead, we'll use the `tf2onnx` package:

1. Load a pre-trained `MobileNetV3` model:

```
import tensorflow as tf
tf_model = tf.keras.applications.MobileNetV3Small(
    weights='imagenet',
    input_shape=(224, 224, 3),
)
```

2. Serialize the model using `tf2onnx`. It follows the same principle as PyTorch, down to the dummy input tensor (`input_signature`), which is necessary for model invocation:

```
import tf2onnx
tf_model_onnx, _ = tf2onnx.convert.from_keras(
    model=tf_model,
    input_signature=[tf.TensorSpec([1, 224, 224, 3])])
onnx.save(tf_model_onnx, 'tf_model.onnx')
```

Once again, we can load the model with ONNX to verify its consistency.

Next, we can use **Netron** (https://netron.app/, https://github.com/lutzroeder/netron) to visualize the NN graph by using the ONNX model file (`torch_model.onnx` or `tf_model.onnx`). This is a graphical viewer for NNs and other ML models. It exists as a web **user interface (UI)** or a standalone app. It supports ONNX, TensorFlow Lite, and PyTorch (experimental), among other libraries. For example, the following figure shows the initial **MobileNetV3** layers in detail, as visualized by Netron (the full model visualization is too large to display within this chapter):

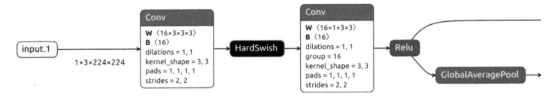

Figure 10.2 – Netron visualization of the MobileNetV3 ONNX model file

Here, the input shape is 3×224×224, **W** is the shape of the convolutional filter, and **B** is the bias. We introduced the rest of the convolution attributes in *Chapter 4*.

Unfortunately, neither PyTorch nor TF comes with the integrated ability to load ONNX models. However, there are open source packages that allow us to do this. Two of them are `onnx2torch` (`https://github.com/ENOT-AutoDL/onnx2torch`) for PyTorch and `onnx2tf` (`https://github.com/PINTO0309/onnx2tf`) for TF.

Next, we'll focus on a tool that will ease the training process.

Introducing TensorBoard

TensorBoard (**TB**, `https://www.tensorflow.org/tensorboard/`, `https://github.com/tensorflow/tensorboard`) is a TF-complement web-based tool that provides visualization and tooling for machine learning experiments. Some of its functions are as follows:

- Metrics (such as loss and accuracy) tracking and visualization
- Model graph visualization (similar to Netron)
- A time series histogram of the change of weights, biases, or other tensors over time
- Low-dimensional embedding projections

TB can work with both TF/Keras and PyTorch, but it has better integration with TF (after all, it is developed by the TF team). In both cases, TB doesn't communicate directly with the models during training. Instead, the training process stores its state and current progress in a special log file. TB tracks the file for changes and automatically updates its graphical interface with the latest information. In this way, it can visualize the training as it progresses. In addition, the file stores the entire training history to be displayed even after it finishes. To better understand how it works, we'll add TB to the transfer learning computer vision examples that we introduced in *Chapter 5*. As a quick recap, we'll start with ImageNet pre-trained MobileNetV3 models. Then, we'll use two transfer learning techniques, **feature engineering** and **fine-tuning**, to train these models to classify the CIFAR-10 dataset. TB will visualize the training.

Let's start with the Keras example. We'll only include the relevant part of the code, and not the full example, as we discussed it in *Chapter 5*. More specifically, we'll focus on the `train_model(model, epochs=5)` function, which takes the pre-trained `model` and the number of training `epochs` as parameters. The following is the function's body (please note that the actual implementation has indentation):

> **Initializing TensorBoard**
>
> This example assumes that TB is initialized and running (although the code works even if it is not available). We won't include the initialization of TB because it differs depending on the environment. However, it is included in the Jupyter Notebook of this example.

Follow these steps:

1. First, we'll configure the training of the pre-trained Keras model with the Adam optimizer, binary cross-entropy loss, and accuracy tracking:

```
model.compile(
    optimizer=tf.keras.optimizers.Adam(
        learning_rate=0.0001),
    loss='categorical_crossentropy',
    metrics=['accuracy'])
```

2. Next, we'll add the special `tensorboard_callback`, which implements the TB connection:

```
tensorboard_callback = tf.keras.callbacks.TensorBoard(
    log_dir='logs/tb/' + datetime.datetime.now().
strftime('%Y%m%d-%H%M%S'),
    update_freq='epoch',
    histogram_freq=1,
    write_graph=True,
    write_images=True,
    write_steps_per_second=True,
    profile_batch=0,
    embeddings_freq=0)
```

The callback parameters are as follows:

- `log_dir`: This instructs `tensorboard_callback` to write the log file in a unique time-stamped folder, `'logs/tb/' + datetime.datetime.now().strftime('%Y%m%d-%H%M%S')`, located in the main `'logs/tb/'` folder. TB will simultaneously pick all training folders under `'logs/tb/'` and display them in its UI as unique training instances.

- `update_freq=1`: Updates the log file once per epoch.

- `histogram_freq=1`: Computes weight histograms once per epoch.

- `write_graph=True`: Generates a graph visualization of the NN architecture.

- `write_images=True`: Visualizes the model weights as an image.

- `write_steps_per_second=True`: Logs the training steps per second.

- `profile_batch=1`: Profiles the first batch to sample its compute characteristics.

- `Embeddings_freq=0`: The frequency (in epochs) at which embedding layers will be visualized (we don't have embedding layers, so it's disabled by default).

3. Finally, we'll run the training with the `model.fit` method:

```
steps_per_epoch=metadata.splits['train'].num_examples // BATCH_
SIZE
validation_steps=metadata.splits['test'].num_examples // BATCH_
SIZE

model.fit(
    train_batches,
    epochs=epochs,
    validation_data=test_batches,
    callbacks=[tensorboard_callback],
    steps_per_epoch=steps_per_epoch,
    validation_steps=validation_steps)
```

We add `tensorboard_callback` to the list of `model callbacks`. The training process notifies each callback for various training events: start of training, end of training, start of testing, end of testing, start of epoch, end of epoch, start of batch, and end of batch. In turn, `tensorboard_callback` updates the log file according to its configuration and the current event.

The TB UI displays all the information in the log file. Although it's too complex to include here, we can still show a snippet with accuracy:

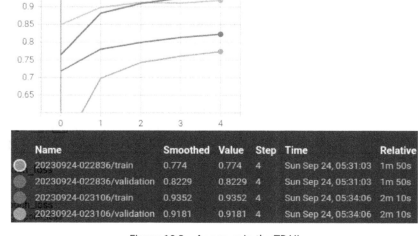

Figure 10.3 – Accuracy in the TB UI

Here, TB displays the accuracy for four different experiments – train/test for feature engineering and train/test for fine-tuning.

Next, let's see how PyTorch integrates with TB. It provides a special `torch.utils.tensorboard.SummaryWriter` class, which writes entries directly to event log files in the `log_dir` folder to be consumed by TB. It follows the same principle as in Keras. The high-level API of `SummaryWriter` allows us to create an event file in `log_dir` and asynchronously add content to it. The main difference with Keras is that we're responsible for adding the content, instead of an automated event listener doing it. Let's see how that works in practice. As with Keras, we'll use the computer vision transfer learning example from *Chapter 5*. We'll only focus on the relevant parts, but you can see the full example in the Jupyter Notebook in this book's GitHub repository.

First, we'll initialize two `SummaryWriter` instances for the feature extractor fine-tuning modes. It doesn't matter where we do it, so long as it happens before we start using them. As with Keras, each training instance has a unique time-stamped folder under `'logs/tb/'` (we're only showing one initialization because they are identical):

```
import datetime
from torch.utils.tensorboard import SummaryWriter
writer = SummaryWriter(
log_dir='logs/tb/' + datetime.datetime.now().strftime('%Y%m%d-
%H%M%S'))
```

For the sake of clarity, we'll include the initialization of the MobileNetV3 pre-trained model:

```
from torchvision.models import (
    MobileNet_V3_Small_Weights, mobilenet_v3_small)
model = mobilenet_v3_small(
    weights=MobileNet_V3_Small_Weights.IMAGENET1K_V1)
```

Next, we'll jump to the training (or testing) loop, where `train_loader`, an instance of `torch.utils.data.DataLoader`, yields pairs of `inputs` and `labels` mini-batches:

```
for i, (inputs, labels) in enumerate(data_loader):
    # Training loop goes here
```

Within the loop, we can add the model graph to the log file. It takes the model and the input tensor as parameters to generate the visualization (hence the need to call `add_graph` in the training loop):

```
writer.add_graph(model, inputs)
```

Finally, at the end of the training loop, we'll add the loss and the accuracy for the current `epoch` as scalar values:

```
writer.add_scalar(
    tag='train/accuracy',
```

```
        scalar_value=total_acc,
        global_step=epoch)
    writer.add_scalar(
        tag='train/loss',
        scalar_value=total_loss,
        global_step=epoch)
```

Each scalar value has a unique `tag` (besides the two tags in the code, we also have `tag='validation/loss'`). Note that `global_step` (equal to the epoch) stores `scalar_value` as a sequence within the same `tag`. In addition to graphs and scalars, `SummaryWriter` can add images, tensors, histograms, and embeddings, among others.

This concludes our introduction to TB. Next, we'll learn how to develop NN models for edge devices.

Developing NN models for edge devices with TF Lite

TF Lite is a TF-derived set of tools that allows us to run models on mobile, embedded, and edge devices. Its versatility is part of TF's appeal for industrial applications (as opposed to research applications, where PyTorch dominates). The key paradigm of TF Lite is that the models run on-device, contrary to client-server architecture, where the model is deployed on remote, more powerful, hardware. This organization has the following implications (both good and bad):

- **Low-latency execution**: The lack of server-round trip significantly reduces the model inference time and allows us to run real-time applications.

- **Privacy**: The user data never leaves the device.

- **Internet connectivity**: Internet connectivity is not required.

- **Small model size**: The devices have limited computational ability, hence the need for small and computationally efficient models. More specifically, TF Lite models are stored in the FlatBuffers (https://flatbuffers.dev/) special efficient portable format, identified by the `.tflite` file extension. Besides its small size, it allows us to access data directly without parsing/unpacking it first.

TF Lite models support a subset of the TF Core operations and allow us to define custom ones:

- **Low power consumption**: The devices often run on battery.

- **Divergent training and inference**: NN training is a lot more computationally intensive compared to inference. Because of this, the model training runs on a different, more powerful, piece of hardware than the actual devices, where the models will run inference.

In addition, TF Lite has the following key features:

- Multi-platform and multi-language support, including Android (Java), iOS (Objective-C and Swift) devices, web (JavaScript), and Python for all other environments. Google provides a TF Lite wrapper API called **MediaPipe Solutions** (`https://developers.google.com/mediapipe`, `https://github.com/google/mediapipe/`), which supersedes the previous TF Lite API.

- Optimized for performance.

- It has end-to-end solution pipelines. TF Lite is oriented toward practical applications, rather than research. Because of this, it includes different pipelines for common ML tasks such as image classification, object detection, text classification, and question answering among others. The computer vision pipelines use modified versions of EfficientNet or MobileNet (*Chapter 4*), and the natural language processing pipelines use BERT-based (*Chapter 7*) models.

So, how does TF Lite model development work? First, we'll select a model in one of the following ways:

- An existing pre-trained `.tflite` model (`https://tfhub.dev/s?deployment-format=lite`).

- Use **MediaPipe Model Maker** (`https://developers.google.com/mediapipe/solutions/model_maker`) to apply feature engineering transfer learning on an existing `.tflite` model with a custom training dataset. Model Maker only works with Python.

- Convert a full-fledged TF model into `.tflite` format.

TFLite model metadata

The `.tflite` models may include optional metadata with three components:

-- **Human-readable part**: Provides additional information for the model.

-- **Input information**: Describes the input data format and the necessary pre-processing steps

-- **Output information**: Describes the output data format and the necessary post-processing steps.

The last two parts can be leveraged by code generators (for example, Android code generator) to create ready-to-use model wrappers in the target platform.

Next, let's see how to use Model Maker to train a `.tflite` model and then use it to classify images. We're only going to show relevant parts of the code, but the full example is available as a Jupyter Notebook in this book's GitHub repository. Let's start:

1. First, we'll create training and validation datasets:

    ```
    from mediapipe_model_maker import image_classifier
    dataset = image_classifier.Dataset.from_folder(dataset_path)
    train_data, validation_data = dataset.split(0.9)
    ```

Here, `dataset_path` is a path to the Flowers dataset (`https://www.tensorflow.org/datasets/catalog/tf_flowers`), which contains 3,670 RGB low-resolution images of flowers, distributed in five classes (one subfolder per class). `data.split(0.9)` splits the dataset (instances of `image_classifier.Dataset`) into `train_data` (90% of the images) and `validation_data` (10% of the images) parts.

2. Next, we'll define the training hyperparameters – train for three epochs with a mini-batch size of 16 and export the trained model in the `export_dir` folder (other parameters are available as well):

```
hparams = image_classifier.HParams(
    export_dir='tflite_model',
    epochs=3,
    batch_size=16)
```

3. Then, we'll define the model parameters (we'll use `EfficientNet`):

```
options = image_classifier.ImageClassifierOptions(    supported_
model=image_classifier.SupportedModels.EFFICIENTNET_LITE4,
    hparams=hparams)
```

4. Finally, we'll create a new model and we'll run the training:

```
model = image_classifier.ImageClassifier.create(
    train_data=train_data,
    validation_data=validation_data,
    options=options,
)
```

This model achieves around 92% accuracy in three epochs. The training process creates a TB-compatible log file, so we'll be able to track the progress with TB (available in the Jupyter Notebook).

5. Next, we'll export the model in `.tflite` format for the next phase of our example:

```
model.export_model('model.tflite')
```

6. Now that we have a trained model, we can use it to classify images. We're going to use the `MediaPipe` Python API (which is different than Model Maker):

```
import mediapipe as mp
from mediapipe.tasks import python
from mediapipe.tasks.python import vision

generic_options = python.BaseOptions(
    model_asset_path='/content/tflite_model/model.tflite')

cls_options = vision.ImageClassifierOptions(
```

```
            base_options=generic_options)

    classifier = vision.ImageClassifier.create_from_options(cls_
    options)
```

Here, `classifier` is the pre-trained model, `generic_options` contains the file path to the `.tflite` model, and `cls_options` contains classification-specific options (we use the default values).

7. We'll load five random flower images (one for each flower class, listed in `labels`) in a list called `image_paths` (not displayed here). We'll classify each image, and we'll compare its predicted label to the real one:

```
    for image_path, label in zip(image_paths, labels):
        image = mp.Image.create_from_file(image_path)
        result = classifier.classify(image)
        top_1 = result.classifications[0].categories[0]
        print(f'Label: {label}; Prediction: {top_1.category_name}')
```

Predictably, the model classifies all images correctly.

Next, we'll learn how to optimize the training with mixed-precision computations.

Mixed-precision training with PyTorch

We discussed mixed-precision training in the context of LLMs in *Chapter 8*. In this section, we'll see how to use it in practice with PyTorch. Once again, we'll use the transfer learning PyTorch example from *Chapter 5* as a base for our implementation. All the code modifications are concentrated in the `train_model` function. We'll only include `train_model` here, but the full example is available as a Jupyter Notebook in this book's GitHub repository. The following is a shortened version of the function definition:

```
def train_model(model, loss_fn, optimizer, data_loader):
    scaler = torch.cuda.amp.GradScaler()
    for i, (inputs, labels) in enumerate(data_loader):
        optimizer.zero_grad()
        with torch.autocast(
            device_type=device,
            dtype=torch.float16):
            # send the input/labels to the GPU
            inputs = inputs.to(device)
            labels = labels.to(device)

            # forward
            outputs = model(inputs)
            loss = loss_fn(outputs, labels)
```

```
# backward with scaler
scaler.scale(loss).backward()
scaler.step(optimizer)
scaler.update()
```

We use a combination of two separate and unrelated mechanisms for mixed-precision training:

- `torch.autocast`: This acts as a context manager (or decorator) and allows a region of the code to run in mixed precision. `device_type` specifies the device that `autocast` applies to. `dtype` specifies the data type with which the CUDA operations work. The PyTorch documentation suggests only wrapping the forward and loss computation with `torch.autocast`. The backward operations automatically run with the same data type as the forward ones.

- `torch.cuda.amp.GradScaler`: When the forward pass uses `float16` precision operations, so does the backward pass, which computes the gradients. However, due to the lower precision, some gradient values will flush to zero. To prevent this, **gradient scaling** multiplies the NN's loss by a scale factor and invokes a backward pass with the scaled value. Gradients flowing backward through the network are also scaled by the same factor. In this way, the entire backward pass uses a larger magnitude to prevent flushing to zero. Before the weight updates, the mechanism *unscales* the scaled gradient values, so the weight updates work with the actual values.

This concludes our introduction to the model development tools. Next, we'll discuss some model deployment mechanisms.

Exploring model deployment

In this section, we'll discuss two basic model deployment examples. They'll help you create simple, yet functional, proof-of-concept apps for your experiments. Let's start.

Deploying NN models with Flask

In our first example, we'll use Google Colab in combination with **Flask** (`https://github.com/pallets/flask`) to create a simple hosted REST API service, which will expose our model to the outside world. For the sake of simplicity, we'll use a **stable diffusion** (**SD**) model: it will accept a textual `prompt` parameter, generate an image with it, and return the image as a result.

According to its home page, Flask is a lightweight **WSGI** (`https://wsgi.readthedocs.io/`) web application framework. It is designed to make getting started quick and easy, with the ability to scale up to complex applications. In our case, it will start a development web server, which will process the requests for the SD model. Although this server will run in the Google Colab environment (think of it as `localhost`), we won't be able to access it. To solve this, we'll need `flask-ngrok` (`https://ngrok.com/docs/using-ngrok-with/flask/`), which will expose the server to the outside world (you'll need a free `ngrok` registration and authentication token to run this example).

To satisfy all dependencies, we'll need the `transformers`, `diffusers`, `accelerate`, and `flask-ngrok` packages. Let's start:

1. First, we'll initialize the SD HF pipeline (`sd_pipe`) in the same way as we did in *Chapter 9*:

```
import torch
from diffusers import StableDiffusionPipeline

sd_pipe = StableDiffusionPipeline.from_pretrained(
    "stabilityai/stable-diffusion-2-1",
    torch_dtype=torch.float16)
sd_pipe.to('cuda')
```

2. Next, we'll initialize our Flask `app`:

```
from flask import Flask
from flask_ngrok import run_with_ngrok

app = Flask(__name__)
run_with_ngrok(app)
```

Here, `run_with_ngrok` indicates that the app will run with `ngrok`, but the actual `app` is not running yet (this will come at the end of this example). Since we don't have access to Colab's `localhost`, `ngrok` will make it possible to access it from our test client.

3. Then, we'll implement our `text-to-image` endpoint, which will process the prompts, which are coming in as web requests, and generate images based on them:

```
import io
from flask import Flask, request, send_file, abort

@app.route('/text-to-image', methods=['POST', 'GET'])
def predict():
    if request.method in ('POST', 'GET'):
        prompt = request.get_json().get('prompt')

        if prompt and prompt.strip():
            image = sd_pipe(
                prompt,
                num_inference_steps=100).images[0]

            image_io = io.BytesIO()
            image.save(image_io, format='PNG')
            image_io.seek(0)
            return send_file(
                image_io,
```

```
                        as_attachment=False,
                        mimetype='image/png'
                )
        else:
                abort(500, description='Invalid prompt')
```

The endpoint's name is /text-to-image and it will process both POST and GET requests (the processing pipeline is the same). The function will parse the textual prompt parameter and it will feed it to sd_pipe to generate an image parameter (in the same way as in the *Chapter 9* example). Finally, the send_file function will return the result of image to the client.

4.   We can now start the Flask app with the app.run() command. It will initialize the Flask development server so that our endpoint will be ready to process requests. In addition, ngrok will expose the app to the outside world with a URL of the http://RANDOM-SEQUENCE. ngrok.io/ type.

5.   We can use this URL to initiate a test request to the text-to-image endpoint (this is outside the Colab notebook):

```
import requests
response = requests.post(
    url='http://RANDOM-SEQUENCE.ngrok.io/text-to-image',
    json={'prompt': 'High quality photo of a racing car on a
track'})
```

6.   We can display the image with the following code:

```
from PIL import Image
import io
image = Image.open(io.BytesIO(response.content))
image.show()
```

This concludes our REST API example. Next, we'll deploy a model in a web environment with a UI.

Building ML web apps with Gradio

Gradio (https://www.gradio.app/) is an open source Python library that allows us to build interactive web-based demos for our ML models. HF Spaces (https://huggingface.co/spaces) supports hosting Gradio apps. So, we can build a Gradio app on top of the HF infrastructure, which includes not only hosting but also has access to all available HF models (https://huggingface. co/models).

We can create an HF space at https://huggingface.co/new-space. The space has a name (which will be its URL as well), a license, and an SDK. At the time of writing, HF Spaces supports Streamlit-based (https://streamlit.io/), Gradio-based, and Static instances. However, you can also deploy custom Docker containers for more flexibility.

Each new HF space has an associated Git repository. For example, the space of this example is located at `https://huggingface.co/spaces/ivan-vasilev/gradio-demo`, which is also the URL of its corresponding Git repository. The Gradio-based space expects a Python module called `app.py` in its root (in our case, the whole example will reside in `app.py`) and a `requirements.txt` file. Every time you push changes to the repository, the app will automatically pick them up and restart itself.

> **Note**
>
> To replicate this example, you'll need an HF account. HF Spaces has different hardware tiers. The basic one is free, but this particular example requires the GPU-enabled tier, which has an hourly cost. Therefore, if you want to run this example, you can duplicate it in your own account and enable the GPU tier.

Gradio starts with a central high-level class called `gradio.Interface`. Its constructor takes three main parameters:

- `fn`: The main function, which will process the inputs and return outputs.

- `inputs`: One or more Gradio input components. These could be textual inputs, file uploads, or combo boxes, among others. You can specify the component as a class instance or via its string label. The number of inputs should match the number of `fn` parameters.

- `outputs`: One or more Gradio components, which will represent the result of the execution of `fn`. The number of outputs should match the number of values returned by `fn`.

Gradio will automatically instantiate and arrange the UI components based on the `input` and `output` parameters.

Next, we'll implement our example. We'll use the same text-to-image SD scenario that we used in the *Deploying NN models with Flask* section. To avoid duplication, we'll assume that the `sd_pipe` pipeline has already been initialized. Let's start:

1. First, we'll implement the `generate_image` function, which uses `prompt` to synthesize an image in a total of `inf_steps` steps:

```
def generate_image(
        prompt: str,
        inf_steps: int = 100):
    return sd_pipe(
        prompt=prompt,
        num_inference_steps=inf_steps).images[0]
```

2. Next, we'll initialize the `gradio.Interface` class:

```
import gradio as gr
interface = gr.Interface(
    fn=generate_image,
    inputs=[
        gr.components.Textbox(label='Prompt'),
        gr.components.Slider(
            minimum=0,
            maximum=100,
            label='Inference Steps')],
    outputs=gr.components.Image(),
    title='Stable Diffusion',
)
```

As we discussed, the `inputs` and `outputs` `gr.Interface` parameters match the input/output signature of the `generate_image` function.

3. Finally, we can run the app with the `interface.launch()` command. Here is what the responsive UI of the app looks like:

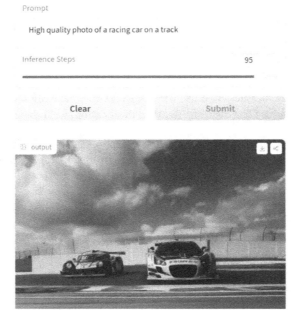

Figure 10.4 – The SD Gradio app's responsive UI, hosted on HF Spaces.
Top: input components; bottom: generated image

This concludes our introduction to Gradio and model deployment.

Summary

In this chapter, we outlined three major components of the ML development life cycle – training dataset creation, model development, and model deployment. We focused on the latter two, starting with development. First, we discussed the popularity of the foundational NN frameworks. Then, we focused on several model development topics – the ONNX universal model representation format, the TB monitoring platform, the TF Lite mobile development library, and mixed precision PyTorch training. Next, we discussed two basic scenarios for model deployment – a REST service as a Flask app and an interactive web app with Gradio.

This concludes this chapter and this book. I hope you've enjoyed the journey!

Index

Symbols

A

www.packtpub.com

Subscribe to our online digital library for full access to over 7,000 books and videos, as well as industry leading tools to help you plan your personal development and advance your career. For more information, please visit our website.

Why subscribe?

- Spend less time learning and more time coding with practical eBooks and Videos from over 4,000 industry professionals

- Improve your learning with Skill Plans built especially for you

- Get a free eBook or video every month

- Fully searchable for easy access to vital information

- Copy and paste, print, and bookmark content

Did you know that Packt offers eBook versions of every book published, with PDF and ePub files available? You can upgrade to the eBook version at packtpub.com and as a print book customer, you are entitled to a discount on the eBook copy. Get in touch with us at customercare@packtpub.com for more details.

At www.packtpub.com, you can also read a collection of free technical articles, sign up for a range of free newsletters, and receive exclusive discounts and offers on Packt books and eBooks.

Other Books You May Enjoy

If you enjoyed this book, you may be interested in these other books by Packt:

Hands-On Graph Neural Networks Using Python

Maxime Labonne

ISBN: 978-1-80461-752-6

- Understand the fundamental concepts of graph neural networks
- Implement graph neural networks using Python and PyTorch Geometric
- Classify nodes, graphs, and edges using millions of samples
- Predict and generate realistic graph topologies
- Combine heterogeneous sources to improve performance
- Forecast future events using topological information
- Apply graph neural networks to solve real-world problems

Enhancing Deep Learning with Bayesian Inference

Matt Benatan, Jochem Gietema, Marian Schneider

ISBN: 978-1-80324-688-8

- Understand advantages and disadvantages of Bayesian inference and deep learning
- Understand the fundamentals of Bayesian Neural Networks
- Understand the differences between key BNN implementations/approximations
- Understand the advantages of probabilistic DNNs in production contexts
- How to implement a variety of BDL methods in Python code
- How to apply BDL methods to real-world problems
- Understand how to evaluate BDL methods and choose the best method for a given task
- Learn how to deal with unexpected data in real-world deep learning applications

Packt is searching for authors like you

If you're interested in becoming an author for Packt, please visit `authors.packtpub.com` and apply today. We have worked with thousands of developers and tech professionals, just like you, to help them share their insight with the global tech community. You can make a general application, apply for a specific hot topic that we are recruiting an author for, or submit your own idea.

Share Your Thoughts

Now you've finished *Python Deep Learning, Third Edition*, we'd love to hear your thoughts! Scan the QR code below to go straight to the Amazon review page for this book and share your feedback or leave a review on the site that you purchased it from.

`https://packt.link/r/1837638500`

Your review is important to us and the tech community and will help us make sure we're delivering excellent quality content.

Download a free PDF copy of this book

Thanks for purchasing this book!

Do you like to read on the go but are unable to carry your print books everywhere?

Is your eBook purchase not compatible with the device of your choice?

Don't worry, now with every Packt book you get a DRM-free PDF version of that book at no cost.

Read anywhere, any place, on any device. Search, copy, and paste code from your favorite technical books directly into your application.

The perks don't stop there, you can get exclusive access to discounts, newsletters, and great free content in your inbox daily

Follow these simple steps to get the benefits:

1. Scan the QR code or visit the link below

https://packt.link/free-ebook/9781837638505

2. Submit your proof of purchase
3. That's it! We'll send your free PDF and other benefits to your email directly

www.ingramcontent.com/pod-product-compliance
Lightning Source LLC
Chambersburg PA
CBHW080615060326
40690CB00021B/4709